ENVIRONMENTAL SYSTEMS & SOCIETIES

IB DIPLOMA PROGRAMME

Andrew Davis
Garrett Nagle

Great Clarendon Street, Oxford, OX2 6DP, United Kingdom

Oxford University Press is a department of the University of Oxford.
It furthers the University's objective of excellence in research, scholarship, and education by publishing worldwide. Oxford is a registered trade mark of Oxford University Press in the UK and in certain other countries

First published in 2020

British Library Cataloguing in Publication Data
Data available

978-0-19-843754-3

10 9 8 7 6 5

Paper used in the production of this book is a natural, recyclable product made from wood grown in sustainable forests. The manufacturing process conforms to the environmental regulations of the country of origin.

Printed and bound by CPI Group (UK) Ltd, Croydon, CR0 4YY

Dedication:

(AJD): For Jenny and John

(GEN): For Angela, Rosie, Patrick and Bethany

Acknowledgements

The authors are grateful to the team at OUP for guiding this project through to completion. Julia Waitring, IB Publisher, gave advice on the framework and coordinated with the IB to provide material for the book. Linda Harvie oversaw the editorial process and was instrumental in keeping the project on track. AD is grateful to Nathan Adams for his help in accessing the Oxford DAM system. In addition, Mandy Ridd carried out the development edit and meticulously copyedited the text. We are also indebted to our reviewer, whose detailed and insightful comments on all parts of the manuscript were invaluable and helped to materially improve the book.

Photo credits:

Cover: Adisa/Shutterstock. All other photos © Garrett Nagle except: **p(iv)**: Adisa/Shutterstock; **p1**: Antonio Guillem/Shutterstock; **p2**: Granger Historical Picture Archive/Alamy Stock Photo; **p3(t)**: Arindambanerjee/Shutterstock; **p3(b)**: Public Health England/Science Photo Library; **p31**: Tom Stack/ Alamy Stock Photo; **p33(l)**: JRJfin/Shutterstock; **p33(m)**: Geogphotos/Alamy Stock Photo; **p33(r)**: RidgebackStudio/Shutterstock; **p34(t)**: Jason Bazzano/ Alamy Stock Photo; **p34(tm)**: Anton Sorokin/Alamy Stock Photo; **p34(bm)**: Nature Picture Library/Alamy Stock Photo; **p34(b)**: Tom Stack/Alamy Stock Photo; **p35(t)**: BSIP SA/Alamy Stock Photo; **p35(bl)**: Ragsltd/Shutterstock; **p35(br)**: Nick Vorobey/Shutterstock; **p58**: Daniel J. Rao/Shutterstock; **p66**: FLHC 24/Alamy Stock Photo; **p67**: Rich Carey/Shutterstock; **p68**: Rich Carey/ Shutterstock; **p78**: FLHC 24/Alamy Stock Photo; **p79(t)**: AfriPics.com/Alamy Stock Photo; **p79(b)**: Nino Marcutti/Alamy Stock Photo; **p88(t)**: Edwin Butter/ Shutterstock; **p88(b)**: Kirsten Wahlquist/Shutterstock; **p90(tl)**: Pedro Moraes/ Shutterstock; **p90(tm)**: Nancy Ayumi Kunihiro/Shutterstock; **p90(tr)**: Tierfotoagentur/Alamy Stock Photo; **p90(bl)**: AGAMI Photo Agency/Alamy Stock Photo; **p90(bm)**: RPFerreira/iStock/Getty Images Plus; **p90(br)**: Pascal Goetgheluck/Science Photo Library; **p150**: Selbst fotografiert von Michael Ertel/Wikimedia/Public Domain (CC BY-SA3.0); **p205, 209**: Adisa/Shutterstock; **p211(c)**: © Killarney House.

Artwork by Q2A Media Services Pvt. Ltd and OUP.

The publisher would like to thank the International Baccalaureate for their kind permission to adapt questions from past examinations and content from the subject guide. The questions adapted for this book, and the corresponding past paper references, are summarized here:

Question practice, **p6**: N17 SP2 TZ0 Q5(b); Question practice, **p11**: M18 SP1 TZ0 Q4(b); Question practice, **p17**: M18 SP1 TZ0 Q11, M17 SP2 TZ0 Q5(a); Question practice, **p23**: N17 SP1 TZ0 Q7; Question practice, **p29**: M18 SP2 TZ0 Q6(a); Question practice, **p37**: M17 SP2 TZ0 Q5(a), N17 SP2 TZ0 Q1(a)(i)(ii); Question practice, **p43**: N17 SP2 TZ0 Q4(a), M18 SP2 TZ0 Q5(b); Question practice, **p51**: N17 SP2 TZ0 Q7(b); Question practice, **p59**: M17 SP1 TZ0 Q1, M18 SP2 Q4(a); Question practice, **p65**: N17 SP2 TZ0 Q2(b); Question practice, **p68**: M17 SP1 TZ0 Q2(a); Question practice, **p74**: M17 SP2 TZ0 Q7(a), N17 SP2 TZ0 Q2(a)(i)(ii); Question practice, **p80**: N17 SP2 TZ0 Q5(c); Question practice, **p90**: M17 SP1 TZ0 Q4, N17 SP2 TZ0 Q5(a), M18 SP1 TZ0 Q12(a)(b); Question practice, **p100**: SPEC SP2 TZO Q6(a); Question practice, **p107**: SPEC SP2 TZO Q6(c); Question practice, **p113**: N17 SP1 TZ0 Q5(d); Question practice, **p116**: M17 SP2 TZ0 Q5(c); Question practice, **p6**: N17 SP2 TZ0 Q5(b); Question practice, **p123**: M18 SP2 TZ0 Q2, N17 SP1 TZ0 Q7(a); Question practice, **p132**: SPEC SP2 TZO Q3, SPEC SP2 TZO Q8(b); Question practice, **p136**: N17 SP1 TZ0 Q3(b)(c); Question practice, **p142**: N17 SP1 TZ0 Q1(b); Question practice, **p146**: M17 SP2 TZ0 Q7(b); Question practice, **p150**: M17 SP2 TZ0 Q7(b), M18 SP2 TZ0 Q6(a), N17 SP2 TZ0 Q3(a)(i)(ii)(b); Question practice, **p155**: M17 SP2 TZ0 Q5(c); Question practice, **p162**: M17 SP1 TZ0 Q7, M18 SP2 TZ0 Q7(b), M18 SP2 TZ0 Q7; Question practice, **p170**: M17 SP2 TZ0 Q3(b)(c), N17 SP1 Q4; Question practice, **p177**: M17 SP2 TZ0 Q3(d)(f), N17 SP2 TZ0 Q4(c); Question practice, **p187–188**: N17 SP1 TZ0 Q2, N17 SP2 TZ0 Q1(b)(i)(ii); Question practice, **p191**: M18 SP2 TZ0 Q7(a), N17 SP2 TZ0 Q6(a)(b); Question practice, **p198**: SPEC SP2 TZO Q1(a), N17 SP2 TZ0 Q6(c); Question practice, **p203**: N17 SP2 TZ0 Q1(a)(i)(ii)(c).

The authors and publisher are grateful to those who have given permission to reproduce the following extracts and adaptations of copyright material:

Aradottir, A. L. et al.: figure adapted from (1992) Hnignun gróðurs og jarðvegs. [A model for land degradation.] (in Icelandic) Groeðum Island (Yearbook of the Soil Conservation Service), (4), pp. 3–82, reproduced by permission.

Barry, Roger G. and Chorley, Richard J.: figure from *Atmosphere, Weather and Climate*, Routledge, © 1968, 1971, 1976, 1982, 1987, 1992, 1998, 2003, 2010 Roger G. Barry and Richard J. Chorley, reproduced by permission of Taylor and Francis, a division of Informa plc, permission conveyed through Copyright Clearance Center, Inc.

Bass, MS., Finer, M., Jenkins, CN., Kreft, H., Cisneros-Heredia, DF., McCracken, SF., et al.: figure from (2010) 'Global Conservation Significance of Ecuador's Yasuní National Park', PLoS ONE 5(1): e8767, https://doi.org/10.1371/journal.pone.0008767.

The Economist: tables from *Pocket World in Figures*, 2017 edition, Profile Books, reproduced by permission.

GRID-Arendal / United Nations Environment Programme: diagram 'Waste Management Hierarchy' adapted from Green Economy Report, 2011 / http://www.grida.no/resources/8247.

Icelandic Institute of Natural History: Fact file on Nootka Lupin and map reproduced by permission.

Nathan, Manjula V.: Figure 3 from *Master Gardener Core Manual* by Manjula V. Nathan, Soil Testing and Plant Diagnostic Service Laboratory, University of Missouri Extension, reproduced by permission.

Melbourne, Barbara: figure 'Life cycle of the malaria protozoan Plasmodium' from *Geofile Online*, September 2007, Issue 553, published by Nelson Thornes 2007, reproduced with permission of the Licensor (Oxford University Press) through PLSClear.

MoFE: table adapted from Table 2. Nepal's Adaptation Pathways, pp. 14-16 of *Nepal's National Adaptation Plan (NAP) Process: Reflecting on lessons learned and the way forward* by the Ministry of Forests and Environment (MoFE), Government of Nepal, the NAP Global Network, Action on Climate Today (ACT) and Practical Action Nepal, retrieved from http://napglobalnetwork.org/, reproduced by permission.

United Nations Framework Convention on Climate Change: excerpt reproduced by permission of UNFCCC.

Sources:

Graph from Energy Agency of Iceland, https://orkusetur.is/
Adapted from https://en.wikipedia.org/wiki/Puffi n#cite_note-BNA_Atlantic-21
Maps from University of Texas Libraries
www.conservation.org
www.nature.org
Maps from wwf.panda.org
www.statice.is.
The World Factbook, www.cia.gov
Living Planet Report 2012, WWF

Contents

Answers to questions and exam papers in this book can be found on your free support website. Access the support website here:
www.oxfordsecondary.com/ib-prepared-support

INTRODUCTION

As an IB student, you are provided with many resources on the path to your final environmental systems and societies (ESS) standard level (SL) exams. This book is just one of the resources to help you prepare.

How to use this book

Welcome to IB Prepared! The main purpose of this book is to help you, the student, prepare for your exams. Using actual student work from past exams (of the new ESS course which had its first exams in 2017) this book provides practical advice on how to approach exam questions, explaining how marks are given, and how marks may be lost, and it explains what the examiners are looking for in the questions. The comments are provided by experienced IB teachers, so what you read is based on knowledge of the marking and teaching of IB ESS.

Your teachers will have provided you with the content material you need to know for the examination; this book is intended to supplement your classroom experiences, not replace them. Active learning is considered to be one of the better ways of learning, so to gain the most benefit from this book we suggest you read through various sections and attempt the past exam questions yourself *before* reading the student answers and the examiner comments.

Get to know your exams

The following table summarizes the assessment for ESS. Examine it carefully, noting that 75% of your marks for the course will be decided by your performance in two formal exams. Be prepared!

There are two externally assessed written papers: paper 1 (one hour) and paper 2 (two hours). Paper 1 is worth 25% of the final marks, and paper 2, 50% of the final marks. The other assessed part of the course (25%) is the internal assessment (practical reports, or IAs) which is marked by your teacher and then moderated by an examiner.

Assessment component	Weighting (%)	Approximate weighting of objectives in each component (%)		Duration (hours)
		1 and 2	3	
Paper 1 (case study)	25	50	50	1
Paper 2 (short answers and structured essays)	50	50	50	2
Internal assessment (individual investigation)	25	Covers objectives 1,2,3 and 4		10

Source: IB environnemental systems and societies guide

Environmental systems and societies SL: grade descriptors

Grade 7: demonstrates comprehensive understanding of relevant ESS concepts and issues; well-structured, consistently appropriate and precise use of ESS terminology; effective use of relevant and well-explained examples that show some originality; some informed appreciation for a range of environmental value systems (EVSs); thorough, well-balanced analysis with detailed evaluations; ability to solve complex and unfamiliar problems; conclusions that are well supported by evidence, and that include some critical reflection. Ability to analyse and evaluate data thoroughly.

Grade 6: demonstrates sound understanding of relevant ESS concepts and issues, showing a wide breadth; well-structured accounts with appropriate and precise use of ESS terminology; effective use of relevant, well-explained examples that may show hints of originality; some informed acknowledgement of a range of EVSs; thorough, well-balanced analysis with valid evaluations; some ability to solve complex and unfamiliar problems; explicit conclusions that are well supported by evidence. Ability to analyse and evaluate data with a high level of competence.

Grade 5: demonstrates several areas of sound understanding of relevant ESS concepts and issues; generally clearly expressed accounts with largely appropriate use of ESS terminology; effective use of relevant examples that include some explanation; some informed awareness of a range of EVSs; clear analysis or argument that shows a degree of balance and attempts at evaluation; some ability to engage effectively with complex or unfamiliar problems; identifiable conclusions that are partially supported by evidence. Ability to analyse and evaluate data competently.

Grade 4: demonstrates one or two areas of sound knowledge and understanding of relevant ESS concepts and issues; sometimes clearly expressed accounts and largely appropriate use of ESS terminology; some use of relevant examples with very limited explanation; some awareness of EVSs; some clear but patchy analysis with a limited attempt at balance; some ability to solve simple or familiar problems; identifiable conclusions that are supported by very limited evidence. Demonstrates some analysis of data.

Grade 3: demonstrates only vague, partial knowledge and understanding of relevant ESS concepts and issues; generally unclear accounts and relevance with some isolated use of ESS terminology; examples that lack relevance and explanation; very limited awareness of EVSs; analysis is lacking or no more than a list of facts; very limited ability to solve simple or familiar problems; conclusions are unclear and not supported by evidence.

Grade 2: demonstrates fragmented or limited knowledge but little understanding of relevant ESS concepts and issues; generally incomprehensible accounts with very little, if any, use of ESS terminology; examples that are incomplete or irrelevant; limited ability to express EVSs; no evidence of real analysis; attempts to solve simple or familiar problems are incorrect or unsuccessful; no clear attempt to make conclusions.

Grade 1: demonstrates very little knowledge of relevant ESS concepts and issues; incomprehensible accounts with no use of ESS terminology; no recognisable use of examples; expresses no clear understanding of EVSs; no analysis or argument; no significant attempts to solve simple or familiar problems; no conclusions.

Command terms for environmental systems and societies

In this section you will gain an understanding of the assessment objectives and the command terms you will find in exam questions.

During the ESS course you are expected to develop a variety of skills. These are summarized in the four assessment objectives that follow:

Objective 1

1. Demonstrate knowledge and understanding of relevant:

- facts and concepts
- methodologies and techniques
- values and attitudes.

Objective 2

2. Apply knowledge and understanding in the analysis of:

- explanations, concepts and theories
- data and models
- case studies in unfamiliar contexts
- arguments and value systems.

Objectives 3 and 4

3. Evaluate, justify and synthesize, as appropriate:

- explanations, theories and models
- arguments and proposed solutions
- methods of fieldwork and investigation
- cultural viewpoints and value systems.

4. Engage with investigations of environmental and societal issues at the local and global level through:

- evaluating the political, economic and social contexts of issues

- selecting and applying the appropriate research and practical skills necessary to carry out investigations
- suggesting collaborative and innovative solutions that demonstrate awareness and respect for the cultural differences and value systems of others.

Papers 1 and 2 test assessment objectives 1–3, while the internal assessment tests AO 1–4.

Assessment objectives	Which component addresses this assessment objective?	How is the assessment objective addressed?	Weighting of paper (% of total marks available)
1–3	Paper 1	Case study; short-answers (35 marks)	25%
1–3	Paper 2	Section A (25 marks) – short-answer and data-based questions; Section B (40 marks) – two structured essays (from a choice of four)	50%
1–4	Internal assessment (IA) – practical work		25%

Command terms for objective 1

Term	Definition	Sample question	How to be prepared
Define	Give the precise meaning of a word, phrase, concept or physical quantity.	Define *biodiversity.* [1]	You should create for yourself a glossary of the key terms in the guide.
Draw	Represent by means of a labelled, accurate diagram or graph, using a pencil.	Draw a flow diagram to show the flows of leaching and decomposition associated with the mineral storage in the 'A' horizon in Figure 2(b). [2]	Be prepared to draw diagrams in both paper 1 and paper 2. Your answer will be electronically scanned so draw the diagram or graph clearly and create labels that can be clearly read.
Label	Add labels to a diagram.	Figure 2 below shows the structure of the Earth's atmosphere. Label the **two** lowest layers of the atmosphere on the diagram. [1]	Each label is likely to be just a word or two – you are labelling rather than annotating (see the definition for annotating within these command terms).
List	Give a sequence of brief answers with no explanation.	List **two** biomes found in Brazil. [1]	A list is likely to consist of just a few words. No explanation is required and you may waste valuable time giving extra detail for which you will receive no credit.
Measure	Obtain a value for a quantity.	Measure the decrease in the thickness of the ice cover on the south coast of Greenland between 1950 and 2018. [2]	Use the scale provided to measure the extent of the decrease in thickness.
State	Give a specific name, value or other brief answer without explanation or calculation.	State the crop that is under the greatest water stress. [1]	The answer is likely to be brief and very specific. You must only use information provided in the question paper/resource booklet.

Command terms for objective 2

Annotate	Add brief notes to a diagram or graph.	Annotate the systems diagram with **two** inputs of water and **two** outputs of water in the marshland ecosystem. [2]	The notes should aid in the description or explanation of the diagram or graph.
Apply	Use an idea, equation, principle, theory or law in relation to a given problem or issue.	Apply the Simpson diversity index to work out the diversity of species in the woodland ecosystem. [3]	Use the application indicated in the question to the specific problem or issue indicated. Your answer should be focused. The formula for the Simpson diversity index will be supplied and does not need to be memorized.
Calculate	Obtain a numerical answer showing the relevant stages of working.	With reference to Figure 1 calculate the doubling time for India. [1]	You should include all the steps involved in calculating the answer, inclusive of any required formula, and substitution of values. The final response should be made clear and have appropriate units.
Describe	Give a detailed account.	Describe the role of primary producers in ecosystems. [4]	Be guided by the number of marks assigned to the question. More marks require a more detailed description. In this example you need to give four points linked to the concept. There is no need to give reasons unless this is also included in the question.

Distinguish	Make clear the differences between two or more concepts or items.	Distinguish between the concept of a "charismatic" (flagship) species and a keystone species using named examples. [4]	Your answer should consist of more than two separate descriptions – it is essential that you emphasize the differences between them, and, in this question, provide named examples.
Estimate	Obtain an approximate value.	With reference to **Figure 7(b)**, estimate the average annual increase in gross domestic product (GDP) between 1990 and 2012. [1]	Use the scale provided on the graph/resource to work out the approximate value of the change.
Identify	Provide an answer from a number of possibilities.	Identify four ways in which solar energy reaching vegetation may be lost from an ecosystem before it contributes to the biomass of herbivores. [4]	Only a brief answer is needed here – sometimes you may have to select from a list.
Interpret	Use knowledge and understanding to recognize trends and draw conclusions from given information.	Interpret the likely impact of the rise of atmospheric CO_2 on global sea levels. [2]	Use knowledge and understanding to outline the effects of changes in the natural and human environment.
Outline	Give a brief account or summary.	Outline how four different factors influence the resilience of an ecosystem. [4]	Your answer should consist of brief statements – it is unlikely that there will be many marks available for an "outline" question.

Command terms for objectives 3 and 4

Term	Definition	Sample question	How to be prepared
Analyse	Break down in order to bring out the essential elements or structure.	With refence to the data presented in the resource booklet, analyse the global ecological value of Mongolia. [6]	From the data provided show the ways in which the subject (Mongolia) has a global ecological significance.
Comment	Give a judgment based on a given statement or result of a calculation.	Comment on the relationship between population growth and food supply. [2]	You should reach a conclusion as to whether there is/is not a relationship between the variables, and whether it is positive or negative.
Compare and contrast	Give an account of similarities and differences between two (or more) items or situations, referring to both (all) of them throughout.	Compare and contrast the impact of humans on the carbon and nitrogen cycles. [7]	For both cycles: • Describe and explain the similarities (compare). • Describe and explain the differences (contrast). • Creating a table (plan) for your compare/contrast response may ensure that you answer all components of the question. • Conclude by reviewing the relative similarities and differences.
Construct	Display information in a diagrammatic or logical form.	Construct a diagram to show how a positive feedback process involving **methane** may affect the rate of global warming. [2]	This may involve boxes and arrows or could be a flow diagram. It is very important that your construction is clear, and that the text is explanatory.
Deduce	Reach a conclusion from the information given.	Deduce, giving a reason, whether the figure below could represent the transfer of energy in a terrestrial ecosystem. [1]	Your answer must state the conclusion reached.
Demonstrate	Make clear by reasoning or evidence, illustrating with examples or practical application.	Demonstrate the higher ecological footprint (EF) associated with MEDCs compared with LEDCs. [2]	The answer requires examples to show the EF from named MEDCs and LEDCs and some reasoning for the greater use of the world's resources by wealthier societies.
Derive	Manipulate a mathematical relationship to give a new equation or relationship.	Using the crude birth rate and crude death rate, derive the natural increase for the selected populations. [1]	Crude birth and death rates are given in rates per thousand, whereas natural increase is given in rates per hundred (percent) so a conversion/manipulation of the data must be conducted to derive the answer. Be careful to read the units carefully.
Design	Produce a plan, simulation or model.	Design a conservation area for the protection of a named species. [3]	Design a reserve using appropriate criteria such as: size, shape, edge effects, corridors and proximity to other reserves.
Determine	Obtain the only possible answer.	With reference to Figure 3, determine the cereal yield in the UK in 1993. [1]	You should have a precise figure – remember to include the units.
Discuss	Offer a considered and balanced review that includes a range of arguments, factors or hypotheses. Opinions or conclusions should be presented clearly and supported by appropriate evidence.	Discuss the implications of environmental value systems in the protection of tropical biomes. [9]	You must consider different environmental value systems so that you cover at least two sides of the issue.

Evaluate	Make an appraisal by weighing up the strengths and limitations.	Pollution management strategies may be aimed at either **preventing** the production of pollutants or **limiting** their release into ecosystems. With reference to **either** acid deposition **or** eutrophication, evaluate the relative efficiency of these two approaches to management. [9]	The response should contain mention of both the advantages and disadvantages of the two strategies and arrive at a conclusion that addresses both viewpoints but favours one more than the other.
Explain	Give a detailed account, including reasons or causes.	Explain the role of **two** historical influences in shaping the development of the environmental movement. [7]	Assume that you will need to describe briefly before you explain, unless that was required in the preceding part of the question. Your response should then go on to outline how and why the two examples shaped the development of the environmental movement.
Examine	Consider an argument or concept in a way that uncovers the assumptions and interrelationships of the issue.	Examine the relationship between energy consumption and air quality. [2]	There are many possible relationships and you should try to look for some contrasts between and within different societies e.g. MEDC/LEDC, rural/urban, fossil fuel, renewables.
Justify	Provide evidence to support or defend a choice, decision, strategy or course of action.	With reference to images in Figure 4, justify your choice of **one** animal as the most suitable to promote conservation. [2]	You must have a balanced discussion that examines the positive and negative reasons for choosing one of the animals rather than the others.
Predict	Give an expected result.	Predict the effect on **nutrient cycling** of increased precipitation over many years in a region that is currently a steppe. [3]	There are three marks available, so you should refer to stores in the biomass, litter and soil, as well as the flows between them and the inputs and outputs to the system. When you are asked to make a prediction, you will be provided with information to assist you. Be sure to use it.
Sketch	Represent by means of a diagram or graph (labelled as appropriate). The sketch should give a general idea of the required shape or relationship and should include relevant features.	Sketch a diagram to show the main features of the water cycle. [3]	Your diagram should show a number of stores and flows. Distinguish between them using boxes/circles for stores and labelled arrows for the flows.
Suggest	Propose a solution, hypothesis or other possible answer.	Suggest a series of procedures that could be used to estimate the net primary productivity of an insect population in kg m-2 yr-1. [7]	The term "suggest" is used when there are several possible answers and you may have to give reasons or a judgment.
To what extent	Consider the merits or otherwise of an argument or concept. Opinions and conclusions should be presented clearly and supported with appropriate evidence and sound argument.	To what extent can different environmental value systems contribute to both causing and resolving the problem of water scarcity? [9]	Your answer should consider at least two environmental systems, and explore how they can cause the problem, as well as offer potential solutions to the problem.

Revision techniques

The techniques for absorbing facts, concepts, case studies and diagrams vary between individuals and what works for one candidate will not work for another. Nevertheless, there are ways of maximizing your revision time – this may be achieved through understanding the techniques which work best for you. They might include mnemonics, word pictures or acronyms and numbered lists. Your teacher could advise you, but ultimately you must decide.

During the exam

Time mismanagement: every year, students lose marks through time mismanagement. The most common tendency is to spend too long on the first question at the expense of the others. Make sure that you are aware of the time allocation for each question and that you stick rigidly to it during the exam. Note that five minutes of reading time is allowed before the start of each exam.

Question choice: in paper 1 all questions are compulsory but in paper 2 you get to choose two extended answers out of four. Again, your revision should be thorough, and no sub-topic omitted.

Generalization: for example, "desertification in Africa". Africa is a massive continent and not all of it suffers from desertification – be precise with the examples you choose.

Lack of evidence: essentially ESS is about the real world and in longer responses there must be plenty of factual support, examples and statistics.

Lack of correct terminology: correct use of terms shows understanding and avoids clumsy description. For example, a situation where "a population keeps on growing due to a lot of young people who are likely to keep the birth rate high for some years into the future" could be described as "population momentum".

Inadequate reading of the question: for example, if you were asked to "outline an example of positive feedback in global climate change" and you described negative feedback, you would inevitably lose marks. Ignoring commands is common and is highly likely to lose marks.

Missing the focus of the question: read the question at least twice; underline the command terms, other key words or the focus of the question. To maximize your marks, you need to ensure that you have given the question its broadest interpretation. It is very easy to forget a category despite having plenty of knowledge about it.

The following summarizes key information when sitting your exam:

Do

- Read the instructions on the cover of your exam paper to remind you of the exam regulations, such as the time allowed and the number of questions you should answer.
- Underline the command terms in the questions and focus on these as you work through each question.
- Write a brief plan for essays, to give your answer a logical structure. This should be written in pencil to allow for secondary thoughts or inspiration that comes to mind later.
- Observe the mark weighting of the sub-parts of structured questions.
- Give sufficient attention to the parts of the question requiring evaluation, discussion or analysis. These carry a heavy mark weighting.
- Complete the correct number of questions.
- Make sure that you all your answers are legible, correctly numbered and in numerical order.

Don't

- Pad your answer with irrelevant content just to make it look better. Examiners are impressed by quality, not quantity.
- Leave the examiner to draw conclusions if you cannot decide.
- Bend the question to fit your rehearsed answer.

- Spend too long on your best question at the expense of others.
- Invent case studies; these will be checked by examiners.
- Use lists or bullet points – these do not help detailed analysis.

Key features of the book

Annotated student answers show you real answers written by previous IB candidates, which mark they achieved and why. The examiner's comments will help you understand how marks may be scored or missed. An example of a question practice section and an accompanying student answer is shown below.

QUESTION PRACTICE

An **exam paper icon** indicates that the question has been taken from a past IB paper.

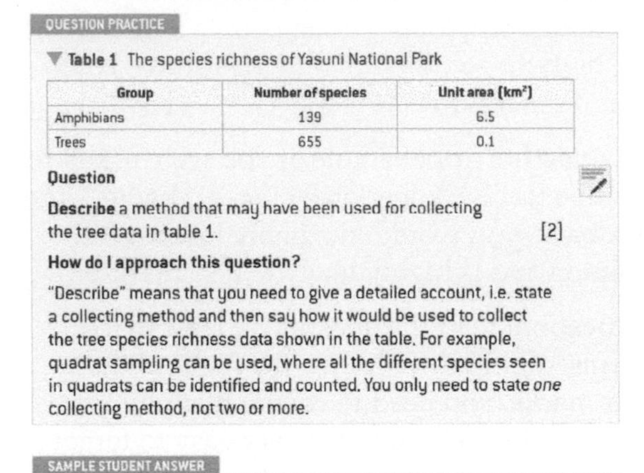

QUESTION PRACTICE

▼ **Table 1** The species richness of Yasuni National Park

Group	Number of species	Unit area (km²)
Amphibians	139	6.5
Trees	655	0.1

Question

Describe a method that may have been used for collecting the tree data in table 1. [2]

How do I approach this question?

"Describe" means that you need to give a detailed account, i.e. state a collecting method and then say how it would be used to collect the tree species richness data shown in the table. For example, quadrat sampling can be used, where all the different species seen in quadrats can be identified and counted. You only need to state *one* collecting method, not two or more.

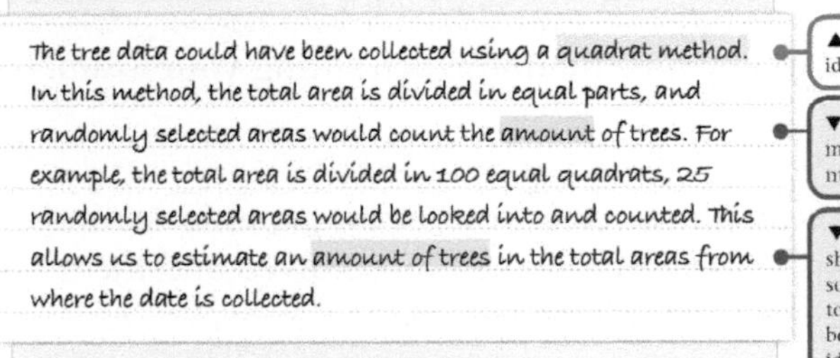

SAMPLE STUDENT ANSWER

The tree data could have been collected using a quadrat method. In this method, the total area is divided in equal parts, and randomly selected areas would count the amount of trees. For example, the total area is divided in 100 equal quadrats, 25 randomly selected areas would be looked into and counted. This allows us to estimate an amount of trees in the total areas from where the date is collected.

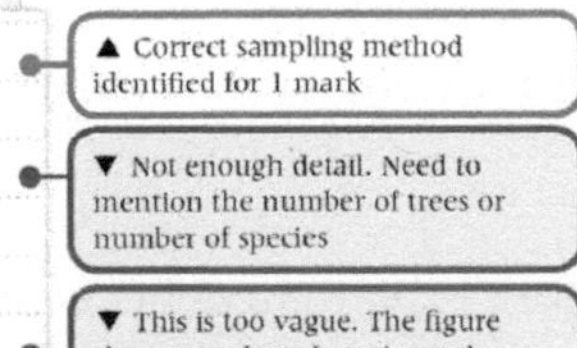

▲ Correct sampling method identified for 1 mark

▼ Not enough detail. Need to mention the number of trees or number of species

▼ This is too vague. The figure shows number of species and so the sampling method needs to refer to how this would have been recorded, for example by identifying and counting all the different species seen in quadrats

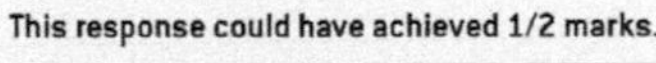

This response could have achieved 1/2 marks.

Assessment tip

Assessment tips give advice to help you optimize your exam techniques, warning against common errors and showing how to approach particular questions and command terms.

Content link

Content links connect different sections that you could revise together, as they offer complementary perspectives on the same topic.

Concept link

Concept links connect to the key concepts discussed in the ESS syllabus.

Test yourself

Test yourself boxes contain exam-style questions relating to the main text, where you can test your knowledge and understanding. The number of marks typically awarded to these questions is also given.

1 FOUNDATIONS OF ENVIRONMENTAL SYSTEMS AND SOCIETIES

This unit introduces the key concepts and environmental principles that are explored throughout the ESS syllabus. Central to the course is the concept of "systems". A system contains components, connected through various processes and interactions, all of which work together to constitute a whole entity. This "holistic" approach enables an appreciation of how environmental and other systems function, and that they do not function in isolation, but rather through interactions with other systems. The systems approach emphasizes that concepts, techniques, and terms can be transferred from one discipline (such as ecology) to another (such as politics), and provides a framework for examining and explaining complex environmental issues.

You should be able to show:

- ✔ an understanding of environmental value systems;
- ✔ how systems and models can help in the study of complex environmental issues;
- ✔ the role of energy and equilibria in the regulation of systems;
- ✔ an understanding of how sustainability and sustainable development can be achieved and monitored;
- ✔ an understanding of how human disturbance produces pollution and ways in which it can be managed.

1.1 ENVIRONMENTAL VALUE SYSTEMS

You should be able to show an understanding of environmental value systems

- ✔ The environmental movement has been affected by significant historical events, originating from the media, environmental disasters, literature, international agreements and through technological developments.
- ✔ Environmental value systems (EVSs) shape the way an individual or society views the environment and approaches environmental issues.
- ✔ EVSs range from ecocentric through anthropocentric to technocentric.
- ✔ Different parts of the biosphere can be given different intrinsic values, depending on a person or society's EVS.

- **Environmental value system (EVS)** – a worldview that affects the way an individual or society perceives and acts on environmental issues, based on their background (e.g. economic, cultural and sociopolitical contexts)
- **Society** – an arbitrary group of individuals who share some common characteristics, such as geographical location, cultural background, historical time frame, religious perspective, and value system
- **Technocentrist** – a person who has a technology-centred EVS
- **Ecocentrist** – a person who has a nature-centred EVS
- **Anthropocentrist** – a person who has a people-centred EVS
- **Deep ecologist** – an ecocentrist who sees humans as subject to nature not in control of it

Historical influences on the development of the environmental movement

The environmental movement promotes the idea of stewardship for planet Earth (i.e. the responsible planning and management of natural resources) both locally and globally. Significant historical influences on the development of the environmental movement have come from:

- literature (e.g. Rachel Carson's *Silent Spring*, 1962)
- the media (e.g. Davis Guggenheim's documentary *An Inconvenient Truth*, 2006)
- major environmental disasters (e.g. Minamata (1956), Bhopal (1984), the Chernobyl disaster (1986))

- **Cornucopian** – a technocentrist who believes that continued progress and providing material items for humanity can be achieved through continued advances in technology
- **Environmental manager** – an anthropocentrist who believes that humans should manage natural systems for economic profit
- **Self-reliance soft ecologist** – an anthropocentrist who believes that communities should play an active role in environmental issues
- **Intrinsic value** – refers to the intangible importance of a species or ecosystem, for example the aesthetic, ethical, spiritual or ecological value

▲ Figure 1.1.1 Rachel Carson

Assessment tip

You may be asked to explain the role of historical influences in shaping the development of the environmental movement. Your answers need to describe the personality or event, explain how it has influenced the movement, and explain exactly what gave rise to the influence.

- international agreements (e.g. the UN Rio Earth Summit in 1992, which produced Agenda 21 and the Rio Declaration)
- technological developments (e.g. the Green Revolution – a time in the mid-20th century when developments in scientific research and technology in farming led to increased agricultural productivity worldwide).

These events led to the development of environmental pressure groups. Media coverage increased and that raised public awareness of the issues.

Rachel Carson's *Silent Spring*

In 1962 American biologist Rachel Carson's (figure 1.1.1) influential book *Silent Spring* was published. Carson wrote about the harmful effects of synthetic pesticides and made a case against the chemical pollution of natural systems. She focused on the activities of chemical companies and explained the impact of use of insecticides on birds of prey. The book led to widespread concerns about the use of pesticides in crop production and the consequent pollution of the natural environment (mainly terrestrial systems). It also contributed to widespread awareness amongst the American public of key environmental issues, and was a focal point for the environmental movements of the 1960s, inspiring many other environmentalists to take action. The book led to a ban on DDT for agricultural uses and inspired the formation of the US Environmental Protection Agency.

The Club of Rome

In 1972 a global body of experts called the Club of Rome published *The Limits to Growth*. The group contained academics, civil servants, diplomats and industrialists and first met in Rome. The report examined the consequences of a rapidly growing world population on finite natural resources. *The Limits to Growth* encouraged scientists, policy-makers and the public to see ecological problems in planetary terms, intrinsically linked to human population growth. It has sold 30 million copies in more than 30 translations and has become the best-selling environmental book in history (source: http://np4sd.org/nominees/club-of-rome/).

Bhopal

On 3 December 1984, the Union Carbide pesticide plant in the Indian city of Bhopal released 42 tonnes of toxic methyl isocyanate gas. The release happened when one of the tanks involved with processing the gas overheated and burst. Some 500,000 people were exposed to the gas. It has been estimated that between 8,000 and 10,000 people died within the first 72 hours following exposure, and that up to 25,000 have died since from gas-related disease (source: https://www.independent.co.uk/news/world/asia/poisoned-legacy-of-bhopal-campaigners-call-on-dow-chemical-to-answer-criminal-charges-31-years-after-a6779231.html) (figure 1.1.2). The disaster showed that enforceable international standards for environmental safety were needed, as well as urgent strategies to prevent similar accidents happening in the future.

Chernobyl

On 26 April 1986, a nuclear reactor at the Chernobyl plant in the Ukraine exploded. A cloud of highly radioactive dust was sent into the atmosphere and fell over an extensive area. Large areas of the Ukraine,

Belarus and Russia were badly contaminated. The disaster resulted in the evacuation and resettlement of over 336,000 people. The fallout caused increased incidence of cancers in the most exposed areas. An area around the plant still remains a no entry area due to radiation. The incident raised concerns over the safety of nuclear power stations.

Fukushima Daiichi nuclear disaster

On 11 March 2011 an earthquake in north-eastern Japan led to a tsunami that flooded the Fukushima Daiichi Nuclear Power Plant (figure 1.1.3). It was the biggest nuclear disaster since Chernobyl. Estimates vary about the number of people affected by the disaster, although it is estimated that around 600 deaths may have been caused by the evacuations following the earthquake and tsunami (source: http://emilkirkegaard.dk/en/wp-content/uploads/Worldwide-health-effects-of-the-Fukushima-Daiichi-nuclear-accident.pdf). The disaster has led to increased public pressure to phase out nuclear power generation. For example, Germany has accelerated plans to close nuclear reactors, and over 90% of Italy's population voted against government plans to expand nuclear power. Switzerland has also decided to phase out nuclear power.

Our Common Future

In 1987, a report by the UN World Commission on Environment and Development (WCED) was published. The report was called *Our Common Future*. It linked environmental concerns to development and aimed to promote sustainable development through international collaboration. It also placed environmental issues firmly on the political agenda.

UN's Earth Summit

The publication of *Our Common Future* and the work of the WCED led to the UN's Earth Summit at Rio in 1992. The summit's message was that a change in our attitudes and behaviour towards environmental issues was required to bring about the necessary changes. The conference led to the adoption of Agenda 21, which is a blueprint for action to achieve sustainable development worldwide.

Environmental value systems (EVSs)

Environmental value systems determine a personal viewpoint and how a person responds to environmental issues. Personal viewpoints depend on many different factors. Factors include social influences, personal characteristics, habits and knowledge of environmental issues.

Personal characteristics and social influences help to determine, for example, personal views of global warming. If somebody is determined, and their parents and friends are environmentally active, they are more likely to take responsibility for solving issues surrounding climate change and try to make a real difference, e.g. trying to reduce energy consumption and using a bike rather than have a lift to college. Education will also determine a person's EVS. They may have read about environmental issues in newspapers and books, such as *Six Degrees* by Mark Lynas, which examines how temperature increase will affect planet Earth.

The options available to a person also affect how they respond to environmental issues. If someone lives in a city that provides a convenient recycling procedure, they are more likely to recycle than someone for whom recycling is inconvenient.

▲ Figure 1.1.2 Victims of the Bhopal tragedy during a rally to mark the 26th year of the Bhopal Gas Disaster, in Bhopal, India on 3 December 2010

▲ Figure 1.1.3 Cleaning up after the Fukushima Daiichi nuclear disaster

Assessment tip

In the range of historical influences selected, it is beneficial to have both local and global examples.

Further possible examples of historical influences could include: James Lovelock's development of the Gaia hypothesis; whaling; Gulf of Mexico oil spill of 2010; Chipko movement; Rio Earth Summit 2012 (Rio+20); Earth Day; Copenhagen Accord. You can also research and find out about any recent or local events that are of interest to you.

Content link

EVSs are explored throughout the course, for example in section 7.2, in relation to the subject of climate change.

Concept link 

EVS: Environmental value systems are a key concept and are discussed in each topic of the ESS syllabus.

EVSs and the systems approach

An environmental value system has inputs (for example education, cultural influences, religious doctrine, media) and outputs (for example decisions, perspectives, courses of action) that are determined by processing these inputs.

The spectrum of EVSs

EVSs range from nature-centred (ecocentrism) through to people-centred (anthropocentrism) and technology-centred (technocentrism) at the other end of the continuum.

The ecocentric worldview is nature-centred and does not trust modern, large-scale technology. **Ecocentrists** prefer to work with natural environmental systems to solve problems, and to do this before problems get out of control. They see a world with limited resources where growth needs to be controlled so that only beneficial forms occur (i.e. those that do not lead to habitat destruction or overuse of natural resources).

At one end of the ecocentrist worldview are the **self-reliance soft ecologists**, who reject materialism and tend to have a conservative view on environmental problem-solving. Self-reliance soft ecologists hold a people-centred (anthropocentric) view that is essentially ecocentric in nature. These environmentalists see humans as having a key role in managing sustainable global systems.

At the other end of the ecocentrist spectrum are **deep ecologists**. The deep ecology movement believes that all species have an **intrinsic value** and that humans are no more important than other species. Deep ecologists put more value on nature than humanity. This EVS rejects the concept of natural resources because it implies that organisms and ecosystems are only important as economic commodities for humans. Deep ecologists argue that an **anthropocentrist** viewpoint (where nature is seen to exist for, and to be used by, humans for human benefit) is at the root of our environmental crisis.

Technocentrists state that technology will provide solutions to environmental problems even when human population growth and its effects (e.g. habitat loss) are pushing natural systems beyond their normal boundaries. At one end of the technocentric spectrum are the **cornucopians**. A cornucopian view is a belief in the never-ending resourcefulness of humans and their ability to control their environment. This view leads to an optimistic view about the state of the world and humanity's impact on it. A more anthropocentric world view is shown by **environmental managers**, who see progress happening within closely defined frameworks to prevent over-exploitation of the Earth's resources.

Advantages and disadvantages of ecocentric and technocentric approaches

Technocentric advantages

- Provides alternatives to individuals that do not inconvenience them
- Substitutes materials or inputs and so avoids costly manufacturing or industrial change
- Allows economic, social and technological development to continue

Assessment tip

It is important to recognize that all EVSs are individual and that there is no "wrong" EVS.

During the ESS course you should develop your own EVSs and be able to justify your decisions on environmental issues based on your EVSs.

Test yourself

1.1 Justify your personal viewpoint on the exploitation of tropical rainforest. [4]

Assessment tip

Prepare arguments examining ecocentric versus technocentric approaches for a range of different environmental issues (e.g. water scarcity, food supply, farming, climate change, etc). This will enable you to comment on how different EVSs influence the way that people respond to environmental issues.

Concept link

SUSTAINABILITY: People's approach to environmental matters; whether it is ecocentric or technocentric, will determine the degree to which the solutions they propose are sustainable.

Technocentric disadvantages

- High cost
- Technological solutions may give rise to further environmental problems
- Promotes and allows for greater resource consumption

Ecocentric advantages

- Approaches may be more sustainable
- Does not have to wait for technological developments to occur
- Raises general environmental awareness in population and communities

Ecocentric disadvantages

- Requires individual change, which can be difficult to encourage
- May hinder economic growth and development

Test yourself

1.2 Distinguish between anthropocentrism and technocentrism. [4]

1.3 State what is meant by the term "environmental value system". [1]

1.4 Sustainable development depends on the interaction between social, environmental, and economic factors. **State** the priority for each of the following sectors of society: self-reliant soft ecologists, conservation biologists, and bankers. [3]

1.5 Describe how a self-reliant soft ecologist, a conservation biologist, and a banker may each support sustainable development. [3]

Decision-making and contrasting EVSs

An individual's or society's EVS will determine the way in which they address environmental issues.

If discussing a technocentric response to biodiversity loss, for example, the role that technological solutions have played in the response could be emphasized.

Web-based monitoring systems have helped to monitor species numbers, and satellite tracking has been used to monitor migrating organisms such as whales and sea birds. ICT systems have been important in enforcing agreements such as the Convention on International Trade in Endangered Species of Wild Fauna and Flora (CITES), and sophisticated technological solutions like seedbanks enable scientists to preserve DNA so that valuable genetic diversity is not lost.

However, it can be argued that technologies such as genetically modified (GM) crops and monocultures are actually responsible for the loss of species diversity. Technocentric solutions would have less concern for intrinsic or ethical rights of biodiversity, and have less concern for local habitats or ecosystems. Many of the causes of species loss, e.g. habitat degradation, are occurring in societies with little access to technology and so it cannot play a role in solving these problems.

An ecocentric approach to biodiversity loss would aim to minimize disruption to ecosystems in the first place, by changing government

Assessment tip

When justifying your personal viewpoint on environmental issues you need to reflect on where you stand on the continuum of EVSs regarding specific issues found throughout the ESS syllabus (e.g. population control, resource exploitation, and sustainable development). The EVS of an individual will inevitably be shaped by cultural, economic, and socio-political context. You should recognize this and appreciate that others may have equally valid viewpoints.

Assessment tip

Examples from the extremes of the EVS spectrum can be used to illustrate contrasting viewpoints. For example, in relation to the exploitation of oil reserves in a pristine (i.e. untouched) environment, deep ecologists would be concerned that nature will be damaged and favour the rights of species to remain unmolested over the rights of humans to exploit resources for economic gain, whereas cornucopians would feel that resources are there to be exploited and used to generate income, believing that with sufficient ingenuity and technical expertise potential environmental obstacles can be overcome.

Content link

CITES is discussed in unit 3.

policies and by persuading individuals of the intrinsic value of all organisms irrespective of monetary worth. Ecocentrists would argue the value of diversity for ecological stability. Ecocentric solutions can focus on localized areas of conservation, including whole habitats or ecosystems, which can generate community support and involvement. However, conservation can be costly, with little economic return, and so can be unpopular with nations seeking economic development.

Assessment tip

It can be argued that technology is a tool which cannot, on its own, solve any problem; there has to be political will to make changes and then technology can help to provide solutions.

Intrinsic values of the biosphere

Nature can be seen as having an intrinsic value. This means that the natural world has integral worth independent of its value (e.g. economic considerations) to anyone or anything else, such as the belief that all life on Earth has a right to exist. Intrinsic values include those based on cultural and aesthetic values.

Assessment tip

You need to be able to discuss the view that the environment can have its own intrinsic value.

Intrinsic value means that something has value in its own right, i.e. inbuilt/inherent worth.

QUESTION PRACTICE

Essay

Explain the role of **two** historical influences in shaping the development of the environmental movement. [7]

How do I approach this question?

There is one mark available for correctly identifying two historical influences, with a maximum of 4 marks for each explanation of how historical influences shaped the development of environmentalism, up to maximum of 7 marks (i.e. if 4 marks are awarded for the first example, up to a maximum of 3 marks can be awarded for the second example). Credit is given for valid statements that describe the event, explaining how it has influenced the movement and exactly what gave rise to the influence.

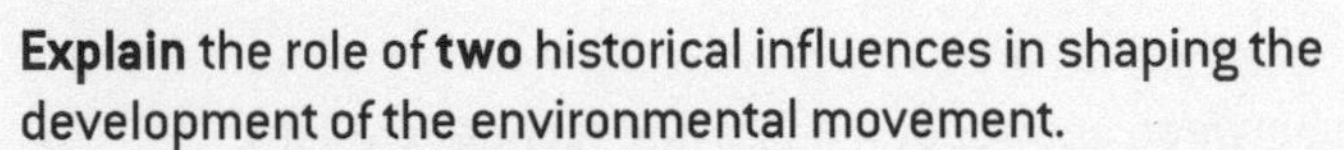

SAMPLE STUDENT ANSWER

Carson's "Silent Spring" (1970s) is a book about the use of DDT, a fertilizer and anti-malaria medicine that greatly harms the environment. Indeed, Carson explains in her book that the pollutant (DDT) builds up in animals' tissues (especially marine animals) through bio-accumulations and biomagnifactes along the food chain. Hence, top predators on the end of the food chain, such as the Bald Eagle, were affected, and its population size decreased significantly. The book raised great awareness about humans' uncontrolled use of pollutants. This led to civil uprisings, which put pressure on countries to ban the use of DDT. The book was so popular that the discussion about DDT even took place in arts and culture (for example, Joni Mitchell's "Big Yellow Taxi"). Finally the USA and most other MEDCs

▲ Correctly identifies one historical influence. One mark is available for naming two historical influences - the second example (the Montreal Protocol) is given later

▼ *Silent Spring* was published in 1962. It is important that factual details of case studies are learnt carefully so they can be accurately reproduced in exams

▼ Poor recall of information, or misconceptions, evident here. DDT is a pesticide, not a fertilizer, and is used to kill malarial mosquitoes (it is not 'anti-malaria medicine')

▲ The concepts of bioaccumulation and biomagnification are referred to here – the reasons why DDT is such a harmful pesticide. This, and subsequent points here, are essential information about factors that led to the publication of *Silent Spring*

▲ This provides background information about what gave rise to *Silent Spring*

▲ A direct link is made to the book's influence on the ban on DDT

(and some LEDCs) banned the use of DDT. Rachel Carson's book had a major influence on environmental movements, giving way to environmental NGOs such as WWF and Greenpeace (other books such as Thoreau's "Walden" also influenced this). Another historical event that influenced the environmental movement is the Montreal protocol in 1987. The protocol is an international agreement signed by 24 countries that aimed to stop the use and production of ozone- depleting substances such as CFCs. Although it was not legally binding, its results were very positive, particularly in MEDCs. By 2000, ODSs had decreased by 99% in MEDCs (merely 13 years after the protocol). This success was of great importance for the environmental movement, as it proved that political (intergovernmental) agreements can change people's and industries' behaviours and slow down the formation of the ozone hole above the Arctic. The Montreal protocol gave hope to a generation of environmentalists that grew up at a time when there was much less awareness about the environment than now.

Moreover, other environmental disasters such as the Bhopal disaster, Chernobyl and the Love Canal and the Exxon oil spill led to many global pro-environmental uprisings that shaped environmental movements.

▲ The wider impacts on the environmental movement are discussed

▲ A second valid example is given, so 1 mark is awarded (two examples given overall). This example is an international agreement

▲ The role of the Montreal Protocol is described

▲ The effectiveness of the agreement is discussed for 1 mark

▲ The importance of the international agreement is clearly stated for 1 mark. At this point in the answer the maximum 7 marks have been awarded. Again, this case study clearly identifies and explains the importance of the event in shaping the environmental movement

▼ This section is not necessary – the questions asks for two examples and this has already been done

This response could have achieved 7/7 marks.

- In sections on both *Silent Spring* and the Montreal Protocol credit has been given for correctly describing the event, explaining what gave rise to the influence, and explaining how this historical influence shaped the development of the environmental movement.
- The syllabus states that you should know about different examples of historical influences from literature, the media, major environmental disasters, international agreements and technological developments. Two valid examples are given here.
- This is an excellent answer which clearly identifies and explains the importance of two distinct influences on the environmental movement. The maximum available marks were awarded.

1.2 SYSTEMS AND MODELS

- **System** – an assemblage of parts and the relationships between them, which together constitute an entity or whole
- **Open system** – a system in which both matter and energy are exchanged with its surroundings (for example, natural ecosystems)
- **Closed system** – a system in which energy, but not matter, is exchanged with its surroundings (for example, the Earth)
- **Isolated system** – a system that exchanges neither matter nor energy with its surroundings (for example, the Universe as far as we know)
- **Model** – a simplified description designed to show the structure or workings of an object, system or concept

>> Assessment tip

A systems approach should be taken for all of the topics covered in the ESS course and you should be able to apply the systems approach to each topic you cover. You should be able to interpret given system diagrams and use data to produce your own, for example to show carbon cycling, food production and soil systems.

>> Assessment tip

Systems diagrams should always be in the same format, with storages (boxes) linked by arrows (flows). They should be kept as visually simple as possible (figure 1.2.1).

You should be able to show how systems and models can help in the study of complex environmental issues

✓ Environmental issues can be studied using the systems approach.

✓ Systems can be shown in diagrammatical form.

✓ Transfers and transformations represent different movements of matter and energy in systems.

✓ Systems can be open, closed or isolated.

✓ Models can be used to understand how systems work.

The systems approach

The systems approach emphasizes the similarities between environmental systems, biological systems and human-made entities such as transport and communication systems. This approach stresses that there are concepts, techniques and terms that can be transferred from one discipline (such as ecology) to another (such as economics).

All **systems** have inputs and outputs. According to the system, these can be inputs and outputs of energy, matter or information. All systems also have storages, flows, processes and feedback mechanisms. The systems method allows different areas of study, such as ESS and economics for example, to be looked at in the same way and for connections to be made between them.

An example of a system is a community of trees in a woodland. A community of trees in a woodland has the following features of a system:

- Individuals or species of trees are the components of the system.
- The components are interrelated.
- The components form an integrated whole, for example they may regulate populations through competition and contribute to succession of community.
- It has flows (i.e. transfers) of matter and energy between components (i.e. storages). For example, leaf fall provides nutrients to other trees through decomposition; glucose is transported from leaves to insects that eat the leaves and to other parts of the forest via litter-fall.
- Components carry out processes. Processes include photosynthesis, respiration and growth.
- It is an **open system** exchanging matter and energy with surroundings.

Systems diagrams

Systems can be shown in diagrams, where boxes indicate storages or matter and energy. Arrows indicate flows between the given storages.

Processes can be labelled on arrows, referring to different transfer or transformation processes. Transfer processes flow through a system

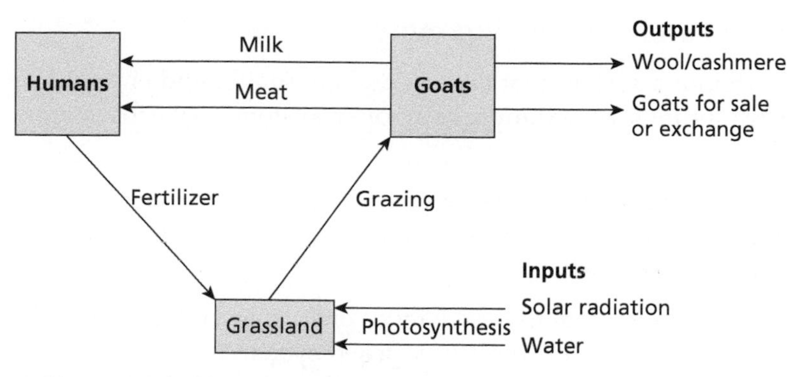

▲ Figure 1.2.1 Diagram of a farming system

>> Assessment tip

You need to be able to construct system diagrams from information provided to you.

and involve a change in location. Transformation processes lead to interaction within a system in the formation of new products or involve a change of state or phase. In figure 1.2.1, photosynthesis (the transformation process of light energy to stored chemical energy in the form of glucose) is a transformation process, with all other flows in figure 1.2.1 indicating transfer processes.

Flows and stores can be drawn proportionately if appropriate quantitative data is provided (i.e. bigger flows have wider arrows, and bigger storages are represented by larger boxes). For example, a quantitative **model** to show the storages and flows in forest carbon cycling can be drawn (figure 1.2.2), using the following data:

- global forest biomass contains 283 gigatonnes of carbon (GtC) (1 gigatonne = 1 billion tonnes)
- dead wood, litter and soil contain 520 GtC
- in the atmosphere there are approximately 750 GtC
- it is estimated that forests release 60 GtC per year into the atmosphere
- worldwide deforestation releases approximately 1.6 GtC per year (most in the tropics)
- some carbon is captured from the atmosphere when other crops are planted in the place of forests.

>> Assessment tip

You must be able to apply the systems concept to a range of scales, from small-scale local ecosystems e.g. a pond, to large ecosystems such as biomes.

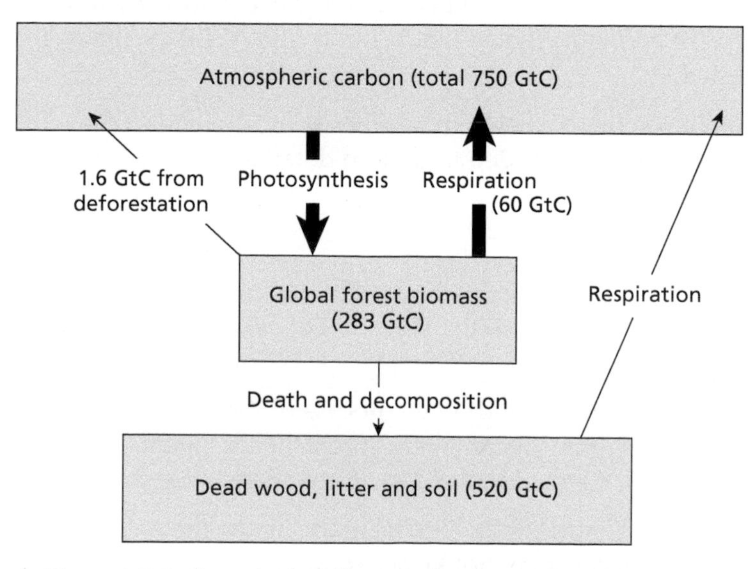

▲ Figure 1.2.2 Quantitative diagram showing carbon flows and storages in a forest ecosystem

Open, closed and isolated systems

An open system is a system that exchanges both matter and energy with its surroundings. An example of an open system is an ecosystem, such as a lake.

A community of trees can be considered an open system because it is supported by the absorption of solar energy and the provision of nutrients for non-tree species.

A **closed system** is a system that exchanges energy, but not matter, with its surroundings. An example of a closed system is the Earth.

An **isolated system** is a system that does not exchange matter or energy with its surroundings. An example of an isolated system is the Universe (figure 1.2.3).

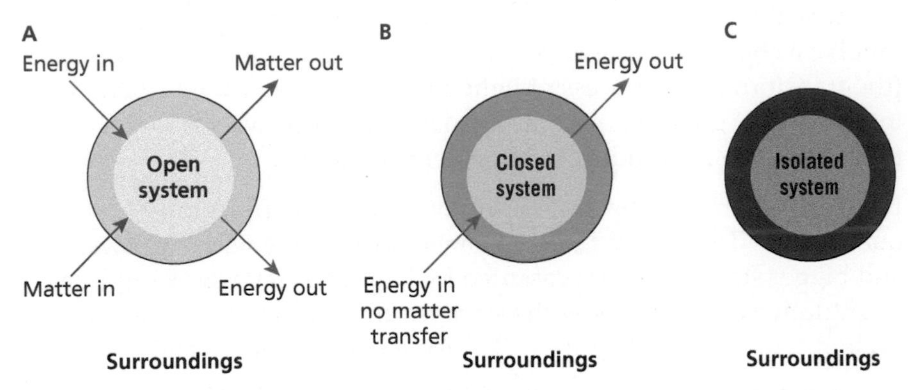

▲ Figure 1.2.3 The exchange of matter and energy across the boundaries of different systems. Open systems (A) exchange both energy and matter; closed systems (B) exchange only energy; and isolated systems (C) exchange neither

Models

>> Assessment tip

You need to be able to construct a model from a given set of information.

Models can be used to show the flows, storages and linkages within systems. While they are unable to show much of the complexity of the real system, they still help in understanding functioning of the system better.

Strengths of models include the following:

- Models allow scientists to predict and simplify complex systems.
- They allow inputs to be changed and outcomes examined without having to wait a long time, as would be the case if real events were being studied.
- Models allow results to be shown to other scientists and to the public, and are easier to understand than detailed information about the whole system.

Limitations of models include:

- Different models may show different effects using the same data. For example, models that predict the effect of climate change may give very different results.
- Systems may be complex, and when models of them are oversimplified they may become less accurate. For example, there are many complex factors involved in atmospheric systems.
- Because many assumptions have to be made about these complex factors, climate models may not be accurate.
- The complexity and oversimplification of climate models has led some people to criticize the limitations of these models.

- Any model is only as good as the data that is used in it. In addition, the data put into the model may not be reliable.
- Models rely on the expertise of the people making them and this can lead to inaccuracies.
- Different people may interpret models in different ways and so come to different conclusions. People who would gain from the results of the models may use them to their advantage.

Assessment tip

You need to be able to evaluate the use of models.

QUESTION PRACTICE

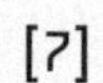

Essay

Explain how a community of trees in a woodland may be considered a system. [7]

How do I approach this question?

The question asks you to apply your knowledge of systems. The answer needs to explain how a community of trees can be considered a system, and not the entire woodland ecosystem. One mark will be awarded for each correct suggestion, up to a maximum of 7 marks. To plan the answer, consider all the different features that make up a system, such as storages and feedback mechanisms, and say how these apply to a tree community. The answer can also define what a system is, and say what *type* of system a community of trees would be.

SAMPLE STUDENT ANSWER

A system is a group of individually moving parts that work together to form a whole. Each system has inputs and outputs. A community of trees in a woodland may be considered a system as it satisfies all the criteria above. The inputs of the community of trees in the woodland are sunlight, carbon dioxide, water and minerals from the soil. They also take in oxygen for respiration. The outputs the trees in the woodland produce include carbon dioxide, oxygen, water (in the form of transpiration), fruits, nuts, leaves, wood etc. A system also consists of transfers and transformations. Transpiration is a transfer in the woodland as there is only a change in location of the water from the plant to the atmosphere. The transformations in the woodland include photosynthesis by plants as they produce oxygen from carbon dioxide and sunlight. It is a transformation due to the change in chemical composition.

As the woodland consists of inputs and outputs, transfers and transformations, they can be considered a system.

▲ A definition of what a system is linked to the example mentioned in the question, i.e. a community of trees. 1 mark

▲ Specific flows into the system (inputs) are given. 1 mark

▲ 1 mark for mentioning that transfers link the storages in systems

▲ 1 mark for saying that transformations are part of a system

▼ Transpiration is a transformation not a transfer because there is a change in state in the water, i.e. it turns from a liquid to a gas. Leaf fall would have been a valid example of a transfer process

▲ 1 mark for correctly identifying a transformation process in woodland

▼ This simply repeats the information already given and is unnecessary

This response could have achieved 5/7 marks.

For this question, a maximum of 4 marks were available for identifying relevant generic features of system and a maximum of 3 marks for examples of these within a tree community. This answer describes the flows between the storages but not what the storages are (i.e. trees). It also does not mention what type of system it is, i.e. an open system exchanging matter and energy with surroundings (e.g. absorption of solar energy and the provision of nutrients for non-tree species). The example chosen for the transformation process is incorrect.

1.3 ENERGY AND EQUILIBRIA

- **Entropy** – a measure of the amount of disorder, chaos or randomness in a system; the greater the disorder, the higher the level of entropy. Entropy increases in a system
- **Equilibrium** – a state of balance among the components of a system
- **Stable equilibrium** – tendency in a system for it to return to a previous equilibrium condition following disturbance (as opposed to unstable equilibrium)
- **Steady-state equilibrium** – the condition of an open system in which there are no changes over the longer term, but in which there may be oscillations in the very short term. There are continuing inputs and outputs of matter and energy, but the system as a whole remains in a more or less constant state (for example, a climax ecosystem)
- **Feedback** – when part of the output from a system returns as an input, so as to affect subsequent outputs
- **Positive feedback** – feedback that increases change; it promotes deviation away from an equilibrium and drives the system towards a tipping point where a new equilibrium is adopted
- **Negative feedback** – feedback that tends to counteract any deviation from equilibrium, and promotes stability
- **Tipping point** – the minimum amount of change within a system that will destabilize it, causing it to reach a new equilibrium or stable state

You should be able to show the role of energy and equilibria in the regulation of systems:

- ✔ The first law of thermodynamics concerns the conservation of energy, and the second law explains the inefficiency and decrease in available energy in systems.
- ✔ Negative feedback stabilizes systems whereas positive feedback moves them further away from equilibrium.
- ✔ The resilience of a system is its tendency to maintain stability, thereby avoiding tipping points where a new state of equilibrium is reached.
- ✔ The size of storages within systems, and their diversity, contributes to the resilience of systems.
- ✔ In a stable equilibrium, there is a tendency in ecological systems to return to the original equilibrium following disturbance, and in steady-state equilibrium there are fluctuations around equilibrium but with no overall change.

Assessment tip

You will need to understand the relationships between resilience, stability, equilibria and diversity – emphasis should be put on these interrelationships as you study this part of the course.

The laws of thermodynamics

The first law of thermodynamics is known as the law of conservation of energy. The first law of thermodynamics states that energy cannot be created or destroyed: it can only be changed from one form into another. This means that the total energy in any system is constant and all that can happen is that energy can change form.

The second law of thermodynamics states that the transfer of energy through a system is inefficient and that energy is transformed into heat. This means that less energy is available to do work and the system becomes increasingly disordered. In an isolated system, entropy increases spontaneously.

Assessment tip

The first law of thermodynamics (the law of conservation of energy) states that energy entering a system equals energy leaving it (energy can neither be created nor destroyed), whereas the second law states that energy in systems is gradually transformed into heat energy due to inefficient transfer, increasing disorder (**entropy**).

The implications of the laws of thermodynamics for ecological systems

Energy is needed in ecosystems to create order, such as to hold complex molecules together. Natural systems cannot be isolated because there must always be an input of energy for work to replace energy that is dissipated.

The maintenance of order in living systems needs a constant input of energy to replace that lost as heat through the inefficient transfer of energy. One way energy enters an ecosystem is as sunlight. Transfer of energy by producers is inefficient due to the inefficient transfer of energy in photosynthesis (the second law of thermodynamics). Sunlight energy is changed into biomass by photosynthesis, i.e. photosynthesis captures sunlight energy and transforms it into chemical energy. Chemical energy in producers may be passed along food chains as biomass, or given off as heat during respiration. The energy entering the system equals the energy leaving it (the first law of thermodynamics).

Available energy is used to do work such as growth, movement, and making complex molecules. The transformation and transfer of usable energy is not 100% efficient (the second law of thermodynamics); whenever energy is converted there is less usable energy at the end of the process than at the beginning. This means that there is a dissipation of energy which is then not available for work. The total amount of energy in a system does not change but the amount of available energy for work does change.

All energy eventually leaves the ecosystem as heat. No new energy has been created; it has simply been transformed and passed from one form to another. Heat is released because of the inefficient transfer of energy. Although matter can be recycled, energy cannot, and once it has been lost from a system in the form of heat energy, it cannot be made available again. Because the transfer and transformation of energy are inefficient, food chains tend to be short.

Equilibrium

There is a tendency in systems to return to the original **equilibrium**, rather than adopting a new one, following disturbance. In forests, for example, insect populations increase and decrease and trees die and grow, but overall the forest remains the same (this is known as **steady-state equilibrium**; see figure 1.3.3). In these cases, disturbance will lead to a return to the original equilibrium (it is said to be "stable"; figure 1.3.1). When a system adopts a new equilibrium following disturbance this is known as **"unstable" equilibrium** (see figure 1.3.2).

- **Resilience** – the tendency of an ecological or social system to avoid tipping points and maintain stability
- **Diversity** – can be defined as "the variety of life", although the meaning depends on the context in which it is used (i.e. can refer to species, habitat or genetic diversity)

Assessment tip

You need to be able to explain the implications of the first and second law of thermodynamics for ecological systems.

Assessment tip

Light energy starts the food chain but is then transferred from producer to consumers as chemical energy.

Assessment tip

You need to be able to apply the first and second laws of thermodynamics to energy transformations and the maintenance of order in living systems.

Concept link

EQUILIBRIUM: This is a key concept and as such it is explored through the ESS syllabus.

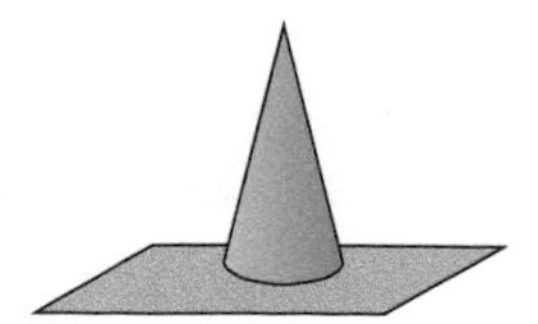

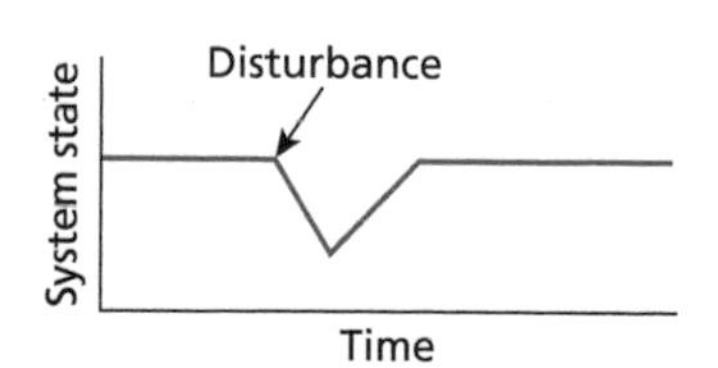

Figure 1.3.1 Stable equilibrium (where disturbance to the system results in it returning to its original equilibrium)

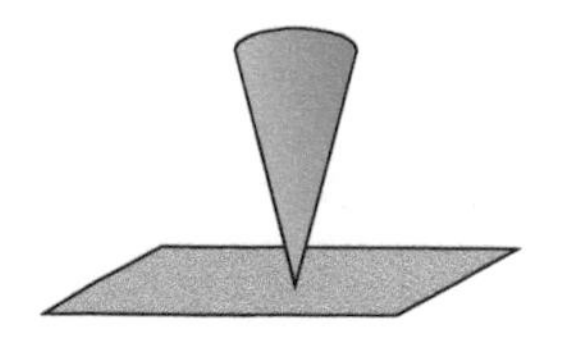

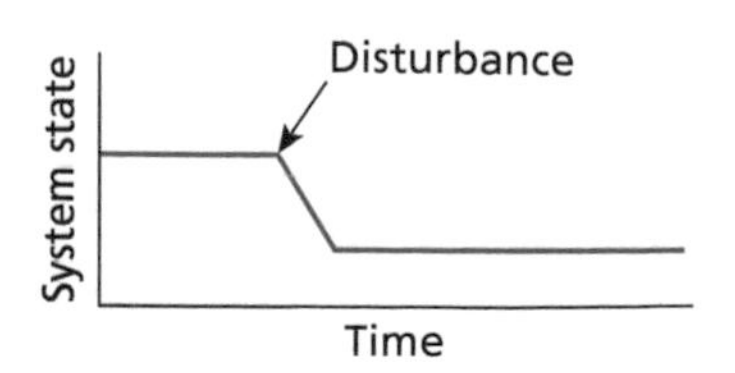

Figure 1.3.2 Unstable equilibrium (where disturbance results in a new equilibrium very different from the first)

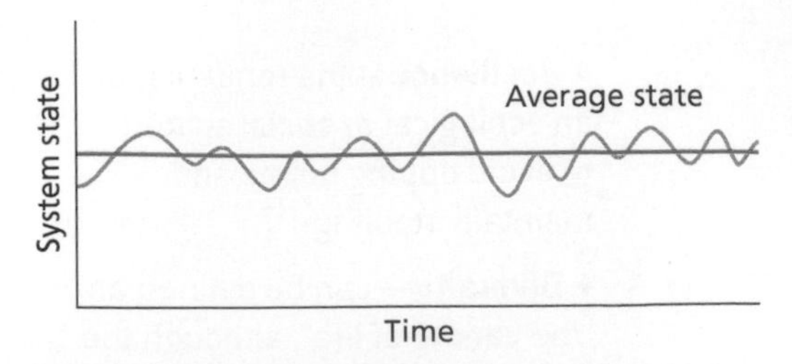

▲ Figure 1.3.3 Steady-state equilibrium, the common property of most open systems in nature. Fluctuations in the system are around a point of equilibrium, and deviation above or below this point results in a return towards it

Negative and positive feedback

Systems have **feedback** mechanisms to maintain equilibrium, thereby balancing inputs and outputs. Such feedback is known as **negative feedback**. For example, in a community of trees in a woodland (see section 1.2): increase in trees (through reproduction) ➔ more competition for light ➔ fewer viable seedlings/more tree deaths ➔ decrease in trees.

Following a move away from equilibrium, negative feedback mechanisms return the system to its original equilibrium.

The concept of negative feedback can be illustrated as an annotated diagram (figure 1.3.4). The diagram shows that equilibrium is promoted by counteracting any deviation from equilibrium, and promotes stability.

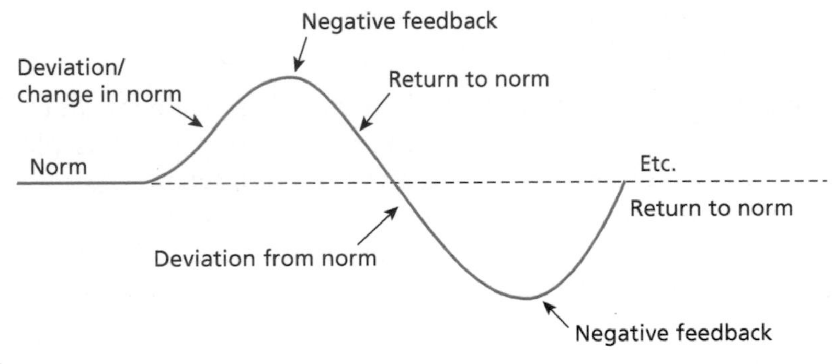

▲ Figure 1.3.4 Negative feedback

Another example of a negative feedback is the predator–prey relationship between snowshoe hare and lynx in the boreal forest of North America, which can be illustrated as shown in figure 1.3.5.

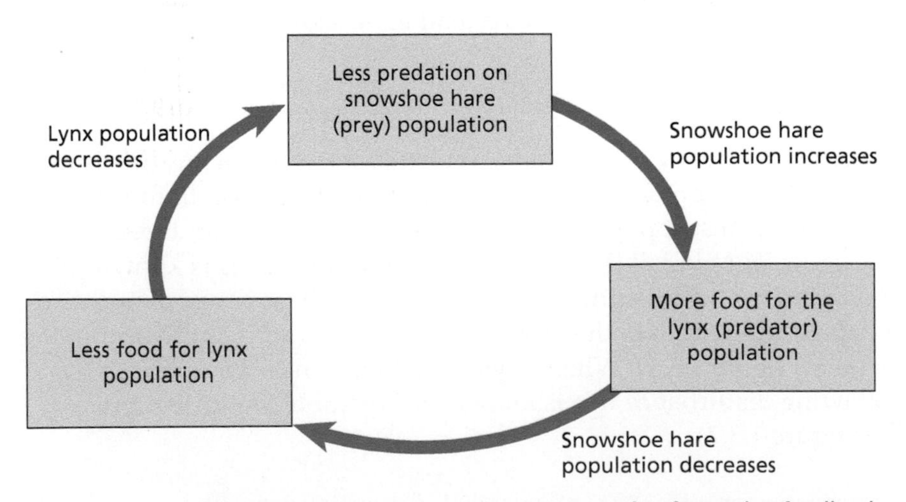

▲ Figure 1.3.5 Predator–prey relationship – an example of negative feedback

In contrast, **positive feedback** mechanisms move the system further away from equilibrium. An example of a positive feedback loop that enhances climate change is shown in figure 1.3.6.

Some systems may show both positive and negative feedback, e.g. human population growth, where more people on the planet produce more children (positive feedback) but at the same time this leads to reduced resource availability, increased mortality and decreased survivorship (negative feedback).

The role of self-regulation in natural systems using feedback systems can be applied to different examples, such as climate change and predator–prey relationships.

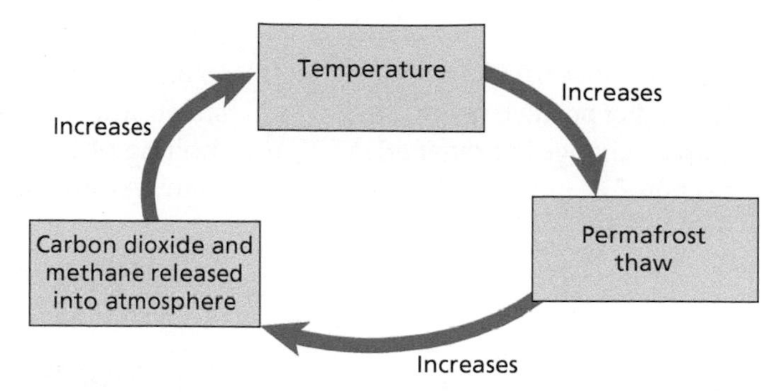

▲ Figure 1.3.6 Enhanced climate change – an example of positive feedback

Test yourself

1.6 Distinguish between negative and positive feedback using examples from environmental systems. [4]

Tipping points

A **tipping point** is a degree of change within a system that will destabilize it, causing it to adopt a new equilibrium. They can be seen as a critical threshold when even a small change can have dramatic effects and completely alter the state of a system, causing extreme outcomes, from sea level rise to widespread drought. Positive feedback loops tend to drive a system towards a tipping point where a new equilibrium is adopted (see above). Most anticipated tipping points are linked to climate change (figure 1.3.7).

Assessment tip

You need to be able to evaluate the possible consequences of tipping points.

Content link

The causes and impacts of climate change are discussed in section 7.2.

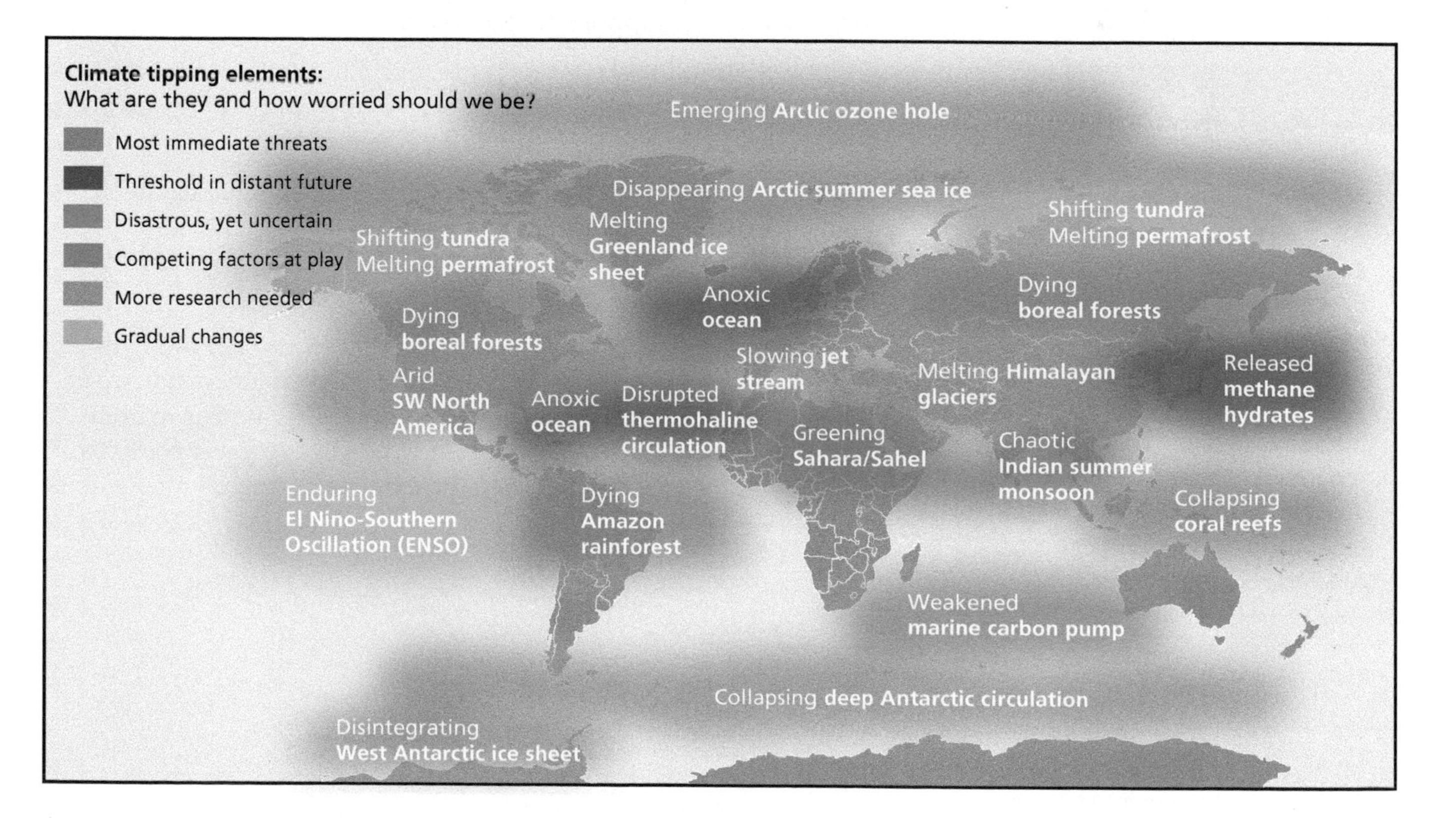

▲ Figure 1.3.7 Potential tipping points and the likelihood of them occurring

Content link

The role of positive feedback in climate change is examined in section 7.2.

For example, increases in CO_2 levels above a certain value would lead to an increase in global mean temperature, causing melting of the ice sheets and permafrost. Reaching such a tipping point could cause the melting of Himalayan mountain glaciers and a lack of freshwater in many Asian societies.

Test yourself

1.7 Outline what is meant by a tipping point. [1]

1.8 Suggest one social and **one** ecological impact that might arise from the equilibrium shift in the Great Barrier Reef system. [2]

Assessment tip

You need to be able to discuss resilience in a variety of systems, both ecological and societal.

Resilience and diversity of systems

Systems that have complex interrelationships between the different components are more stable and therefore less likely to reach a tipping point than simpler systems (figure 1.3.8). For example, when considering an ecosystem:

- If there are several food chains that contain many interconnected food webs then it is inherently more stable than one with fewer food chains because loss of individual species will leave many more food chains to support the overall structure of the system.
- The greater the **diversity**, the greater the **resilience** of the system. Greater diversity implies more complexity (i.e. many niches), which leads to a greater resistance to change.
- Similarly, systems with larger storages, e.g. natural resources (biomass, nutrients, water, etc) will be more resilient than those with smaller storages. Large storages confer abundant supplies of key resources to support the ecosystem.
- Mature communities, such as those found in climax communities, will have larger storages and more developed nutrient cycles and food webs, adding to the resilience of the ecosystem.
- The presence of negative feedback mechanisms increases the resilience of systems and returns them towards equilibrium. Steady-state equilibrium, where there is a balance of inputs and outputs, will lead to stability.

Content link

Food chains, and how they can be used as a model to indicate flows of energy and matter in ecosystems, are discussed in section 2.2.

Content link

Climax communities are discussed in section 2.4.

Assessment tip

Human threats themselves do not affect resilience directly, because resilience is the inherent property of the system to resist threats (tied up in its storages, diversity and so on) and the degree of threat will not change this. However, human activities that diminish the inherent resilience of the system by reducing its storages and diversity will change the resilience of the system.

Human activities which reduce the size of storages (e.g. removing timber from forest ecosystems) or reducing the complexity of interactions, e.g. food webs, will reduce the resilience of a system. This applies to both ecological and social systems. Social systems that have fewer complex connections are less stable than those that have robust social interactions.

Assessment tip

You should be able to refer to examples of human impacts and relate these to possible tipping points.

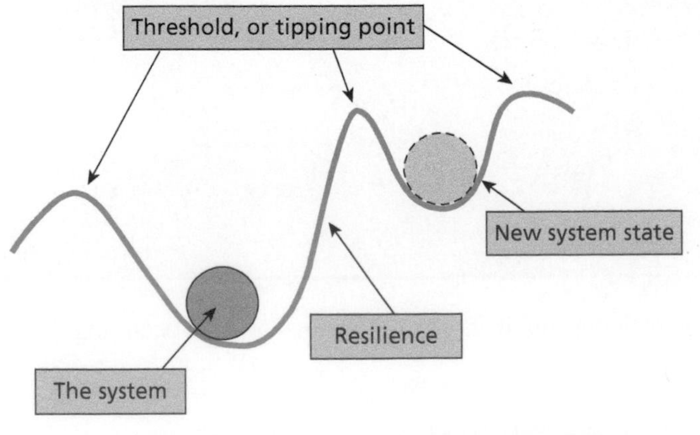

▲ Figure 1.3.8 The role of resilience in avoiding tipping points

QUESTION PRACTICE

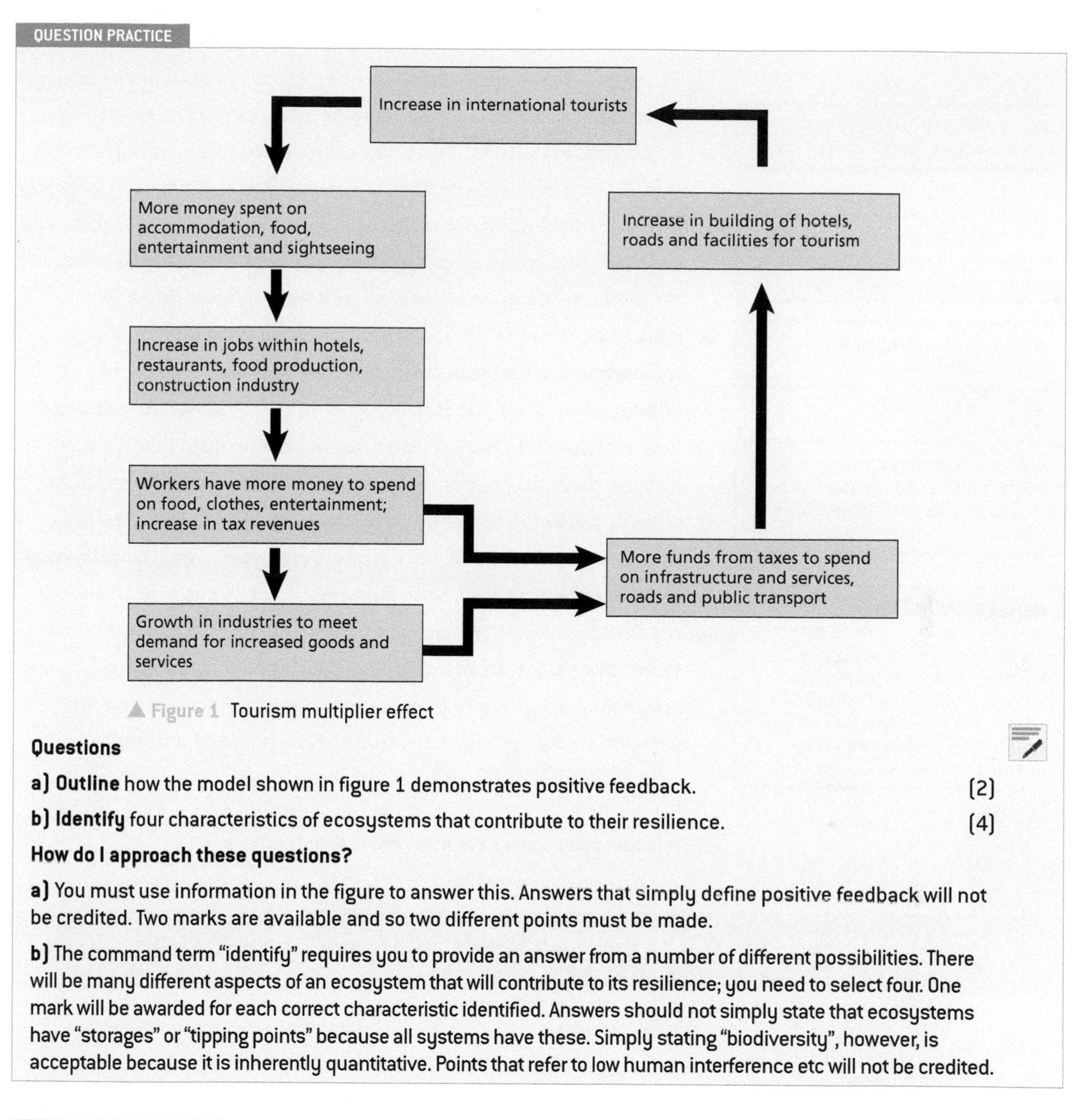

▲ Figure 1 Tourism multiplier effect

Questions

a) **Outline** how the model shown in figure 1 demonstrates positive feedback. [2]

b) **Identify** four characteristics of ecosystems that contribute to their resilience. [4]

How do I approach these questions?

a) You must use information in the figure to answer this. Answers that simply define positive feedback will not be credited. Two marks are available and so two different points must be made.

b) The command term "identify" requires you to provide an answer from a number of different possibilities. There will be many different aspects of an ecosystem that will contribute to its resilience; you need to select four. One mark will be awarded for each correct characteristic identified. Answers should not simply state that ecosystems have "storages" or "tipping points" because all systems have these. Simply stating "biodiversity", however, is acceptable because it is inherently quantitative. Points that refer to low human interference etc will not be credited.

SAMPLE STUDENT ANSWER

a) It shows positive feedback as the increase in tourists continues to grow each time, as more money and jobs mean more industry growth and tax revenue. This revenue is used to improve infrastructure and attracts more tourists, leading to more tourists coming rather than a return to the original number of tourists (equilibrium). It results in a new number of tourists (each cycle/time) and so a new equilibrium is reached.

▲ A valid comment, based on the figure, about how positive feedback is illustrated

▲ A second point is made, leading on from the first, and so a second mark is awarded. This shows good exam technique - by breaking the answer down into two separate aspects of the positive feedback illustrated in the figure, full marks are achieved

This response could have achieved 2/2 marks.

b) The resilience of an ecosystem refers to its ability to return to its original state after a change or disturbance and thus avoid tipping points. First, the stage of succession influences this ability, as mature climax communities have more complex food webs and nutrient recycling, and thus are more stable and resilient than pioneer ecosystems in early stages of succession. Second, the presence of limiting factors is important for resilience, since if limiting factors such as water, sunlight, temperature, nutrients and space are abundant, even if one of them decreases, the ecosystem can still maintain stability and resilience as many other resources remain. Third, a high habitat diversity providing many habitats and niches leads to high species diversity, meaning even if one prey species is removed, predators can still feed on other prey species, allowing for more complex food webs and resilience of the ecosystem to the disappearance of a species. Lastly, high genetic diversity from large populations also contributes to ecosystem resilience, since diversity in gene pools means a lower chance of a whole population being wiped out due to the spread of a disease.

▲ 1 mark awarded for referring to the maturity of the community

▲ 1 mark for referring to the complexity of interactions and linkages between components

▲ 1 mark for referring to abundance of resources

▲ 1 mark for mentioning that biodiversity is an important factor in determining the resilience of a system

▲ This point has already been covered by the discussion on diversity. There is only one mark available for mentioning biodiversity (genetic/species/habitat diversity), and the maximum 4 marks have been awarded above

This response could have achieved 4/4 marks.

1.4 SUSTAINABILITY

- **Sustainability** – the use and management of resources that allows full natural replacement of the resources exploited and full recovery of the ecosystems affected by their extraction and use
- **Natural capital** – a term used for natural resources that can produce a sustainable natural income of goods or services
- **Natural income** – the yield obtained from natural resources
- **Goods** – marketable commodities exploited by humans

You should be able to show an understanding of how sustainability and sustainable development can be achieved and monitored

✔ Sustainability refers to the use of resources at a rate that allows for natural regeneration and minimizes damage to the environment.

✔ Ecosystems provide life-supporting services and goods.

✔ Natural capital is natural resources that produce a sustainable natural income, where natural income refers to the yield obtained from natural capital.

✔ Factors such as pollution, biodiversity, population or climate can be used as environmental indicators to assess sustainability.

✔ The Millennium Ecosystem Assessment (MA) plays a role in assessing the world's ecosystems and the services, and ways in which they can be conserved and used sustainably.

✔ Environmental Impact Assessments (EIAs) provide information on the environmental effects of development projects.

✔ Ecological footprints (EFs) provide an index to measure sustainability.

Sustainability

Sustainability is the responsible use and management of resources that allows natural regeneration and minimizes environmental damage. It involves the use of renewable resources and the sustainable use of **natural capital**. Whether a country has sustainable development policies is usually evident in how it deals with its energy requirements.

Concept link

SUSTAINABILITY is a key concept and as such it is addressed throughout the course.

A country could be seen as pursuing sustainable development if:

- renewable energy accounts for a very high percentage of energy consumption and there is a reduced reliance on fossil fuels that produce greenhouse gases
- the needs of the present are met without compromising the needs of future generations
- it is taking steps to address soil degradation
- it is attempting to remove invasive species
- it is promoting sustainable fishing techniques that lead to an increase in wild fish populations
- it is using the profits made from tourism to protect wildlife
- areas of wilderness are promoted.

A country could be seen as **not** pursuing sustainable development if:

- wild species are harvested, whose numbers are in decline
- hunting of species in decline or under threat is legalized
- it has a very high **ecological footprint** (see page 22)
- it is home to industries that produce lots of greenhouse gases
- there are high rates of soil degradation
- there has been a large loss of original ecosystems
- invasive species are not fully controlled and so they are a threat to native species and habitats
- mass tourism is leading to stress to wildlife, and that is reducing reproductive success or causing injury
- the government has approved oil exploration, increasing the risk of oil pollution and the further emissions of greenhouse gases
- the population has grown significantly, resulting in greater use of natural resources, resulting in overfishing and overgrazing, for example
- the energy industry and industrial processes are emitting substantial amounts of greenhouse gases.

Natural capital and natural income

Natural resources, such as forest ecosystems, can provide both **goods** and **services**. Goods are tangible products, such as timber. Services include water replenishment, flood and erosion protection, and climate regulation, which are life-supporting but often difficult to value economically.

- **Services** – natural processes that provide a benefit to the human environment
- **Environmental Impact Assessment (EIA)** – a detailed survey that provides decision-makers with information in order to consider the environmental impact of a project. The assessment should include a baseline study to measure environmental conditions before development commences, and to identify areas and species of conservation importance. Monitoring should continue for some time after the development
- **Ecological footprint (EF)** – the area of land and water required to sustainably provide all resources at the rate at which they are being consumed by a given population

Assessment tip

The concept of sustainability should be used throughout the course, where appropriate.

Content link

Ecological footprints are used to give you a sense of your own impact at the start of the course and are addressed in more detail in section 8.

> **Assessment tip**
>
> You need to be able to explain the relationship between natural capital, **natural income** and sustainability.

For example, the potential ecological services and goods provided by a coral reef ecosystem in Australia would include the following.

Goods:

- Sustainably sourced coral rubble provides building materials.
- Reef fish provide food to local communities.
- Plants and animals provide medicines such as cancer treatments.

Services:

- Coral reefs help to protect the shore from erosion.
- They are a nursery for many ocean species, helping to maintain global biodiversity.
- Coral reefs provide amenity value and act as an ecotourism attraction which can be used to fund conservation efforts and to provide a sustainable income to local communities.
- Coral reefs surround many small island nations containing indigenous peoples and so provide cultural and spiritual value.

The goods and services from a tropical forest ecosystem in Borneo include those shown in table 1.4.1.

▼ Table 1.4.1 Goods and services from a tropical forest ecosystem in Borneo

Goods	Services
Timber/wood	Vegetation and trees to prevent soil erosion
Plants and animals for food	Absorption of CO_2 (i.e. carbon sink)
Plant extracts	Aesthetic quality/value
Medicines	

Natural capital is resources that can be utilized to produce a sustainable natural income. Figure 1.4.1 shows the sustainable and unsustainable use of a forest ecosystem to produce goods (i.e. timber). Part (a) shows that forest growth is harvested whilst the remaining storage (or stock) remains. The original stock (natural capital) is not depleted and so continued harvesting is sustainable. Part (b) shows the storage (i.e. the forest) being depleted year on year, and this leads to a reduction in natural capital.

> **Assessment tip**
>
> If asked to discuss the potential ecological services and goods provided by a named ecosystem, make sure that you give a valid, specific, named example, such as a tropical forest in Amazonia. The range of goods and services provided should be discussed in the context of the named ecosystem. The goods and services must be for human use, and so "biodiversity" would be too vague unless linked to medicines, food or tourism for example.

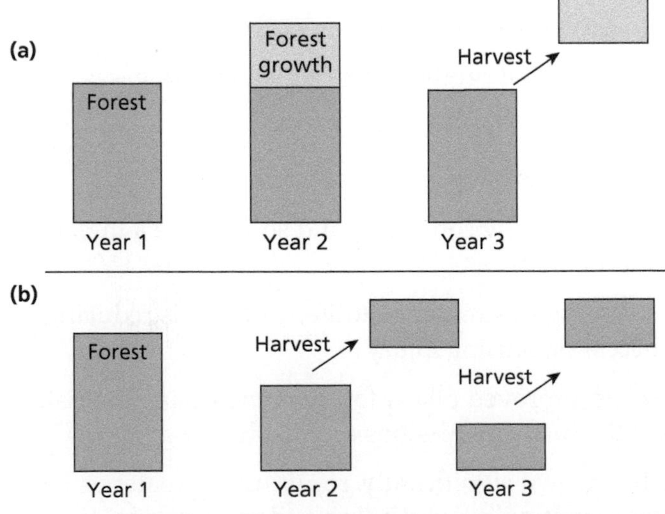

▲ Figure 1.4.1 (a) The sustainable use of a forest ecosystem to provide natural income (i.e. timber). (b) Unsustainable harvesting

> **Test yourself**
>
> **1.9 Explain** how the concepts of natural capital and natural income are useful models in managing the sustainable exploitation of a resource. [4]

Ecosystem services

There are three types of ecosystem services.

Supporting services:

- These are the essentials for life and include primary productivity, soil formation and the cycling of nutrients.
- All other ecosystem services depend on these.

> **Assessment tip**
>
> You need to be able to discuss the value of ecosystem services to a society.

Regulating services:

- These are a diverse set of services and include pollination, regulation of pests and diseases, and production of goods, such as food, fibre and wood.
- Other services include climate and hazard regulation and water quality regulation.

Cultural services:

- These are derived from places where people interact with nature, providing cultural goods and benefits.
- Open spaces – such as gardens, parks, rivers, forests, lakes, the seashore and wilderness – provide opportunities for outdoor recreation, learning, spiritual well-being and improvements to human health.

Test yourself

1.10 State two regulating services that mountains, moorlands and heaths provide. [2]

1.11 Suggest two likely cultural services supplied by urban ecosystems. [2]

1.12 State supporting services from enclosed farmland. [2]

Indicators of sustainability

It is possible to assess whether a country or society is functioning in a sustainable way by using specific indicators. These indicators provide a measure by which a society can be compared to others. They can also be used to assess whether the country or society is following a pathway of sustainable development.

The Millennium Ecosystem Assessment (MA)

In 2000 the UN initiated a consultation exercise that ultimately led to the launch of the Millennium Ecosystem Assessment (MA) in 2001. It was a large study to assess knowledge in this area and to reach agreement, involving both social and natural scientists.

The aims of the MA were to improve the decision-making process relating to ecosystem management, with a view to improving human well-being. The MA was also developed to inform and improve future scientific assessments of this kind.

The main findings of the MA were as follows:

- Humans have changed ecosystems more rapidly and extensively in 50 years between 1950 and 2000 than in any similar period in human history. The changes are primarily due to meeting rapidly growing demands for resources such as food, fresh water, timber and fuel. The effect of this disturbance has led to a large, irreversible loss of biodiversity.
- Economic development and human well-being resulting from ecosystem change have been achieved at the cost of degradation of many ecosystem services, increased risk of tipping points being reached, and the increase of poverty for some individuals or societies. Unless these problems are addressed, there will be a significant reduction in the benefits future generations can obtain from ecosystems.

Content link

The United Nations identified the need to coordinate countrywide approaches to sustainable development at the 1992 Rio Earth Summit. The Rio Earth Summit is discussed in section 1.1.

- Significant changes to institutions, practices and policies will be needed to address the challenges highlighted by the MA (i.e. reversing ecosystem degradation while meeting the increased needs for global resources). There are many options for enhancing or conserving ecosystem services in ways which need to be considered and implemented.

Assessment tip

You need to be able to discuss how environmental indicators such as MA can be used to evaluate the progress of a project to increase sustainability.

Environmental Impact Assessment (EIA)

Assessment tip

You need to be able to evaluate the use of EIAs in assessing sustainable development.

An **Environmental Impact Assessment (EIA)** is carried out before any major development project. An EIA is an evaluation of the current ecosystem or environment and likely impacts from the development. An EIA estimates change to the environment that occurs as a result of a project, and helps to decide whether the advantages outweigh the disadvantages.

An EIA is designed to protect the local environment in the following ways.

Baseline study:

- This provides an inventory of social and cultural aspects, keystone and red-listed species, and unique habitats that are of particular value.
- It can help to focus, organize and prioritize protective strategies.

Content link

The concepts of red-listed species and keystone species are covered in sections 3.3 and 3.4 respectively.

Assessment of potential social and ecological impacts and benefits:

- This provides a holistic evaluation, i.e. taking all factors into account.
- It allows input from all stakeholders.
- By comparing the development to similar projects (already executed), the validity of the EIA is increased.

Assessment tip

You are not expected to explore an environmental impact assessment (EIA) in depth, but rather to focus on the principles of their use.

Recommendations/mitigations:

- These are designed to limit impact and protect the environment.
- The EIA report is a public document, so the public can respond to the report, leading to recommendations.
- This recommends changes to the development, with modified construction techniques, which mitigate potentially damaging environmental impacts, for example.

Test yourself

1.13 Outline possible limitations of an Environmental Impact Assessment (EIA). [4]

In some countries the EIA is advisory while in others it is compulsory and so may determine the implementation of mitigation strategies. It is dependent on effective enforcement.

Ecological footprint (EF)

Content link

The ecological footprint is discussed in detail in section 8.4

An ecological footprint is the area of land and water required to support a defined population at a given standard of living. Ecological footprints greater than the biocapacity of a country (i.e. the ability of a biologically productive area to generate sustainable supply of resources) indicate unsustainability.

Assessment tip

You need to be able to explain the relationship between ecological footprint and sustainability.

Test yourself

1.14 Compare ecocentric and technocentric approaches to reducing an ecological footprint. [4]

1.15 Explain how developments in technology may increase or decrease the ecological footprint of a human population. [4]

QUESTION PRACTICE

- Land area of 103 000 km^2.
- Terrain is mountainous and volcanic.
- Isolated island so biological diversity is low, and there are few endemic species.
- Only 0.7% of land is suitable for growing crops, and harsh climate means farming is limited to livestock and geothermally heated greenhouses.
- 60% of population lives in the capital city Reykjavik.
- Total fertility rate is two children per woman.
- Important industries include fishing, aluminium smelting and tourism
- Ecological footprint is 7.4 GHa compared to a world average of 2.6 GHa.
- A representative democracy and high income country, ranked 13th highest on the human development index.
- Badly affected by the global financial crisis in 2008.
- Hydroelectric and geothermal power sources provide 85% of primary energy.
- Expects to be energy-independent, using 100% renewable energy by 2050.
- Government recently approved oil exploration in Icelandic waters by oil companies.

▲ Figure 1 Fact file on Iceland

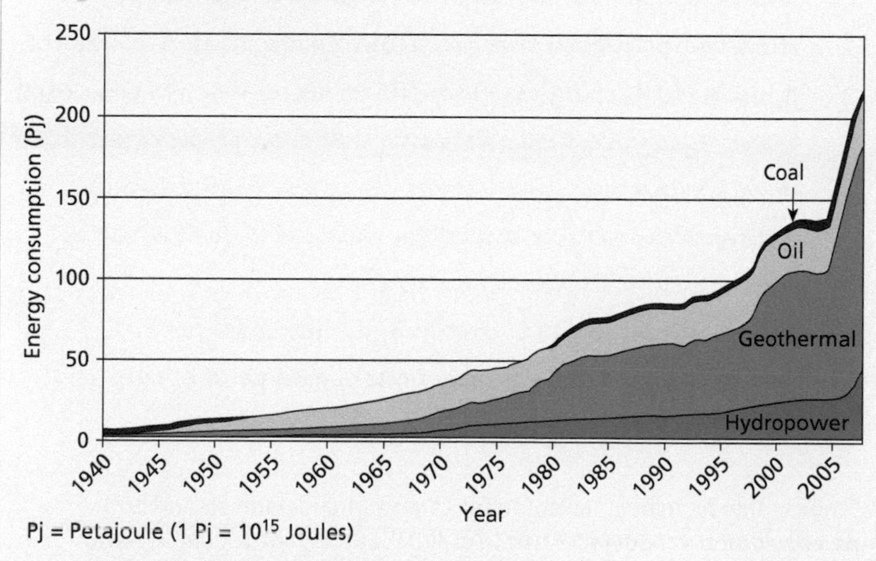

◀ Figure 2 Graph showing primary energy consumption in Iceland 1940–2008

- Estimated worldwide population of twelve million.
- 60% of the world's puffins live in Iceland.
- Puffins lay their eggs in burrows on cliffs in June–July, one egg per year.
- Adult puffins bring small fish to their young.
- Classified as "vulnerable" on International Union for Conservation of Nature (IUCN) red list.
- Current population in decline.
- Threats to puffins include overfishing, native predators such as foxes and gulls, introduced predators such as cats, hunting and egg collection by humans, oil spills, extreme weather, and disturbance from tourists.
- Puffins can be hunted legally in Iceland in April by a technique called "sky fishing", which involves catching low-flying birds with a big net. Their meat and eggs are commonly featured on hotel menus.
- Puffin populations are affected by extreme weather events and changes in availability of food.

▲ Figure 3 Fact file on the Atlantic puffin

Question

With reference to figures 1, 2 and 3, **to what extent** might Iceland be viewed as a role model for sustainability by other countries? [6]

How do I approach this question?

The command term "to what extent" means that evidence needs to be considered that is in favour of, and against, an argument. A conclusion should be given that is supported by the evidence provided in the answer. An answer could begin "It is a model for sustainability because" followed by valid points, and then a section that begins "It is not a model for sustainability because" followed by relevant points. A conclusion at the end of the answer should give a balanced view of what the evidence presents (i.e. whether or not Iceland can be viewed as a role model for sustainability). If there is no conclusion, then a maximum of 5 marks can be awarded.

SAMPLE STUDENT ANSWER

Iceland can be viewed as a role model to some extent because:

1. Its primary energy sources mainly come from hydroelectric and geothermal power sources (85%), which are renewable.
2. It is energy independent, expecting to use 100% of renewable energy by 2050.
3. Not all areas for hydroelectric power stations are being used, because energy is meeting the demand, and so there is less disturbance to the environment.

However, some of its practices are less sustainable, as follows:

1. Recently approved oil exploration in Icelandic waters, releasing pollution and disturbing natural environments.
2. It has a high ecological footprint at about ~~7.4GHa~~ compared to average of 2.6GHa, meaning that their people create a lot of waste and consume a lot of resources unsustainably.
3. It hunted vulnerable species for meals and food which is non-ethical and unsustainable to do.

In conclusion, Iceland is potentially a role model for sustainability, but some of its practices and policies can be improved.

▲ 1 mark awarded for correctly saying that renewable energy is a major component of energy generation

▼ This is not saying anything in addition to the first point, and so no further marks are awarded here

▲ The question asks "to what extent" which means providing arguments for and against the proposal (i.e. in this case that Iceland can be considered a role model for stainability). Putting "however" here clearly indicates that both points of view are being provided

▲ Valid point why Iceland may not be considered a role model of sustainability

▲ A second reason against the proposal in the question. It is good practice to quote figures provided in data etc. to support your case, as shown here

▲ A third valid point about why Iceland is not a model for sustainability

▲ In a "to what extent" question you need to provide an overall conclusion supported by evidence given in the answer. Here a short but reasonable conclusion is provided. 1 mark awarded

Three marks have been awarded for saying why Iceland should not be considered a model for sustainability, but only one for saying why it should be considered a model. One mark has been awarded for a valid conclusion. Either one further point in favour of the statement, or one against, would have secured full marks for this question.

This response could have achieved 5/6 marks.

1.5 HUMANS AND POLLUTION

- **Pollution** – the addition of a substance or an agent to an environment through human activity, at a rate greater than that at which it can be rendered harmless by the environment, and which has an appreciable effect on the organisms in the environment

You should be able to show an understanding of how human disturbance produces pollution and ways in which it can be managed

- ✔ Pollution is the addition of a harmful substance to an environment at a rate greater than that at which it can be removed, and which has a noticeable effect on the organisms within the environment.
- ✔ Non-point pollution is emitted from a variety of disparate sources whereas point source pollution is released from a clearly definable source.
- ✔ Persistent pollution is not broken down naturally and remains in food chains, whereas biodegradable pollution can be broken down and removed from the ecosystem.

- Acute pollution operates in the short term and has immediate serious effects whereas chronic pollution acts over a longer period of time.
- Primary pollution is active on emission whereas secondary pollution arises from primary pollutants that have undergone change.
- Systems diagrams can be used to show the causes and effects of pollution.
- There are three different levels of intervention that are used to manage pollution: altering human activity, controlling the release of a pollutant, and cleaning up pollution and restoring damaged systems.
- Dichlorodiphenyltrichloroethane (DDT) is an example of how the effects of a "pollutant" on the environment can be in conflict with beneficial uses.

- **Non-point source pollution** – pollution that arises from numerous widely dispersed origins
- **Point source pollution** – pollution that arises from a single clearly identifiable site
- **Persistent organic pollutants (POPs)** – organic compounds that are difficult to break down
- **Biodegradable** – capable of being broken down by natural biological processes
- **Primary pollution** – a pollutant that is active on emission
- **Secondary pollution** – pollution arising when primary pollutants undergo physical or chemical change
- **Acute pollution** – pollution that produces its effects through a short, intense exposure, where symptoms are usually experienced within hours
- **Chronic pollution** – pollution that produces its effects through low-level, long-term exposure, and where disease symptoms develop up to several decades later

Assessment tip

The terms "pollutant" and "contaminant" in environmental chemistry mean the same thing.

Introduction to pollution

Pollution is the addition of a harmful substance to an environment at a rate greater than that at which it can be removed, and which has a noticeable effect on the organisms within the environment. It may be used quantitatively as an environmental indicator of sustainability.

Types of pollution are related to human activities, including acidification of forests and buildings, and eutrophication of streams and ponds.

Content link

The causes, effects and management of pollution are important principles that are explored in several sections, such as 3.3, 4.4, 6.3, 6.4 and 7.3.

Assessment tip

Pollutants only become pollutants when there is too much. For example, not all fertilizer or manure causes pollution. Spreading manure on fields does not cause pollution if the amount spread can be used by plants.

Content link

Acid deposition is explored in section 6.4.

Non-point and point source pollution

Non-point source pollution is the release of pollutants from numerous, widely dispersed origins. In contrast, **point source pollution** is the release of pollutants from a single, clearly identifiable site. Point source pollution is easier to manage and clean up than non-point source pollution because its origin can be identified. An example of point source pollution would be pollution from an outflow pipe, whereas an example of non-point source pollution might be emissions from vehicles.

Content link

Eutrophication is covered in section 4.4.

Persistent and biodegradable pollution

POPs are **persistent organic pollutants** that are resistant to environmental degradation. They can therefore accumulate in food chains. Once chemicals enter food chains, the top predators are often at extra risk because of the biomagnification effects of some chemicals. In contrast, **biodegradable** pollution is able to be broken down by organisms and so does not persist in food chains.

Most modern pesticides, used to treat crops so as to ensure maximum yield, are biodegradable (e.g. Bt proteins, which are toxic to insects, are rapidly decomposed by sunlight), although earlier chemicals were persistent (e.g. DDT).

Acute and chronic pollution

The effect of, for example, UV (ultraviolet) radiation, due to a reduction of the ozone layer as a result of pollution from CFCs, may be acute or chronic.

- The effects of UV radiation on the eye may be acute, and temporary blindness can occur. Chronic effects may be irreversible, leading to the development of cataracts and eventually blindness.
- Acute exposure of the skin to UV radiation can cause mutations during cell division and sunburn. In the long term, skin cancers can result when mutated or damaged cells begin uncontrolled division and invade other areas (an effect called metastasis). Chronic exposure of the skin to UV-B radiation also causes wrinkling, thinning and loss of elasticity.

In terms of air pollution, acute effects include asthma attacks. Chronic effects include lung cancer, chronic obstructive pulmonary disease (COPD) and heart disease.

Primary and secondary pollution

A **primary pollutant** is one which is active on emission and directly impacts the environment. For example, CO_2 is released from burning fossil fuels and actively contributes to global warming; CFCs are released from aerosols and actively contribute to ozone depletion.

A **secondary pollutant** is one that is formed from a primary pollutant through physical or chemical change. For example CO_2 combines with seawater to form carbonic acid that has an impact on calciferous shelled organisms or corals. NO_x combines with water to form acid precipitation. NO_2 forms ozone that contributes to photochemical smog.

> **Assessment tip**
>
> When providing examples of primary pollutants you need to include their direct impact. For example, NO_x can be either primary or secondary without such specification. Examples of secondary pollutants need to include the process leading to their pollutionary impact.

Systems diagrams: the impact of pollutants

Systems diagrams can be used to show the effect of pollution on natural systems. For example, the impact of nitrates leaching into a body of water is shown in figure 1.5.1.

> **Assessment tip**
>
> You need to be able to construct systems diagrams to show the impact of pollutants.

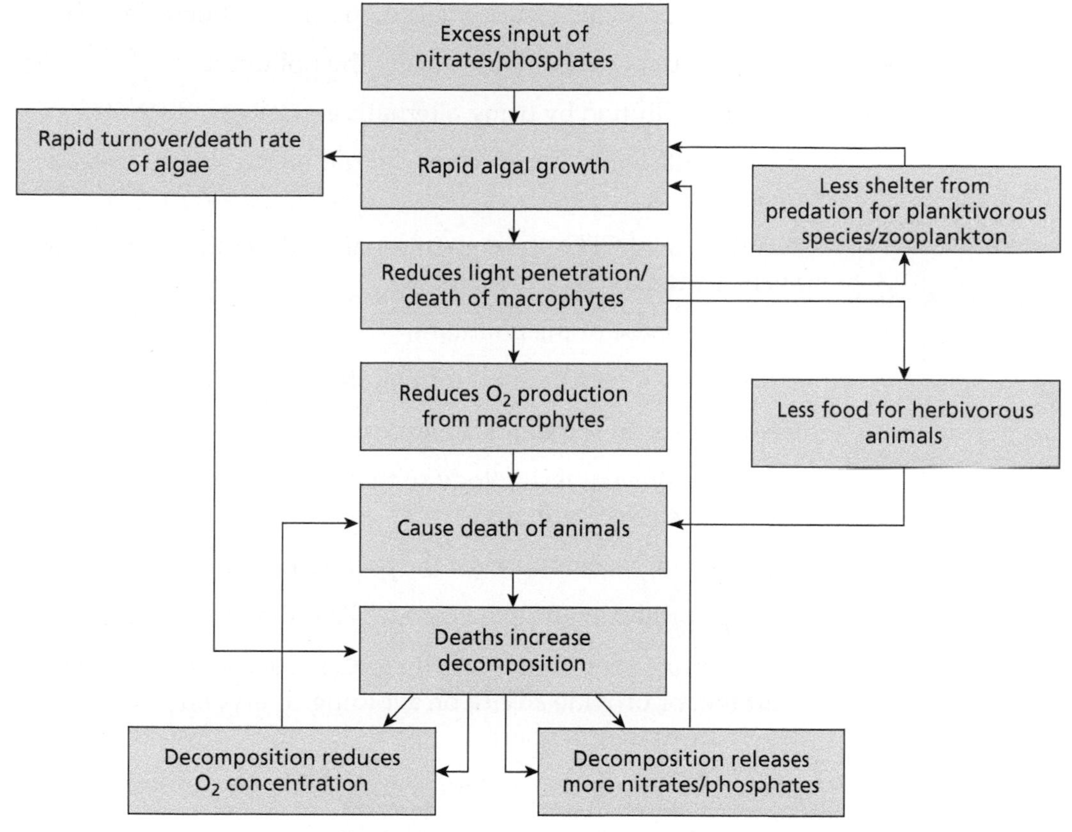

▲ Figure 1.5.1 System diagram showing the role of positive feedback mechanisms in affecting the equilibrium of an aquatic ecosystem during the process of eutrophication

Levels of intervention

There are different ways to approach the management of a pollutant. The three-step pollution management model (table 1.5.1) summarizes these different approaches.

▼ Table 1.5.1 Pollution management targeted at three different levels

Process of pollution	Level of pollution management
Human activity producing pollutant ↓	**Altering human activity** The most fundamental level of pollution management is to change the human activity that leads to the production of the pollutant in the first place, by promoting alternative technologies, lifestyles and values through: • campaigns • education • community groups • governmental legislation • economic incentives/disincentives.
Release pollutant into environment ↓	**Controlling release of pollutant** Where the activity/production is not completely stopped, strategies can be applied at the level of regulating or preventing the release of pollutants by: • legislating and regulating standards of emission • developing/applying technologies for extracting pollutant from emissions.
Impact of pollutant on ecosystems	**Clean-up and restoration of damaged systems** Where both the above levels of management have failed, strategies may be introduced to recover damaged ecosystems by: • extracting and removing pollutant from ecosystem • replanting/restocking lost or depleted populations and communities.

Concept link 

STRATEGY: Clear management strategies are needed to tackle the environmental problems caused by pollution.

Assessment tip

The principles of pollution, particularly relating to pollution management (see table 1.5.1), should be used throughout the course when addressing issues of pollution. There are clear advantages of employing the earlier strategies of pollution management over the later ones. Collaboration is an important factor in pollution management.

For example, if approaching the management of pollution from CFCs:

- Manage the human activity producing the pollutant:
  - stop the pollution by using alternative gases or substitutes
  - provide alternative technology e.g. roll-on deodorant instead of aerosol
  - ban the use of CFCs through international treaties or protocols such as the Montreal Protocol.
- Manage the release of the pollutant:
  - recycle CFCs from disused refrigerators
  - create emission standards, laws or regulations
  - use more efficient technology so that less ozone-depleting substances (ODS) are used.
- Manage the long-term impact of the pollutant on the ecosystem:
  - provide protection from increased UV radiation if necessary
  - for example, protect human skin with sunscreen or protective clothing, or provide advice on avoiding times of day when UV levels are at their highest
  - protect buildings and materials using UV-resistant technologies
  - use improved crops with more UV resistance, e.g. using genetically modified organisms (GMOs) that have UV resistance.

Assessment tip

You need to be able to evaluate the effectiveness of each of the three different levels of intervention given a specific example of pollution.

Test yourself

1.16 Explain how economic factors affect a country's approach to pollution management. [4]

DDT: conflict between utility and environmental effects

Dichlorodiphenyltrichloroethane (DDT) is a man-made pesticide that has both advantages and disadvantages. Its main advantages are in the control of diseases such as malaria and in improving crop yields. During the 1940s and 1950s, it was used extensively to control lice and mosquitoes. Lice spread the disease typhus and mosquitoes spread the disease malaria (Figure 1.5.2). Today there are about 250 million cases of malaria each year. DDT was also used as a pesticide in farming, which helped to increase agricultural yields.

An economic benefit of controlling malarial mosquitoes is that time off work due to malaria is reduced and the productivity of workers increases. It also reduces health care costs to the government, employer, family and the individual. Overall, it is cheaper to kill malarial mosquitoes than to treat malaria.

Costs of DDT

In the 1960s, public opinion turned against DDT due to the publication of the book *Silent Spring* by Rachel Carson (see section 1.1). Carson claimed that the large-scale spraying of pesticides was killing top predators due to bioaccumulation and biomagnification. An example of the biological effect of DDT is the thinning of eggshells in birds, such as the peregrine falcon which is at the top of the food chain. DDT can cause cancer in humans. There are also links between DDT and premature births. DDT has also been linked to low birth weight and reduced mental development.

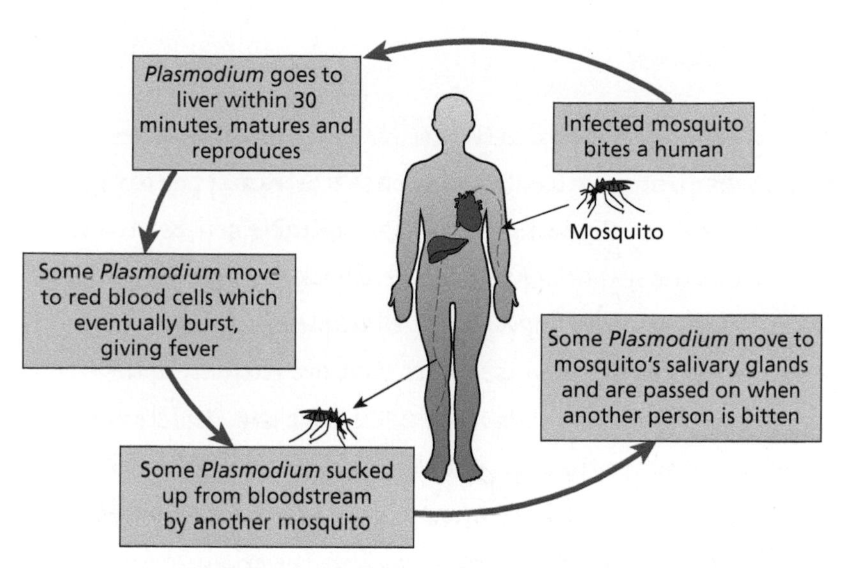

▲ Figure 1.5.2 Malaria is a disease caused by *Plasmodium*, a parasitic infection spread by mosquitoes

Developments in the management of DDT

In the 1970s and 1980s, the use of DDT in farming was banned in many MEDCs. In 2001, the Stockholm Convention on persistent organic pollutants regulated the use of DDT – it was banned from use in farming but was permitted for disease control. Cases of malaria increased in South America after countries stopped spraying DDT. Cases of malaria in Ecuador decreased by over 60% following an increase in the use of DDT.

The plan is to find alternatives for disease control by 2020. In 2006, the World Health Organization (WHO) changed its recommendations. It recommended the use of DDT for regular treatment in buildings and in areas with a high incidence of malaria. The WHO still aims for a total phase out of the use of DDT by the early 2020s.

DDT use: evaluation

There are many advantages and disadvantages of using DDT. As the WHO is suggesting the total phase out of DDT by the early 2020s, it suggests that it believes the disadvantages outweigh the advantages.

Assessment tip

You need to be able to evaluate the use of DDT, i.e. advantages and disadvantages, with an overall conclusion about its use.

QUESTION PRACTICE

With reference to named examples, **distinguish** between a primary and a secondary pollutant. [4]

How do I approach this question?

The command term "distinguish" means making clear the differences between two or more concepts or items. One mark will be awarded for a valid example of a primary pollutant and one mark for an example of a secondary pollutant. The other two marks are for saying why these examples are primary and secondary pollutants. Examples of primary pollutants need to include their direct impact and examples of secondary pollutants need to include the process leading to their pollutionary impact. A maximum of two marks can be awarded if no examples are given.

SAMPLE STUDENT ANSWER

A primary pollutant refers to the source of its origin, where pollution was first produced. In this case a primary pollutant is pollution that is released from an identifiable source and it brought about the impact of positive feedback in the environment that may cause climate change. An example of a primary pollutant would be greenhouse gases that are released to the atmosphere with the composition of CFCs that can deplete the ozone layer, thereby causing global warming. On the other hand, a secondary pollutant refers to the pollutions that are released from non-point sources, meaning the origin could not be identified as it has already been dispersed and mixed with other pollutants. A secondary pollutant can be determined by its impact such as acid rain.

▼ The candidate confuses point source pollution with primary pollution

▲ A correct example of a primary pollutant is given. Even though an incorrect definition of a primary pollutant has been given, 1 mark can be awarded for a correct example

▼ The sentence here makes it sound as if the depletion of the ozone layer causes global warming, which is not the case. CFCs are also a greenhouse gas, which is a separate issue to the hole they have caused to the ozone layer

▼ The candidate confuses secondary pollutants with non-point source pollution

More thorough learning of key definitions would have helped the candidate tackle this question more successfully.

This response could have achieved 1/4 marks.

2 ECOSYSTEMS AND ECOLOGY

This topic examines the structure, functioning and changes that occur in ecosystems. You need to understand and be able to evaluate different ecological sampling techniques. Practical fieldwork in marine, terrestrial, freshwater or urban ecosystems will help you to achieve this. Ecological sampling techniques are used to obtain qualitative (i.e. descriptive) and quantitative (i.e. numerical) data for abiotic (non-living) and biotic (living) factors in ecosystems. You can select ecosystems to study that are local to you. By using various practical measurements to study different aspects of the same ecosystem, techniques are not studied in isolation, but can be used to build a complete picture of that ecosystem, so that they can be better understood. These methods can then be used to study the effects of human disturbance on ecosystems.

You should be able to show:

- ✔ an understanding of the species concept and factors that affect populations;
- ✔ what is meant by "community" and "ecosystem", and different models that can be used to represent them;
- ✔ how matter cycles and energy flows through ecosystems;
- ✔ an understanding of how climate determines ecosystem distribution, and how ecosystems develop though time and space and are affected by human disturbance;
- ✔ how ecological investigations can be carried out, using biotic and abiotic sampling methods.

2.1 SPECIES AND POPULATIONS

You should be able to show an understanding of the species concept and factors that affect populations:

- ✔ A species is the smallest unit into which organisms can be divided, and the habitat is the place where a species lives.
- ✔ The fundamental niche is a complete description of where a species can theoretically exist whereas a realized niche is the actual role a species plays in an ecosystem through biotic interactions.
- ✔ "Abiotic" refers to non-living factors in an environment and "biotic" refers to living factors.
- ✔ Interactions between different populations in an ecosystem include predation, herbivory, parasitism, mutualism, disease, and competition, with each affecting the carrying capacity of organisms.
- ✔ Population growth can be exponential, leading to a J-shaped growth curve, or sigmoidal, leading to an S-shaped curve controlled by limiting factors.

- **Species** – a group of organisms that share common characteristics and that interbreed to produce fertile offspring
- **Habitat** – the environment in which a species normally lives
- **Niche** – the particular set of abiotic and biotic conditions and resources to which an organism or population responds
- **Fundamental niche** – the full range of conditions and resources in which a species could survive and reproduce
- **Realized niche** – the actual conditions and resources in which a species exists due to biotic interactions
- **Abiotic factors** – non-living, physical factors that influence the organisms and ecosystem, such as temperature, sunlight, pH, salinity, and precipitation

Species, habitat and niche

A **species** can be defined as a group of organisms that interbreed and produce fertile offspring. An example of a species is the lion, *Panthera leo*.

- **Biotic factors** – interactions between the organisms, such as predation, herbivory, parasitism, mutualism, disease, and competition
- **Competition** – the demand by two or more species for limited resources
- **Parasitism** – interaction where one organism gets its food from another organism that does not benefit from the relationship
- **Disease** – an illness or infection caused by a pathogen
- **Mutualism** – an interaction between two species where both species benefit
- **Predation** – interaction where one organism hunts and eats another animal
- **Herbivory** – interaction where an animal feeds on a plant
- **Carrying capacity** – the maximum number of individuals of a species that can be sustainably supported by a given area
- **Limiting factors** – circumstances that restrict the growth of a population or prevent it from increasing further

Habitat is the kind of biotic and abiotic environment in which a species normally lives. For example, lions are found in grassland, savanna, dense scrub, and open woodland. In contrast, a **niche** refers to the biotic and abiotic environment with which a species interacts, for example the prey that it eats, its vulnerability to parasites, access to fresh water, and so on. A habitat may be shared by many species whereas a niche is limited to a single species. For example, different cat species inhabit tropical grasslands but only lions hunt in groups and so they tend to take larger prey.

The theoretical range of conditions in which a species can exist is called its **fundamental niche**. Species interact with other species in their environment, through **competition** and other ecological relationships. Because species interact with their environment in this way, individuals in a species cannot exist in all possible conditions determined by their fundamental niche – the range of conditions and environments where a species is actually found is called its **realized niche** (see figure 2.1.1).

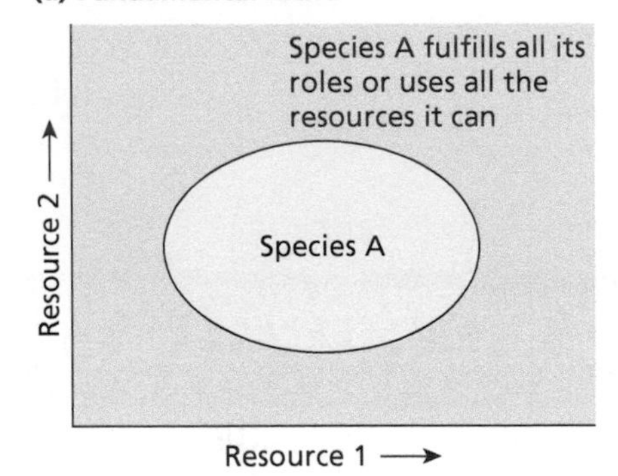

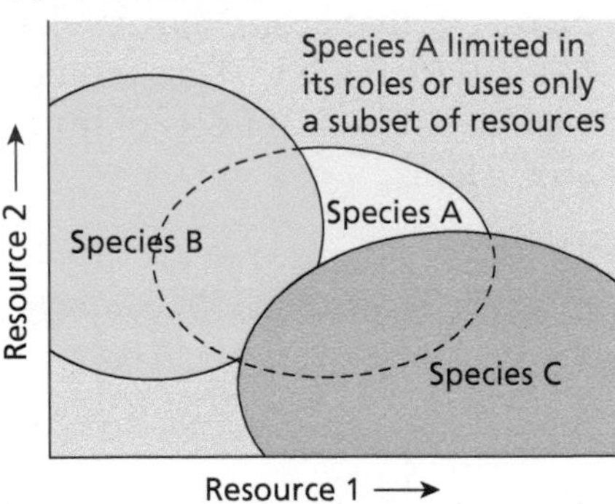

▲ Figure 2.1.1 The fundamental and the realized niche. In (a), species A occupies all conditions within its fundamental niche, whereas in (b) biotic interactions with two other species limit the conditions and resources that species A can utilize. This is its realized niche

Assessment tip

Valid named species should be used in your answers. For example, use "Australian plague locust" rather than "insect", "marram grass" rather than "grass" and "oak tree" rather than "tree".

Assessment tip

If you are asked to distinguish between the terms "niche" and "habitat" with reference to a named species, make sure you give valid examples. "Role of species within ecosystem" would be acceptable as the definition of niche, but not "job", which is a human-centred term and only addresses the impact of species on systems, not the mutual relationship.

Abiotic factors

Non-living, physical factors that influence organisms are known as **abiotic factors**. These include temperature, sunlight, pH, salinity, and precipitation. During fieldwork it will be necessary to measure abiotic factors and evaluate the methods used (table 2.1.1).

▼ Table 2.1.1 Methods used to measure abiotic factors and an evaluation of each technique

Abiotic factor	Equipment for measuring the factor	Methodology	Evaluation
Wind speed	Anemometer (figure 2.1.2a)	The anemometer is hand-held and is pointed into the wind. It is held at the same height for each measurement.	Gusty conditions may lead to large variations in data. Care must be taken not to block the wind with your body when you are holding the anemometer.
Temperature	Digital thermometer	The digital thermometer can be used to measure temperature in air and water, and at different depths of soil. The digital thermometer is held at the same depth or height for each measurement.	Data will vary if temperature is not taken at the same depth/height each time. Temperature is measured for a short period of time. Data-loggers can be used to measure temperature over long periods of time.
Light intensity	Light meter (figure 2.1.2b)	The light meter is hand-held with the sensor facing upwards. The light meter is held at the same height above the ground for each measurement. The reading is taken when there is no fluctuation in the reading.	Cloud cover will affect the light intensity. Shading from plants or the person operating the light meter will also affect the light intensity. Care must therefore be taken when taking readings using a light meter.
Flow velocity	Flow meter (figure 2.1.2c)	The impeller is put into the water just below the surface. The impeller is pointed into the direction of the flow. A number of readings are taken to ensure accuracy.	Velocity varies according to the distance from the surface, so readings must be taken at the same depth. Results can be misleading if only one part of a stream is measured. Water flows can vary over time because of rainfall or ice melting events.
Turbidity	Secchi disc	The Secchi disc is mounted on a pole or line and is lowered into the water until it is just out of sight. The depth is measured using the scale of the line or pole. The disc is raised until it is just visible again and a second reading is taken. The average depth calculated is known as the Secchi depth.	Reflections off water will reduce visibility and make it difficult to take turbidity measurements. Measurements are subjective and depend, to some extent, on the technique used by the person taking the measurements.
Dissolved oxygen	Oxygen-sensitive electrodes are attached to an oxygen meter	Hold probe at a set distance beneath the surface of the water.	There may be spatial variation in concentration due to mixing of water with air through turbulence, for example.
Soil moisture	A sample of soil is placed in an oven and heated so that water evaporates	Weigh soil, then heat in an oven. Heat the soil until there is no further loss in weight. Loss of weight can be calculated as a percentage of the starting weight. Soil moisture probes can also be used.	If the oven is too hot when evaporating the water, organic content can also burn off.

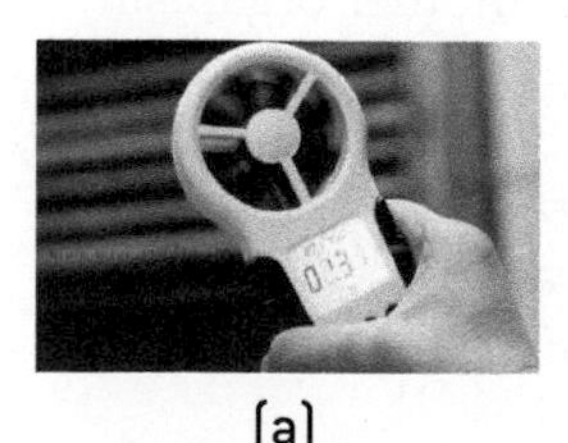

(a)

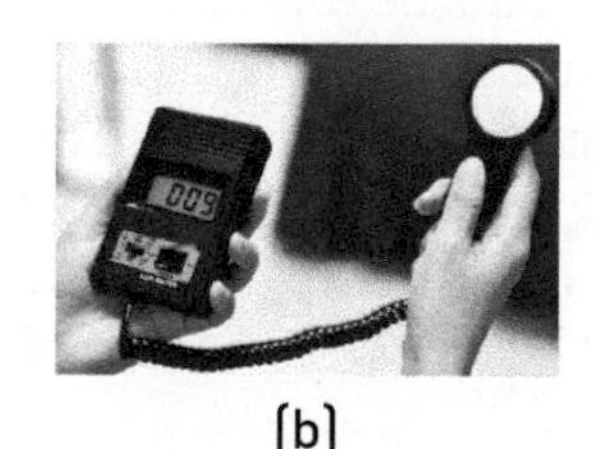

(b)

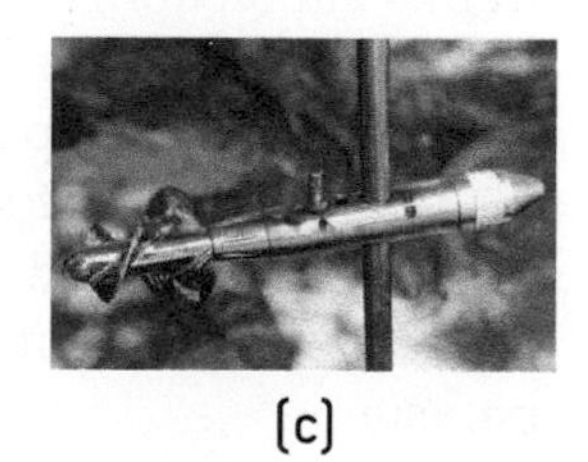

(c)

▲ Figure 2.1.2 Apparatus for measuring abiotic factors: (a) anemometer, (b) light meter, (c) flow velocity meter

Sampling must be carried out carefully so that an accurate representation of the study area can be obtained. An inaccurate representation of a study area may be obtained if errors are made in sampling. Short-term and limited field sampling (i.e. small sample sizes taken over short periods of time) reduce how effective sampling methods are because abiotic components may vary from day to day and from season to season. Most abiotic components can be measured using data-logging devices. The advantage of data-loggers is that they can provide continuous data over a long period of time, and this makes results more representative of the area. Results can be made more reliable by taking many samples.

Assessment tip

You need to be able to evaluate methods to measure at least three abiotic factors in an ecosystem.

Biotic factors

Biotic factors are the species, populations and communities present in an ecosystem. There are several different types of interaction that occur between populations: **predation, herbivory, parasitism, mutualism, disease,** and competition.

Content link

Negative feedback is discussed in section 1.3.

Predation

Predation is an interaction where one animal hunts and eats another animal. The predator is the animal that hunts and kills the other animal. The animal that is hunted and killed is called the prey.

The **carrying capacity** of the prey is affected by the predator because the number of prey is reduced by the predator. The carrying capacity of the predator is affected by the prey because the number of predators is reduced when prey become fewer. These predator–prey interactions are controlled by negative feedback mechanisms.

▲ Figure 2.1.3 Predation – an orb-weaving spider eating an alate (winged ant)

Test yourself

2.1 Explain how predation may lead to:

a) stability in a population of the prey species; [3]

b) long-term population decrease or extinction of the prey species. [3]

Herbivory

Herbivory is an interaction where an animal feeds on a plant (figure 2.1.4). The animal that eats the plant is called a herbivore. An example of herbivory is a caterpillar eating a leaf. The carrying capacity of herbivores is affected by the quantity of the plant they feed on. An area with more abundant plant resources will have a higher carrying capacity than an area that has less plant material available as food for a consumer.

▲ Figure 2.1.4 Herbivory – a Galápagos giant tortoise (*Chelonoidis nigra*) feeding on grass

Parasitism

In this interaction, one organism gets its food from another organism that does not benefit from the relationship. The organism that benefits from the relationship is called the parasite. The organism from which the parasite gets its food is called the host. The parasite benefits from the interaction, but the host is harmed by the interaction. The carrying capacity of the host may be reduced because of the harm caused by the parasite.

An example of a parasite is *Rafflesia* (figure 2.1.5). *Rafflesia* are plants that have giant flowers and no leaves. Because they have no leaves they cannot carry out photosynthesis and so they cannot make sugar. *Rafflesia* flowers get the sugars they need from a vine on which they live (i.e. they parasitize the vine).

▲ Figure 2.1.5 Parasitism – a *Rafflesia* flower growing on a Tetrastigma vine on the slopes of Mt Kinabalu, Sabah, Malaysia

Mutualism

Mutualism is an interaction in which both species benefit. An example of a mutualistic interaction is coral. In a coral, an animal polyp makes a hard structure from calcium carbonate. Single-celled algae live inside the polyp. These algae are called zooxanthellae. The zooxanthellae photosynthesize and make sugar for the polyp, and in return the polyp creates a hard structure that protects the zooxanthellae (figure 2.1.6).

▲ Figure 2.1.6 Mutualism – a coral polyp containing photosynthetic algae within its tentacles

Disease

A disease is an illness or infection causes by a pathogen. Pathogens are harmful organisms that include bacteria, viruses, fungi and some single-celled organisms (protoctistans, such as *Plasmodium* that causes malaria – see figure 2.1.7). A pathogen will reduce the carrying capacity of the organism it is infecting. Changes in the incidence of the pathogen, and therefore the disease, can also cause a decrease in population size below the carrying capacity. When the population recovers it will increase again towards the carrying capacity.

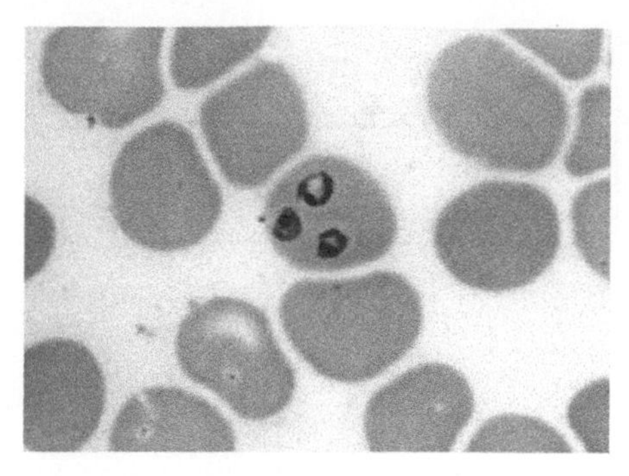

▲ Figure 2.1.7 Photo of *Plasmodium falciparum* in a blood sample – a single-celled organism that causes malaria

Competition

When resources are limited, individuals must compete in order to survive. This competition can be either within a species or between individuals of different species. When competition is within a species it is called intraspecific competition. When competition is between different species it is called interspecific competition (figure 2.1.8). No two species can occupy the same niche (i.e. they cannot be identical in all ways) and so interspecific competition occurs when the niches of different species overlap. In this interaction, the stronger competitor (i.e. the one better able to survive) will reduce the carrying capacity of the other's environment.

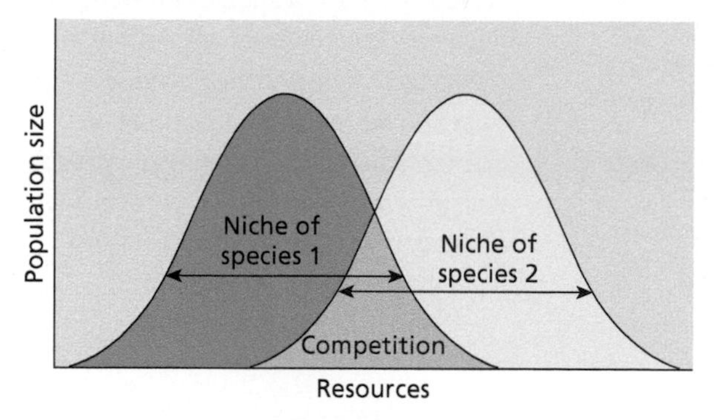

▲ Figure 2.1.8 Competition between two species

>> **Assessment tip**

You need to be able to interpret graphical representations (or models) of factors that affect an organism's niche, such as competition and organism abundance over time.

Figure 2.1.9 shows the difference between intraspecific competition and interspecific competition:

(a)

(b)

▲ Figure 2.1.9 (a) Intraspecific competition – gazelles competing (*Gazella* sp.), and (b) Interspecific competition – competition between individuals of different species (a Eurasian tree sparrow and great tit competing at a birdfeeder)

Test yourself

2.2 Outline the similarities and differences between predation and competition. [4]

2.3 Distinguish between biotic and abiotic factors. [4]

Content link

The carrying capacity in human populations is explored in section 8.4.

Assessment tip

If asked to define carrying capacity, you need to make it clear that carrying capacity is associated with a single species, i.e. "The maximum number of a species that can..." rather than "the maximum number of species that can...", which is incorrect. You should also refer to *sustainable* support.

Carrying capacity

The population of a species will fluctuate around its carrying capacity due to changes in resources and the abiotic environment (figure 2.1.10). Competition and other biotic interactions will lead to changes in the carrying capacity.

Carrying capacity may be difficult to estimate.

- There are many different potential **limiting factors** for natural populations.
- The population's needs may alter over time due to genetic changes and evolution.
- Environmental conditions may change, e.g. climate change, introduced species.
- It takes extensive and long-term studies to identify a precise relationship between a species and given environmental factors.

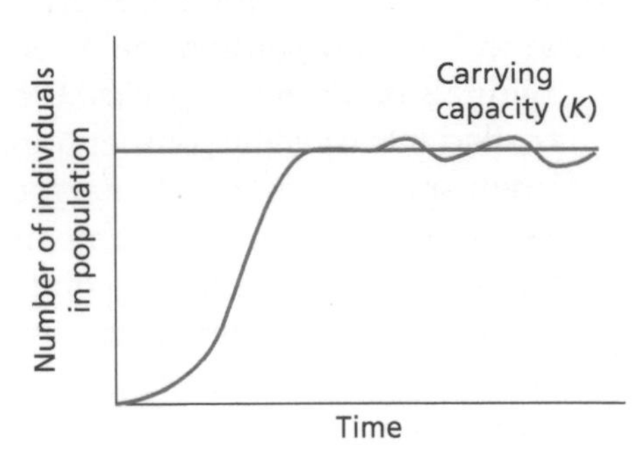

▲ Figure 2.1.10 A species introduced into a new environment will start with a low population, and as individuals meet and reproduce, the population will increase rapidly. Limiting factors, such as water, food and other resources will lead to an upper limit, or carrying capacity (*K*) for the population

Population growth

Population growth of populations can be either exponential, leading to a J-shaped population curve (figure 2.1.11 (a)) or logistic, leading to an S-shaped population curve (figure 2.1.11 (b)). The initial gradient of both S- and J-curves is low due to limiting factors, with low numbers of individuals reproducing at sub-optimum gender ratios. Limiting factors in plants include light, nutrients, water, carbon dioxide, and temperature. Limiting factors in animals include food, mates, and water. When these limiting factors are overcome, as biotic potential exceeds environmental resistance, exponential increase occurs with positive feedback in both S- and J-curves (i.e. more individuals are available to reproduce, and they produce more offspring which in turn reproduce). In S-curves, new limiting factors eventually slow the population growth and maintain equilibrium, leading to a plateau. New limiting factors include, for example, limited food, increased predation, disease, accumulation of wastes, and competition. Limiting factors may be density-dependent, providing negative feedback mechanisms to maintain this equilibrium.

In an S-curve, the population reaches its carrying capacity (figure 2.1.11 (b)). Populations may fluctuate around the carrying capacity due to seasonal changes, changes in resource availability, predation, disease or other factors.

In J-curves, limiting factors such as seasonal climate change, disease and overexploited food resources eventually lead to rapid decline (i.e. a population crash).

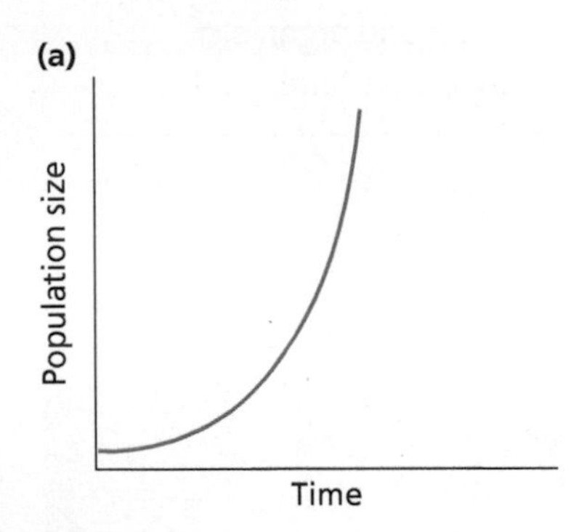

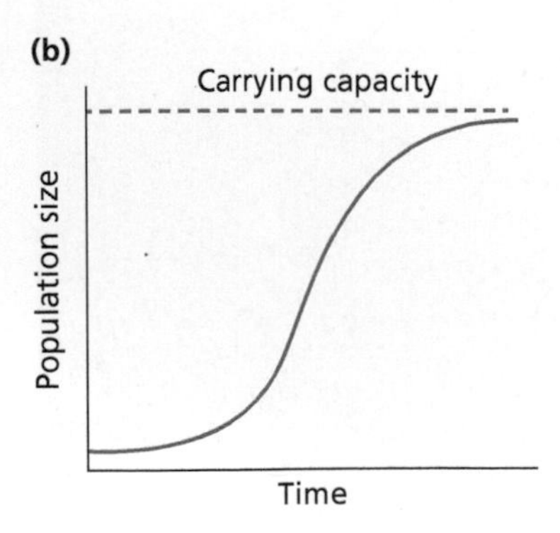

▲ Figure 2.1.11 (a) A J-shaped population growth curve; (b) An S-shaped (logistic) population growth curve

Concept link 

EQUILIBRIUM: Limiting factors act on populations so that they reach a point of equilibrium about which they can fluctuate.

Assessment tip

You need to be able to explain population growth curves in terms of numbers and rates.

QUESTION PRACTICE

a) **Distinguish** between the terms "niche" and "habitat" with reference to a named species. [4]

b) (i) **Define** the term "carrying capacity". [1]
(ii) **Identify three** reasons why carrying capacity can be difficult to estimate. [3]

How do I approach these questions?

a) The command term "distinguish" means pointing out the differences between two or more concepts or items. You need to apply your answers to valid named examples. Use the definitions of a niche and a habitat to answer this question. One mark will be awarded for each correct answer, up to a maximum of 4 marks. A maximum of 2 marks can be awarded if no examples are given.

b) (i) "Define" means that you need to give the precise meaning of "carrying capacity".
(ii) "Identify" means selecting answers from a range of possibilities. There are several reasons why carrying capacity may be difficult to estimate – you need to write down three valid points.

SAMPLE STUDENT ANSWER

a) Niche and habitat are terms that can be distinguished based on how they relate to a species. A habitat is the location in which a species resides and is essentially where the species specifically lives. A niche, however, is a species' role within an ecosystem. For example, a tucan's habitat may be in the canopy of the rainforest of Brazil, however, its niche is its function within the ecosystem, such as collecting food to eat.

▲ 1 mark for describing what a habitat is. A more sophisticated answer would have referred to the kind of (biotic and abiotic) environment in which a species normally lives

▲ Correct definition of "niche" given for 1 mark. The answer could have added more detail by referring to "all its interactions with its (biotic and abiotic) environment"

▲ Uses the example chosen (i.e. a toucan) to illustrate the meaning of "habitat"

▲ An example of what is meant by "niche" is given for 1 mark. A more specific example of the toucan's role within the rainforest would have been eating fruits of certain forest tree species and dispersing their seeds

This response could have achieved 4/4 marks.

b) (i) The maximum population size that an environment/ ecosystem can sustainably sustain.

▲ A valid definition for 1 mark, although a bit awkward. Better to say "the maximum number of individuals of a species that can be sustainably supported by a given area"

This response could have achieved 1/1 marks.

(ii)

- because it is difficult to estimate a population (only an estimate)
- because limiting factors are unpredictable (limiting factors prevent an ecosystem from reaching its full carrying capacity)
- due to technological development, humans can push to Earth's carrying capacity; one does not know when technology can no longer extend the Earth's carrying capacity.

▼ This simply repeats the question, so no mark awarded

▲ 1 mark for giving a valid reason. There are many different potential limiting factors for natural populations and so their effect on carrying capacity is difficult to predict

▲ 1 mark for identifying a second reason. For human populations, technological developments cause, over time, changes in resources which are required and available

This response could have achieved 2/3 marks.

2.2 COMMUNITIES AND ECOSYSTEMS

- **Community** – a group of populations living and interacting with each other in a common habitat
- **Ecosystem** – a community and the physical environment with which it interacts
- **Respiration** – the conversion of organic matter into carbon dioxide and water in all living organisms, releasing energy
- **Photosynthesis** – the conversion and storage of sunlight energy as organic matter (new biomass), releasing oxygen in the process
- **Bioaccumulation** – this occurs when non-biodegradable toxins absorbed by a given trophic level are not broken down or excreted, and so accumulate in tissues
- **Biomagnification** – the increasing concentration of a non-biodegradable toxin as it passes along a food chain due to the loss of other biomass

You should be able to show what is meant by "community" and "ecosystem", and different models that can be used to represent them

✔ Photosynthesis supports communities and ecosystems by changing sunlight energy into chemical energy for new growth; respiration uses this to provide energy for life processes.

✔ Producers provide the foundation for food chains.

✔ Feeding relationships can be divided into different trophic levels, which can be modelled using food chains, food webs, and ecological pyramids.

✔ The second law of thermodynamics determines the length of food chains and the shape of ecological pyramids.

✔ Persistent or non-biodegradable pollutants bioaccumulate and biomagnify in food chains.

Community and ecosystem

An **ecosystem** can be defined as a **community** of organisms that depend on each other and the environment they live in. All ecosystems, such as tropical rainforest, include both living (e.g. animals, plants and fungi) and non-living components (e.g. rocks, water and climatic conditions). They are open systems, where both matter and energy are exchanged with the outside environment.

Test yourself

2.4 Outline the basic components of an ecosystem using the systems approach. [4]

Content link

The systems approach is covered in section 1.2.

Respiration and photosynthesis

Assessment tip

When discussing photosynthesis and respiration, details of chloroplasts, enzyme-mediated reactions, mitochondria, carrier systems, adenosine triphosphate (ATP), and specific intermediate biochemicals do not need to be known.

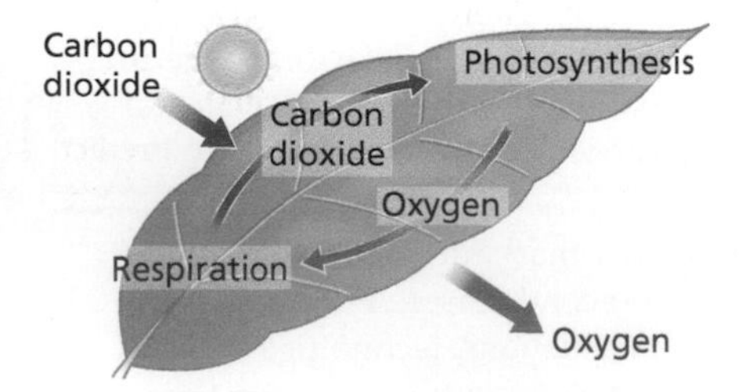

▲ Figure 2.2.1 The relationship between photosynthesis and respiration

Respiration is a transformation process which releases energy from organic molecules such as glucose, producing water and carbon dioxide.

Photosynthesis is a transformation process which takes energy captured by plant pigments (e.g. chlorophyll) and stores it as chemical energy in the form of glucose. Oxygen is produced as a by-product.

Photosynthesis takes place in producers and is the essential reaction that provides energy and matter in food chains. All organisms carry out respiration to provide energy for metabolic processes in cells and life processes.

Assessment tip

Photosynthesis involves the transformation of light energy into the chemical energy of organic matter. Respiration is the transformation of chemical energy into energy necessary for biological activity with, ultimately, heat lost from the ecosystem. All organisms respire: bacteria, algae, plants, fungi, and animals.

The role of producers

Plants are producers that convert light energy into chemical energy by photosynthesis.

- Plants convert carbon dioxide and water into glucose and oxygen.
- Glucose forms the raw material of biomass and is the basis of food chains.
- Producers therefore provide energy in a form that can be passed along food chains.
- The production of oxygen by producers is vital for the majority of ecosystems.

The absorption of CO_2 maintains a balance of CO_2 in the atmosphere and reduces global warming.

Plants may also provide other resources, such as services for ecosystems, e.g. habitats, soil conservation, and the cycling of matter.

Primary producers may alternatively generate biomass through chemosynthesis (using energy released by chemical reactions to produce glucose). Chemosynthetic bacteria use chemical energy to produce food in the absence of sunlight (e.g. those found in subterranean caves and deep ocean trenches).

Assessment tip

You need to be able to explain the transfer and transformation of energy as it flows through an ecosystem.

Food chains and food webs

Trophic levels are feeding levels. The first trophic level contains producers; these organisms produce their own food by photosynthesis. All other trophic levels contain consumers. These are animals that eat other organisms to obtain their food. The second trophic level contains primary consumers. These organisms eat plants and are also known as herbivores. The third trophic level contains secondary consumers. These organisms eat other animals and are also known as carnivores. Some secondary consumers may eat both animals and plants and are known as omnivores. The fourth level contains tertiary consumers.

A food chain is a simple diagram that shows feeding relationships in an ecosystem (see figure 2.2.2). Arrows from one organism to the next represent energy flow.

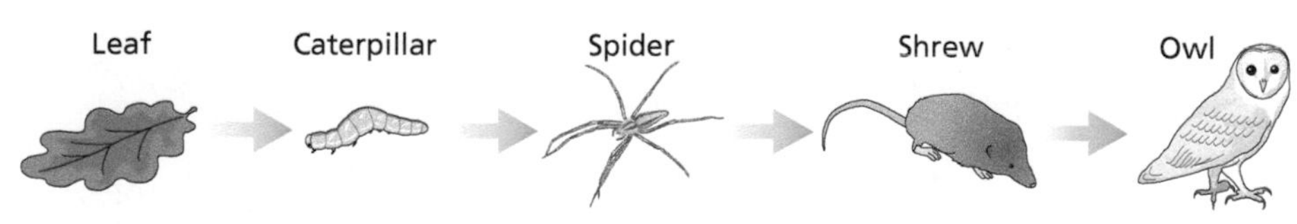

▲ Figure 2.2.2 A food chain

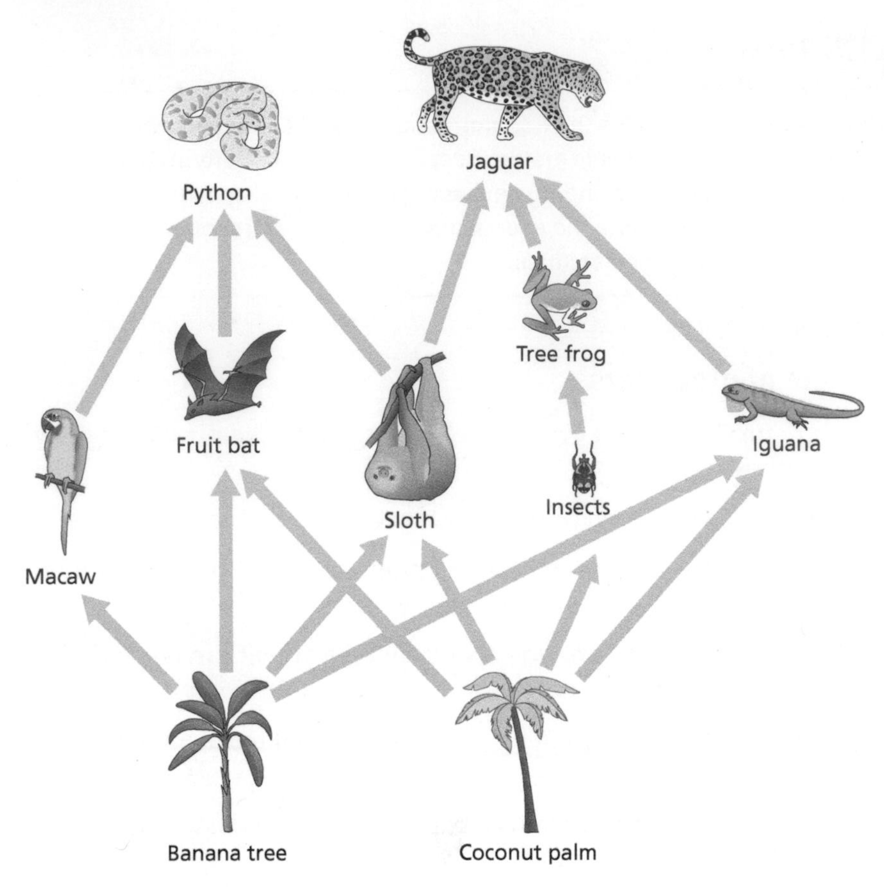

▲ Figure 2.2.3 A food web

A food web is a diagram that shows interconnected food chains in an ecosystem. Figure 2.2.3 shows a food web from a local environment with its different trophic levels.

One species may occupy several different trophic levels depending on which food chain is being considered. In Figure 2.2.3, the jaguar is both a secondary and a tertiary consumer depending on which food chain is being considered. Decomposers feed on dead organisms at each trophic level.

Assessment tip

A food chain is linear, showing energy flow through an ecosystem, whereas a food web shows the complex interactions between different food chains. One species may occupy several different trophic levels in a food web depending on which food chain it is present in.

Assessment tip

Be aware of the distinction between storages of energy illustrated by boxes in energy-flow diagrams (representing the various trophic levels), and the flows of energy or productivity often shown as arrows (sometimes of varying widths).

Assessment tip

You should be able to use your knowledge of thermodynamics to explain how energy flows in ecosystems.

Energy loss in food chains

Unit 1 described how quantities of energy decrease as it passes along a food chain. Energy is lost through respiration, excretion, egestion and inedible parts of organisms that are not passed on in a food chain. This issue is explored further in section 2.3 on page 45.

Test yourself

2.5 Explain how the first and second laws of thermodynamics are demonstrated as energy from the Sun flows through the primary producers in a food chain. [4]

Ecological pyramids

Pyramids are graphical models of the quantitative differences that exist between the trophic levels of a single ecosystem. Quantitative data for each trophic level are drawn to scale as horizontal bars arranged symmetrically around a central axis.

Assessment tip

You need to be able to create models of feeding relationships, such as food chains, food webs and ecological pyramids, from given data.

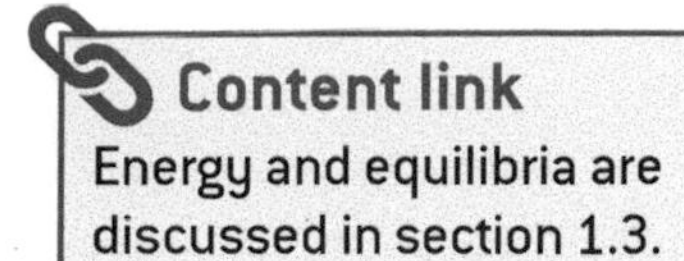

Content link

Energy and equilibria are discussed in section 1.3.

The shape of ecological pyramids

A pyramid of numbers represents the number of organisms (producers and consumers) coexisting in an ecosystem. A pyramid of biomass represents the stores of energy and matter at each trophic level measured in units such as grams of biomass per square metre (g m^{-2}). Because energy decreases along food chains (the second law of thermodynamics) there is a tendency for biomass to decrease along food chains, and so pyramids become narrower as the trophic levels are ascended.

Pyramids of numbers can sometimes have a narrower base (producers) than the second trophic level (primary consumers), when producers, such as trees, are relatively large at lower trophic levels and so few in number. Similarly, pyramids of biomass can show greater quantities at higher trophic levels because they represent the biomass present at a given time (there may be marked seasonal variations) – see figure 2.2.4. When they are inverted they show more biomass at higher trophic levels than lower trophic levels.

The amount of biomass present at a certain moment in time is called the standing crop biomass. The standing crop biomass does not show the amount of productivity in the trophic level over time. For example, a field may have a lower standing crop biomass of grass than the herbivores that feed on it. This results in an inverted pyramid of biomass. Inverted pyramids may also be the result of seasonal variations in biomass. In terms of productivity over time, however, as the grass generates new biomass at least as fast as it is being eaten, the productivity of the grass will be higher than the productivity of the herbivores that feed on it (see pyramids of productivity below).

Both pyramids of numbers and pyramids of biomass represent storages, and are "snap-shots" of the community in time. The third form of

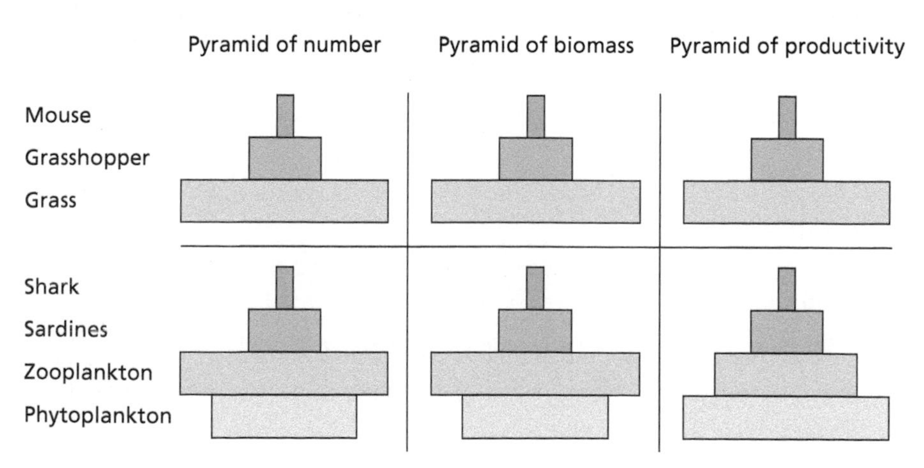

▲ Figure 2.2.4 Contrasting ecological pyramids

pyramid, pyramids of productivity or energy, refer to the flow of energy through trophic levels and invariably show a decrease in energy along the food chain. For example, the turnover of two retail outlets cannot be compared by simply comparing the goods displayed on the shelves; the rates at which the shelves are being stocked and the goods sold also need to be known. Similarly, a business may have substantial assets but cash flow may be very limited. In the same way, pyramids of numbers and biomass simply represent the stock at a given moment, whereas pyramids of productivity show the rate at which that stock is being generated.

Assessment tip

"Pyramids of biomass" refers to a standing crop (a fixed point in time) and "pyramids of productivity" refers to the rate of flow of biomass or energy. The units of measurement are different in each. Biomass, measured in units of mass (for example, $g\ m^{-2}$), should be distinguished from productivity, measured in units of flow (for example, $g\ m^{-2}\ yr^{-1}$ or $J\ m^{-2}\ yr^{-1}$).

Pyramids of productivity represent the flow of energy through a food chain from one trophic level to the next trophic level. The second law of thermodynamics explains the inefficiency and decrease in available energy along a food chain. Because the rate of flow of energy decreases along a food chain, pyramids of productivity always become narrower towards the top. There are no inverted pyramids of productivity.

Assessment tip

Pyramids of productivity cannot be inverted. Measurements are taken over one year; this takes into account any seasonal changes. Over a full year, energy will decrease as it flows from one trophic level to the next. Pyramids of productivity therefore demonstrate energy loss in food chains, where there is a reduction in the flow of energy from producers through to consumers.

There are several implications of this pyramidal structure in ecosystems, for example:

- biomagnification (see below)
- relative vulnerability of top carnivores
- minimum size of conservation areas to support top carnivores (see unit 3)
- efficiency in food choice and production for human populations (see unit 5).

The pyramid structure of an ecosystem may be changed by human activities, for example:

- crop farming increases the producer base and decreases higher trophic levels
- livestock farming increases the primary consumer bar in the pyramid and decreases secondary and tertiary consumers
- trophy hunting reduces top carnivores
- deforestation reduces the producer bar in a biomass pyramid.

Content link

Terrestrial food production systems are covered in section 5.2.

Test yourself

2.6 For an ecosystem you have studied, **draw** a food chain of at least four named species. [2]

2.7 State one other type of pyramid used to show trophic levels. [1]

2.8 Evaluate pyramids of numbers as a method of representing the biotic components of an ecosystem. [3]

2.9 Distinguish between a pyramid of numbers and a pyramid of productivity. [4]

Bioaccumulation and biomagnification

Non-biodegradable toxins are not broken down as they pass along food chains (Figure 2.2.5). The pyramid structure of food chains can influence the impact of non-biodegradable toxins on an ecosystem:

- Many non-biodegradable toxins are fat-soluble and so accumulate in the fatty tissue of organisms. Toxins therefore accumulate over time within the biomass of each trophic level (i.e. **bioaccumulation** takes place).
- Biomass is also lost through respiration at each trophic level, which increases the concentration of these toxins (i.e. **biomagnification** takes place).
- The concentration of toxins becomes greatest and most lethal in the top carnivores.
- Populations of top carnivores are often the least stable and so most prone to decline.
- Loss of the top carnivores will lead to an imbalance in the lower populations, which may lead to disruption of the entire food chain, food web or ecosystem, e.g. DDT in bald eagles; mercury in swordfish.

Human activities, such as mining and the combustion of fossil fuels, have led to widespread mercury pollution: mercury emitted into the air settles on land where it can be washed into water. Fish and shellfish concentrate mercury in their bodies, and as the mercury passes along the food chain it magnifies in top predators such as tuna, marlin, shark and swordfish.

Content link

The role of DDT as a pollutant is discussed in section 1.5.

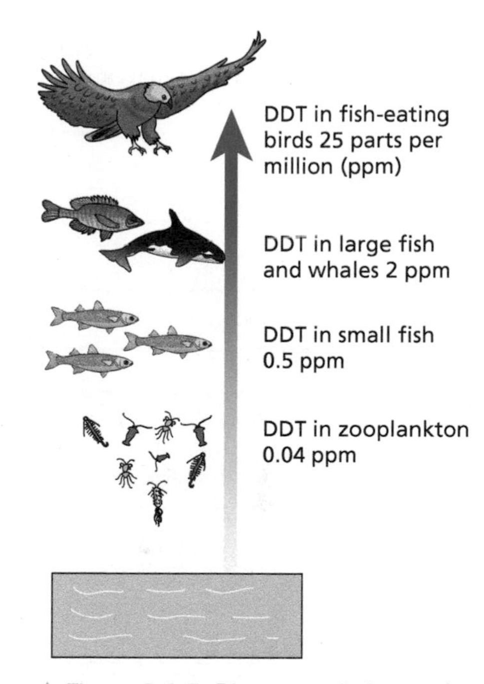

▲ Figure 2.2.5 Bioaccumulation and biomagnification in a food chain

QUESTION PRACTICE

1) **Describe** the role of primary producers in ecosystems. [4]

Essay

2) **Suggest** the procedures needed to collect data for the construction of a pyramid of numbers for the following food chain:

Plants ⟶ Snails ⟶ Birds [7]

How do I approach these questions?

1) The question asks you to describe, not explain, the role of primary producers, which means that you need to give a detailed account but not include reasons or causes. You need to say what role they play in ecosystems. One mark will be awarded for each correct role described, up to a maximum of 4 marks.

Essay

2) This question requires you to apply your knowledge of sampling techniques to the organisms listed in the food chain. Different sampling strategies are applied to different organisms; for example, the techniques used to sample plants may be different to those used to sample animals. The sampling techniques need to supply data to construct a pyramid of numbers, which means that population sizes have to be calculated – what techniques are available to do this? You may also want to say how the pyramid would be constructed using the data obtained.

SAMPLE STUDENT ANSWER

▲ 1 mark for stating that producers are the basis of food chains. A second point here could be that producers are plants that convert light energy into chemical energy by photosynthesis

▲ Correctly states that they provide food for consumers, i.e. they provide energy in a form that can be passed along food chains

1) Primary producers are at the bottom of the food chain. They are essential as they provide food for consumers and are the main source of energy for the rest of the food chain.

Only 2 different points are offered, even though there are 4 marks available. This demonstrates poor exam technique – candidates need to consider the number of marks that are awarded to a question and ensure that enough information has been included to earn full marks.

This response could have achieved 2/4 marks.

▲ Correct method given – 1 mark

▼ Needs to say how the quadrats will be randomly positioned to attain a mark

▲ 1 mark for describing how the total population can be estimated by multiplying the mean of samples by total sample area

▲ A method for estimating the populations of mobile species is correctly identified

▲ The way in which individuals are sampled is given. 1 mark

▲ It is made clear that the marking method must not harm the animals. 1 mark awarded

▲ Individuals are released and allowed to redistribute before resetting traps. 1 mark

▼ Unclear how the population size is estimated. Needs to say here that the ratio of marked : unmarked in recapture is recorded and used to estimate the total population

2) A pyramid of numbers shows the population of the organisms at each trophic level in a food chain. To collect population data for plants, one might use the quadrat method. To use this method one first lays out a grid of squares in the area in which the population is to be determined. Several squares are then randomly selected from the grid and the number of organisms in each selected square is counted. One then finds the average population of the selected squares (for a single square) and multiplies by the total number of squares. To collect population data from mobile organisms such as snails and birds, one might use the capture-mark-release method. To use this method, one first places non-harmful traps in the area to be studied. After a few days (to allow the organisms to become used to the traps), the organisms in them are captured and marked (in a way that will not impact their ability to survive). The marked organisms are then released. After an amount of time sufficient for reintegration of the marked organisms into their population, a group of individuals from the population is captured and the population is determined based on the amount of marked organisms in the new sample.

This response could have achieved 6/7 marks.

2.3 FLOWS OF ENERGY AND MATTER

You should be able to show how matter cycles and energy flows through ecosystems:

- ✔ Solar radiation passes through the Earth's atmosphere and is absorbed by producers.
- ✔ Energy flows through ecosystems when consumers eat producers, and when consumers eat consumers.
- ✔ Primary productivity refers to the gain in energy and biomass in producers, and secondary productivity refers to the creation of new biomass in consumers.
- ✔ "Gross productivity" is the gain in energy in an organism prior to respiratory and other losses, and "net productivity" refers to the quantity of energy that is stored in new biomass following respiration.
- ✔ The carbon and nitrogen cycles illustrate the flow of matter through ecosystems.
- ✔ Human activities, such as deforestation, agriculture, urbanization, and burning fossil fuels, affect energy flows and matter cycles.

- **Primary productivity** – the gain by producers in energy or biomass per unit area per unit time. This term could refer to either gross or net primary productivity
- **Gross productivity (GP)** – the total gain in energy or biomass per unit area per unit time, which could be through photosynthesis in primary producers or absorption in consumers
- **Gross primary productivity (GPP)** – the total gain in energy or biomass per unit area per unit time fixed by photosynthesis in green plants
- **Net productivity (NP)** – the gain in energy or biomass per unit area per unit time remaining after allowing for respiratory losses (R)
- **Net primary productivity (NPP)** – the gain by producers in energy or biomass per unit area per unit time remaining after allowing for respiratory losses (R). This is potentially available to consumers in an ecosystem
- **Secondary productivity** – the biomass gained by heterotrophic organisms, through feeding and absorption, measured in units of mass or energy per unit area per unit time
- **Gross secondary productivity (GSP)** – the total gain by consumers in energy or biomass per unit area per unit time through absorption
- **Net secondary productivity (NSP)** – the gain by consumers in energy or biomass per unit area per unit time remaining after allowing for respiratory losses (R)

Pathways of radiation through the atmosphere

Not all sunlight emitted by the Sun is used by plants for photosynthesis.

- The light does not reach the Earth: it is reflected away from the Earth into space or reflected by clouds.
- It is absorbed in the atmosphere.
- It is reflected by the Earth's surface.
- It falls on non-photosynthetic surfaces.
- It is reflected by leaf surfaces or transmitted by leaves.

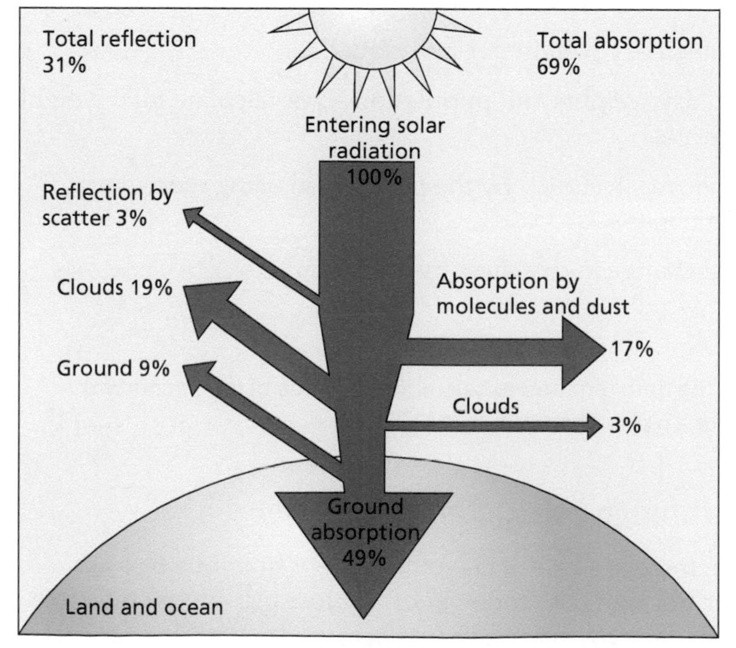

◀ **Figure 2.3.1** Pathways of radiation through the atmosphere involve a loss of radiation through reflection and absorption

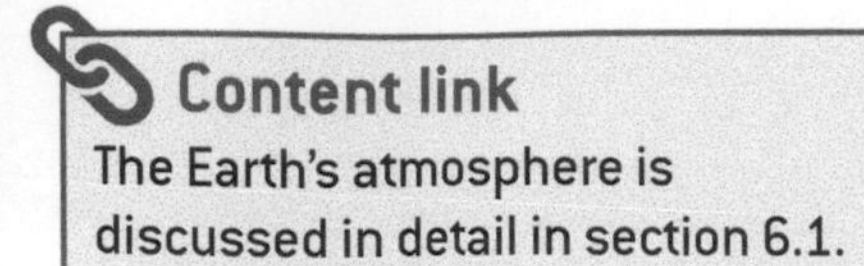

The Earth's atmosphere is discussed in detail in section 6.1.

- The light is not red or blue wavelengths and is not used in photosynthesis (i.e. it is of an incorrect wavelength).
- It is absorbed or reflected by water so does not reach aquatic plants.

Primary productivity

Solar energy reaching vegetation may be lost from an ecosystem before it contributes to the biomass of herbivores.

- Chemical energy is respired by vegetation (see below).
- Chemical energy is not eaten or harvested by consumers: dead material is consumed by decomposers.
- Biomass is eaten but not absorbed by herbivores and is lost in faeces.
- Biomass is absorbed by herbivores, but lost through respiration.

Assessment tip

You need to be able to calculate the values of both GPP and NPP from given data.

Gross primary productivity (GPP) is the total chemical energy produced by a plant. **Net primary productivity (NPP)** is productivity once respiratory losses have been taken into account (i.e. gross primary productivity minus respiration). NPP therefore represents the amount of energy converted to new biomass, available for the next trophic level. The definitions for productivity must include units. For example, for NPP, the gain by producers in energy or biomass per unit area per unit time remaining after allowing for respiratory losses (R).

Secondary productivity

Gross secondary productivity (GSP) is the gain in energy or biomass minus faecal loss per unit area per unit time. Gross secondary productivity is **net secondary productivity** plus respiration.

Estimating secondary productivity

The following procedure could be carried out to estimate the net secondary productivity of, for example, an insect population in kg m^{-2} yr^{-1}.

- Measure change in population size over year, using capture-mark-release-recapture technique (Lincoln Index – see 2.5) for example.
- Weigh a sample of insects to find their wet weight.
- Use a conversion factor to calculate dry weight from wet weight.
- Calculate mean dry weight per individual.
- From mean dry weights and population sizes calculate total weight change over year.
- Estimate the area occupied by the population using measuring tapes or scale maps.
- Divide total change in dry mass by area in m^2.

Test yourself

2.10 Identify the data required to calculate the value of net secondary productivity for a named population. [3]

Energy flow through ecosystems

In a food chain, there is a loss of chemical energy from one trophic level to another through respiration and heat loss. All energy therefore ultimately leaves the ecosystem as heat energy.

Ecological efficiency can be defined as the percentage of energy transferred from one trophic level to the next. Ecological efficiency is low, with an average of one-tenth of the energy available to one trophic level becoming available to the next trophic level. Energy is lost through respiration, inedible parts and faeces. All energy is ultimately lost as heat through the inefficient energy conversions of respiration. This heat is re-radiated to the atmosphere. Overall there is a conversion of light energy to heat energy by an ecosystem.

Figure 2.3.2 shows the movement of energy flow through an ecosystem. Such energy-flow diagrams show the productivity of the different trophic levels.

>> **Assessment tip**

You need to be able to analyse and construct quantitative models of flows of energy and matter.

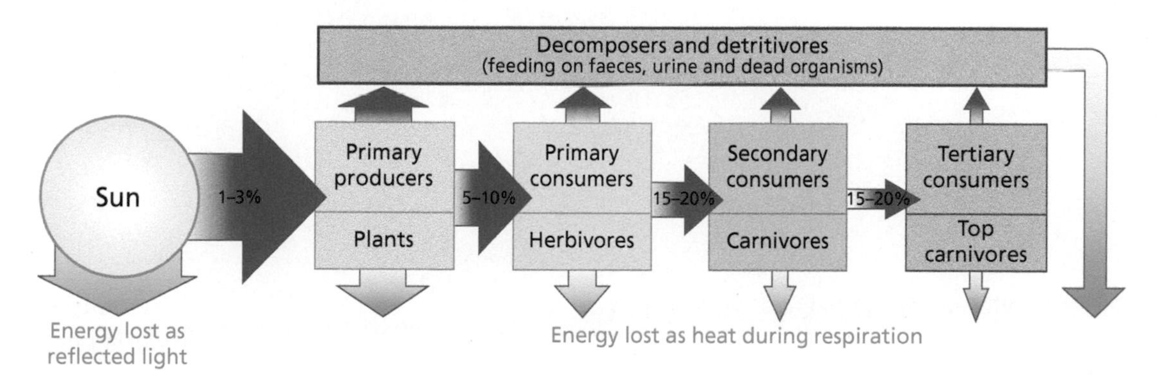

▲ Figure 2.3.2 An energy-flow diagram showing the flow of energy through an ecosystem

The **net productivity** of secondary consumers is much smaller than that of primary consumers:

- Energy is used for respiration by primary consumers and ultimately lost as heat, meaning that there is less energy stored in biomass for the next trophic level
- Greater activity of secondary consumers (carnivores) means more energy is used for respiration than by primary consumers (herbivores).

>> **Assessment tip**

In food chains, energy flows from producer to consumer, and then from consumer to consumer. Transfer processes pass on the energy without a change in state, whereas transformation processes pass on the energy with a change in state (for example chemical energy into heat energy).

Test yourself

2.11 Explain the transfer of energy through an ecosystem. [3]

Test yourself

2.12 An aquatic food chain contains the following storages: algae (4200 kJ), small aquatic animals (630 kJ), small fish (125 kJ), trout (25 kJ) and human (5kJ).

Calculate the ecological efficiency of energy transfer between:

a) Algae and small aquatic organisms

b) Trout and humans [2]

>> **Assessment tip**

You need to be able to analyse the efficiency of energy transfers through a system.

Carbon cycle

Unlike energy, matter cycles in ecosystems. Organic molecules (e.g. glucose) are continually being broken down into simpler inorganic molecules, such as carbon dioxide and water. Processes inside organisms take the inorganic molecules and build them into complex molecules, and so the cycle continues.

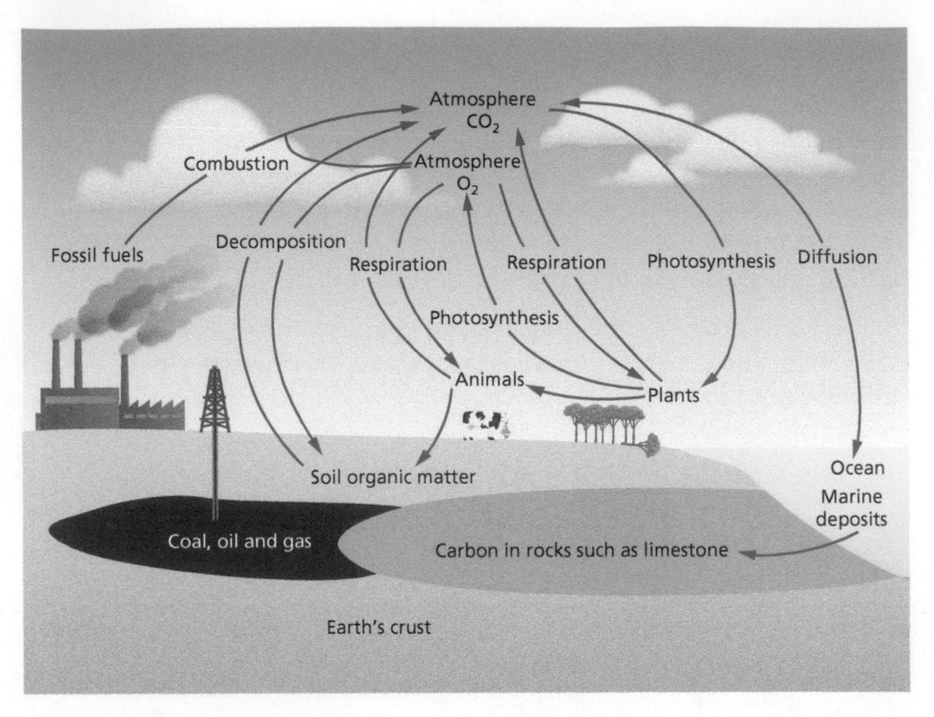

▲ **Figure 2.3.3** Carbon cycle

The transfer processes in the carbon cycle are as follows:

- herbivores feeding on producers
- carnivores feeding on herbivores
- decomposers feeding on dead organic matter
- carbon dioxide from the atmosphere dissolving in rainwater and oceans.

The transformation processes in the carbon cycle are as follows:

- photosynthesis transforms carbon dioxide and water into glucose using sunlight energy trapped by chlorophyll
- respiration transforms organic matter such as glucose into carbon dioxide and water, releasing energy in the process
- combustion transforms biomass into carbon dioxide and water
- biomineralization transforms carbon dioxide into calcium carbonate in shellfish and coral
- fossil fuels are made from the sedimentation of organic matter, incomplete decay and pressure.

Test yourself

2.13 Identify the processes described below:

a) The process that transforms carbon in fossil fuels to carbon dioxide in the atmosphere

b) The process that transforms carbon in plants to carbon dioxide in the atmosphere

c) The process that transfers carbon in plants to carbon in animals

d) The process that transforms carbon dioxide in the atmosphere to carbon in plants. [4]

Nitrogen cycle

The transfer processes in the nitrogen cycle are:

- herbivores feeding on producers
- carnivores feeding on herbivores
- decomposers feeding on dead organic matter
- plants absorbing nitrates through their roots.

The transformation processes in the nitrogen cycle involve four different types of bacteria. The processes are:

- nitrogen-fixing bacteria transform nitrogen gas in the atmosphere into ammonium ions
- nitrogen-fixing bacteria include bacteria that live in the soil and other bacteria that live in plant root nodules

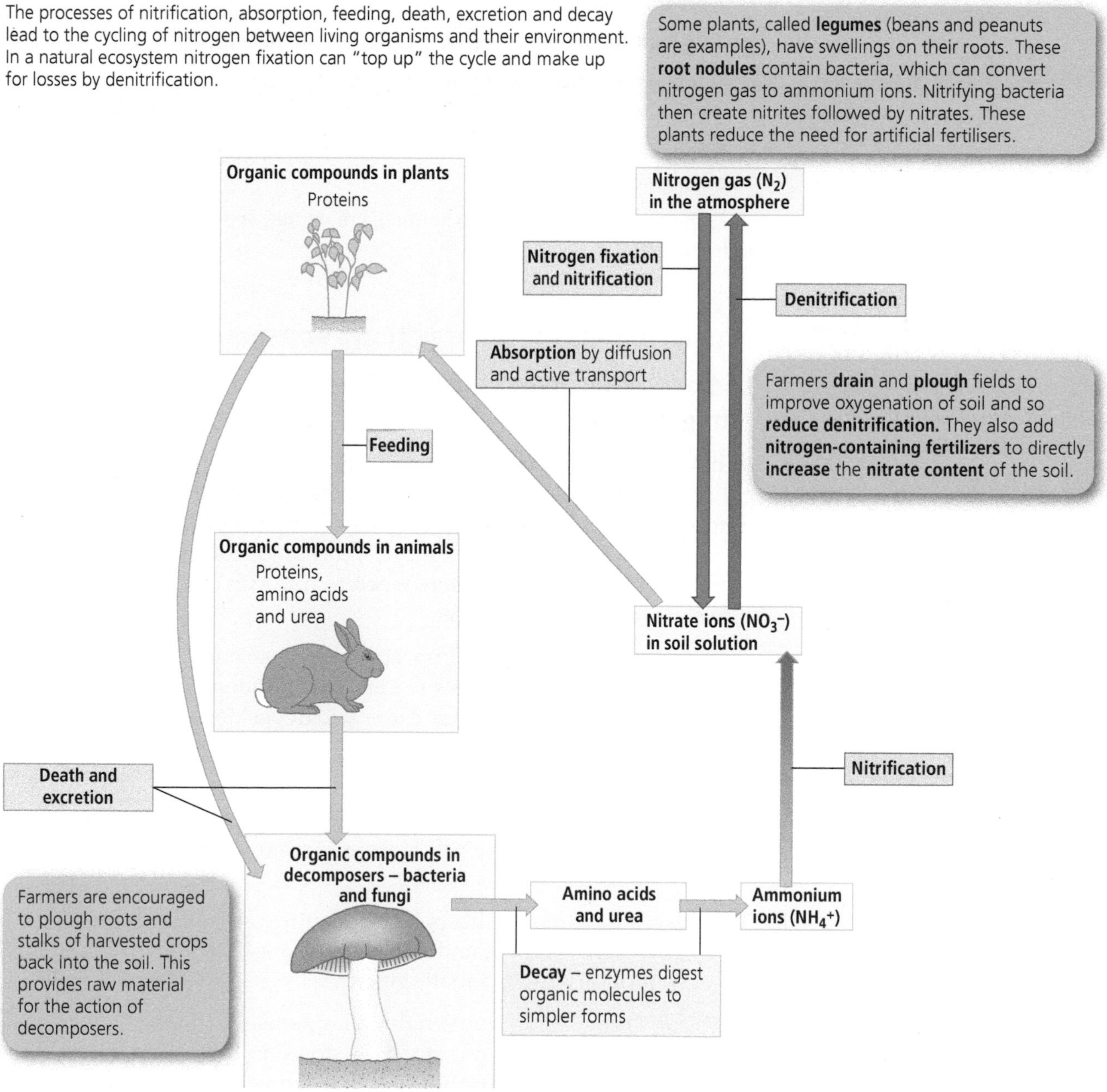

The processes of nitrification, absorption, feeding, death, excretion and decay lead to the cycling of nitrogen between living organisms and their environment. In a natural ecosystem nitrogen fixation can "top up" the cycle and make up for losses by denitrification.

Some plants, called **legumes** (beans and peanuts are examples), have swellings on their roots. These **root nodules** contain bacteria, which can convert nitrogen gas to ammonium ions. Nitrifying bacteria then create nitrites followed by nitrates. These plants reduce the need for artificial fertilisers.

Organic compounds in plants
Proteins

Nitrogen gas (N_2) in the atmosphere

Nitrogen fixation and nitrification

Denitrification

Absorption by diffusion and active transport

Farmers **drain** and **plough** fields to improve oxygenation of soil and so **reduce denitrification.** They also add **nitrogen-containing fertilizers** to directly **increase** the **nitrate content** of the soil.

Feeding

Organic compounds in animals
Proteins, amino acids and urea

Nitrate ions (NO_3^-) in soil solution

Nitrification

Death and excretion

Organic compounds in decomposers – bacteria and fungi

Amino acids and urea

Ammonium ions (NH_4^+)

Farmers are encouraged to plough roots and stalks of harvested crops back into the soil. This provides raw material for the action of decomposers.

Decay – enzymes digest organic molecules to simpler forms

▲ Figure 2.3.4 Nitrogen cycle

- nitrifying bacteria transform ammonium ions into nitrite and then nitrate
- denitrifying bacteria transform nitrates into nitrogen
- decomposers break down organic nitrogen into ammonia (deamination)
- producers convert inorganic materials into organic matter in the nitrogen cycle
- producers use nitrogen from nitrates to make amino acids and then protein
- decomposers convert organic storage into inorganic matter in the nitrogen cycle
- decomposers transform protein and amino acids into ammonium ions.

Content link

The soil system is covered in section 5.1.

Assessment tip

You need to be able to discuss human impacts on energy flows, and on the carbon and nitrogen cycles.

Test yourself

2.14 Identify the processes X, Y and Z described below:

X: The process that transforms nitrogen in the air to nitrogen compounds in green plants

Y: The process that transforms nitrate in soil to nitrogen gas in the atmosphere

Z: The process that converts ammonia to nitrites and then nitrates in the soil. [3]

Human impacts on energy flows and matter cycles

Comparing the impacts of humans on the carbon and nitrogen cycles

Content link

The effects of human activities on the carbon cycle are discussed further in section 7.2.

The effects of oxides of nitrogen on the production of photochemical smog are discussed in section 6.3, and in contributing to climate change in section 7.3.

Content link

Acid deposition is discussed further in section 6.4.

Content link

Soil degradation and conservation is examined in detail in section 5.3.

In both cycles combustion of forests and fossil fuels increases the concentration of oxides in the atmosphere, and in both cycles deforestation and agriculture lead to decomposition that also releases oxides. However, in the carbon cycle, carbon dioxide is released by respiration into the atmosphere whereas nitrogen oxides are released into soil water by nitrification.

Both oxides will increase the impact of global warming, although NO_x to a smaller degree. Both oxides result in the acidification of water and aquatic bodies, although only NO_x may cause acid deposition and acidify soils.

Deforestation removes organic storages of both nitrogen and carbon (stored in plant biomass) and reduces the absorption of carbon from the atmosphere (via photosynthesis) but not nitrogen. Deforestation also causes soil erosion, which reduces inorganic nitrogen storages in soil (not carbon).

The use of inorganic fertilizers increases nitrogen in the soil but does not increase carbon, and run-off may cause excessive inorganic nitrogen (not carbon) in aquatic systems.

Herbicide use in agriculture might kill organisms, thus reducing both carbon and nitrogen organic storages (stored in their biomass). In turn this will reduce the nitrification, denitrification and decomposition process, whereas the effect on the carbon cycle is limited to reducing respiration by soil animals.

The extraction of fossil fuels (oil, coal and gas) reduces ancient underground carbon storages and transfers carbon storages to the Earth's surface, for human use, whereas the effect on the nitrogen cycle is limited to a few organic compounds found in oil.

Human impacts on energy transfers

Energy trapped by plants millions of years ago, in the form of fossil fuels, is being released. The amount of energy available to humans has increased enormously, enabling industry and agriculture to be powered. Population growth, through increased food output, has therefore increased rapidly.

This change in the Earth's energy budget has led to many environmental issues, such as habitat destruction, climate change and the reduction of non-renewable resources.

Increased carbon dioxide levels and the corresponding increase in temperatures have led to the reduction in Arctic land and sea ice, leading to a potential tipping point.

Content link

Climate change causes and impacts are discussed in section 7.2.

QUESTION PRACTICE

Essay

Compare and contrast the impact of humans on the carbon and nitrogen cycles. [7]

How do I approach this question?

"Compare and contrast" means that you must give an account of both similarities *and* differences between the two cycles, referring to both of them throughout your answer. The danger in this question is that you give a simple account of each cycle without comparing and contrasting them. If only similarities or only differences are identified, a maximum of 5 marks can be awarded.

When thinking about your answer, you may want to do a rough plan, outlining both cycles and their key components, then arranging these in a table, with the carbon cycle in one column and the nitrogen cycle in the other – this will allow you to think about similarities between them, and also the differences.

SAMPLE STUDENT ANSWER

In the case of the CO_2 cycle, humans have contributed by releasing millions of tonnes by burning fossil fuels and by deforestation. This means that a lot of CO_2 has been released into the atmosphere and now there are fewer plants for turning it into O_2 again. By contributing to the CO_2 cycle in this way, the greenhouse effect has increased.

▲ A valid point about how humans have affected the carbon cycle for 1 mark

On the other hand, humans haven't contributed to the nitration cycle by burning fossil fuels, but by creating nitrogen-based fertilizers humans have added nitrogen to the soil of fields and in that way we should be contributing to the cycle. But the cycle is broken when the fertilized crops are taken to supermarkets and houses far away from the soil where nitrogen and other nutrients have been taken. By doing this, we prevent nitrogen from returning to the soil. This is why fertilizers are constantly added. In other cases, some plants with a high population of

▲ A contrasting point about the nitrogen cycle. The question requires the candidate to "compare and contrast the impact of humans on the carbon and nitrogen cycles"

▲ A valid point about how humans have affected the nitrogen cycle for 1 mark

▲ A further point about human impacts on the nitrogen cycle. Nitrogen is taken away from the original site of origin to supermarkets where the biomass containing nitrogen is sold. One further mark awarded

▼ Nitrogen fixation by bacteria is a natural process. Although an increase in nitrogen fixation by agriculture will increase levels of fixation, it is not on the same scale as nitrogen added to fields by fertilizer

▼ This paragraph repeats the points already made, and does not add any new marking points

nitrogen-fixing bacteria had been placed in the fields. This is a big contribution to the cycle because it allows the fixation of atmospheric nitrogen into the soil.

To sum up, humans have increased the amount of CO_2 in the atmosphere by releasing CO_2 present in fossil fuels kept away from the cycle. And we have interfered with the cycle itself by deforestation. While in the nitrogen cycle we have interfered by taking away nitrogen from the soil and we have been forced to add it in the way of fertilizer or with more nitrogen-fixing bacteria.

This response could have achieved 3/7 marks.

Some knowledge of the carbon and nitrogen cycles is shown but only a limited discussion of the human impacts on them. Only one contrasting point is made and no similarities are discussed. A similarity between the two cycles would be, for example, that in both cycles combustion of forests or fossil fuels increases concentration of oxides in the atmosphere. A difference would be that carbon dioxide is released by respiration into the atmosphere whereas nitrogen oxides are released into soil water by nitrification.

Better planning of the answer would have ensured that more of the marking points were covered.

2.4 BIOMES, ZONATION, AND SUCCESSION

- **Biome** – collections of ecosystems sharing similar climatic conditions, each having characteristic limiting factors, productivity and biodiversity
- **Zonation** – changes in community along an environmental gradient due to factors such as changes in altitude, latitude, tidal level or distance from shore (coverage by water)
- **Succession** – the process of change over time in an ecosystem involving pioneer, intermediate and climax communities

You should be able to show an understanding of how climate determines ecosystem distribution, and how ecosystems develop though time and space and are affected by human disturbance

✔ Biomes have different limiting factors, productivity and biodiversity depending on their geographical location.

✔ The distribution of biomes is being altered by climate change.

✔ Communities change both spatially and temporally to develop different ecosystems: "succession" refers to the change in communities through time and "zonation" refers to spatial changes, along environmental gradients.

✔ *r*-strategist species are found in pioneer communities and *K*-strategist species are found in climax communities.

✔ Human activities affect the development of ecosystems.

✔ An ecosystem's capacity to survive change depends on its diversity and resilience.

Biomes

Biome distribution depends on levels of insolation (sunlight), temperature and precipitation (rainfall). For example, the distribution, structure and relative productivity of tropical rainforest can be explained as follows:

- Distribution: they are found between the tropics of Cancer and Capricorn (23.5° N and S), where warm conditions exist throughout the year, rainfall is high, and levels of insolation are high throughout the year due to lack of seasons.
- Structure: high levels of photosynthesis support a complex ecosystem structure. There are a number of layers from ground level to canopy to emergent plants; trees of up to 50 metres with lower layers of shrubs and vines.
- Relative productivity: there are high levels of productivity due to high levels of photosynthesis.

The following factors will limit productivity:

- cold temperatures in autumn and winter will limit the growing season
- limited annual rainfall (e.g. less than 250mm per year) will provide insufficient water
- uneven seasonal distribution of rainfall throughout the year
- frozen soil and snow cover during winter months.

Climate change is altering the distribution of biomes and causing biome shifts. For example, it is thought that tundra, rainforest and desert will have different vegetation types in the year 2100 than they do now, due to changes in climate (source: www.pnas.org/content/104/14/5738).

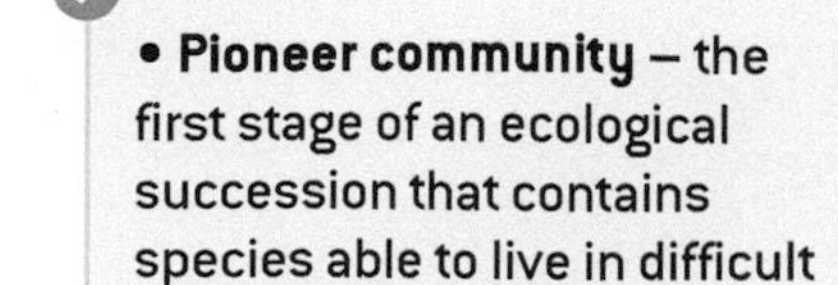

- **Pioneer community** – the first stage of an ecological succession that contains species able to live in difficult environmental conditions
- **Climax community** – the final stage of a succession that is more or less stable/balanced

Assessment tip

You need to be able to explain the distributions, structure, biodiversity and relative productivity of contrasting biomes, and analyse data for a range of biomes.

Assessment tip

You need to be able to discuss the impact of climate change on biomes.

Test yourself

2.15 Explain the role of climate in the distribution and relative productivity of a named biome. [4]

Zonation

Zonation refers to changes in communities in space, along an environmental gradient. Zonation is due to factors such as altitude, latitude, tidal level or distance from shore. Zonation occurs at different spatial scales, from small-scale changes along rocky shores (see below), to larger-scale zonation on mountains due to changes in climate (with different communities found in different zones due to abiotic factors such as temperature and precipitation), to planet-wide zonation due to latitude (i.e. different communities found at different latitudes due to variation in insolation, precipitation and temperature).

An example of zonation is the rocky shore (see Figure 2.4.1), which can be divided into zones from the lower to upper shore. On a rocky shore each zone can be defined by the spatial patterns of animals and plants. Seaweeds in particular show distinct zonation patterns. Seaweed species that are more resilient to water loss, such as channel wrack, are found on the upper shore. Seaweed that is less resilient to water loss, such as kelp, is found on the lower shore where it is not out of the water for long.

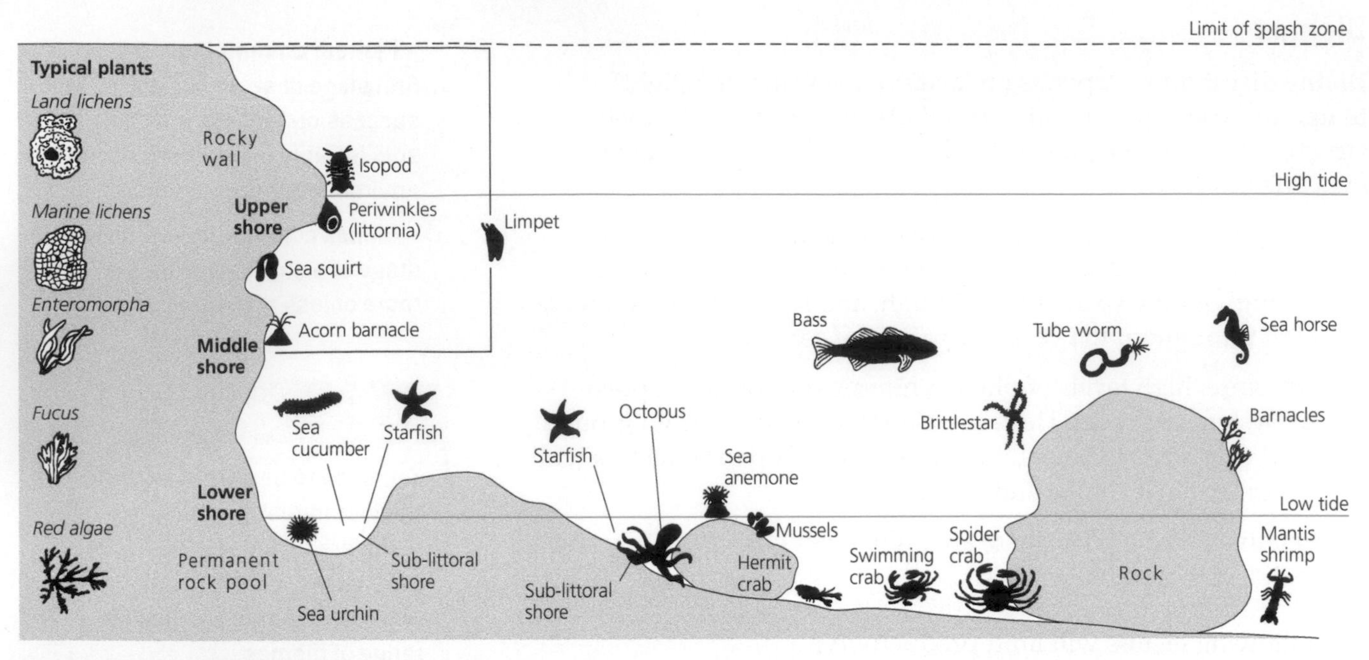

▲ Figure 2.4.1 Zonation on a rocky shore

Succession

Succession happens when species change the habitat they have colonized and make it more suitable for new species. Changes in the community of organisms cause changes in the environment they live in. These changes in the environment allow another community to become established and replace the one before through competition.

Each stage of a succession is called a seral stage. The first seral stage of a succession is called the **pioneer community**. The later communities in a sere are more complex than those that appear earlier. The final seral stage of a succession is called the **climax community**.

Table 2.4.1 shows the differences between pioneer and climax communities.

>> Assessment tip

You need to be able to describe the process of succession using a given example, and explain the general patterns of change in communities undergoing succession.

▼ Table 2.4.1 Comparing pioneer and climax communities

Pioneer community	Climax community
The first seral stage of a succession	The final seral stage of a succession
r-strategists are abundant	*K*-strategists are abundant
Simple in structure with low diversity	Complex in structure with high diversity
Species can tolerate harsh conditions such as strong light and low nutrient levels	Characteristics of climax community are determined by climate and soil
For example: a community of lichens covering bare rock	For example: a community of trees and shrubs

Succession causes changes in the biotic and abiotic conditions:

- It increases levels of vegetation.
- It causes an increase in soil aeration.
- Decaying vegetation improves the soil fertility.
- The soil and vegetation provide habitats for other organisms.
- This leads to increased biodiversity.

Concept link 

BIODIVERSITY: Climax communities generally have higher levels of biodiversity than earlier communities in a succession.

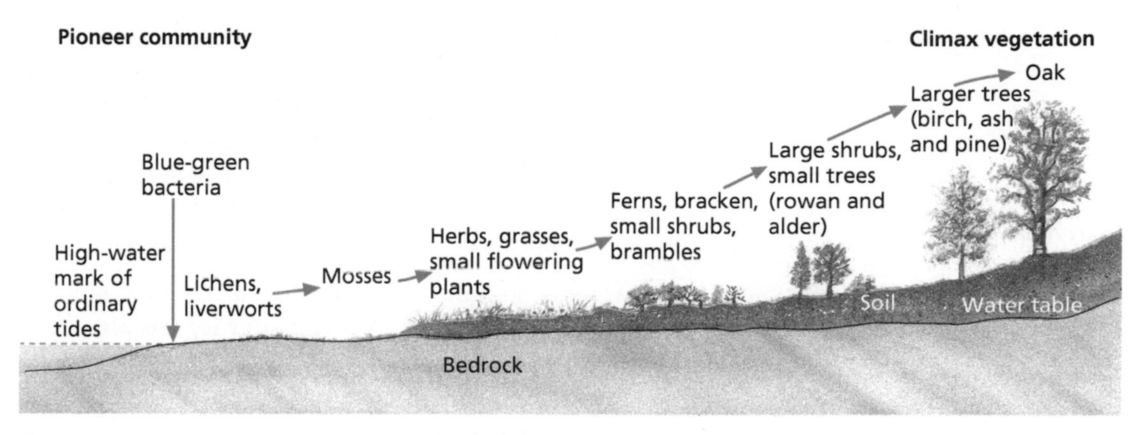

▲ Figure 2.4.2 A succession on rock from pioneer to climax community

An example of a succession is the change that occurs from bare rock to climax temperate forest community (figure 2.4.2):

- Lichens and mosses are pioneer species. Very few species can live on bare rock as it contains little water and has few available nutrients. Lichens can photosynthesize and are effective at absorbing water. Lichens therefore do not need soil to survive and are excellent pioneers. Once established, lichens and mosses trap particles blown by the wind. When the lichens and mosses die and decompose they form a soil in which grasses can germinate. The growth of pioneers helps to weather parent rock, adding still further to the soil.
- Grasses and ferns that grow in thin soil can now colonize the area. These new species are better competitors than the pioneer species. Grasses grow taller than mosses and lichen, and so get more light for photosynthesis. Grass roots trap soil and stop erosion. Grasses have a larger photosynthetic area and so can grow faster.
- The next stage involves the growth of herbaceous plants. Herbaceous plants include dandelions and goose-grass, which need more soil to grow but which outcompete the grasses. These herbaceous plants have wind-dispersed seeds and rapid growth, and so become established before larger plants arrive.
- Shrubs then appear, such as bramble, gorse and rhododendron. Shrubs are larger plants than the ones in earlier seral stages. The larger plants can grow in good soil and are better competitors than the slower-growing plants of the earlier seral stages.
- The final stage of a succession is the climax community. Here, trees that have grown produce too much shade for the shrubs. The shrubs are replaced by shade-tolerant forest floor species (species that can survive in shady conditions). The amount of organic matter increases as succession progresses because as pioneer and subsequent species die out, their remains contribute to a build-up of litter from the biomass. Soil organisms move in and break down litter, leading to a build-up of organic matter in the soil, making it easier for other species to colonize. Soil also traps water and so increasing amounts of moisture are available to plants in the later stages of the succession.

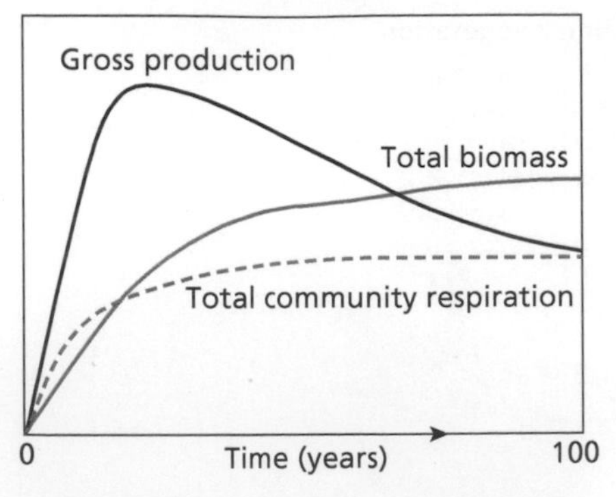

▲ Figure 2.4.3 Due to high NPP and relatively low community respiration in the pre-climax successional stages, gross productivity remains high and the P : R ratio therefore remains more than 1. There is high community respiration in the climax stage and so the P : R ratio equals 1 (P : R = 1)

In early stages, **gross productivity** is low due to the low density of producers. The proportion of energy lost through community respiration is relatively low too, so net productivity is high, i.e. the system is growing and biomass is accumulating. In later stages, with an increased consumer community, gross productivity may be high in a climax community. However, this is balanced by respiration, so net productivity approaches zero and the production : respiration (P : R) ratio approaches one.

Primary succession happens on bare substrate such as rocks or shingle ridges. Secondary succession occurs where a community has already existed in the past and soil exists (figure 2.4.4).

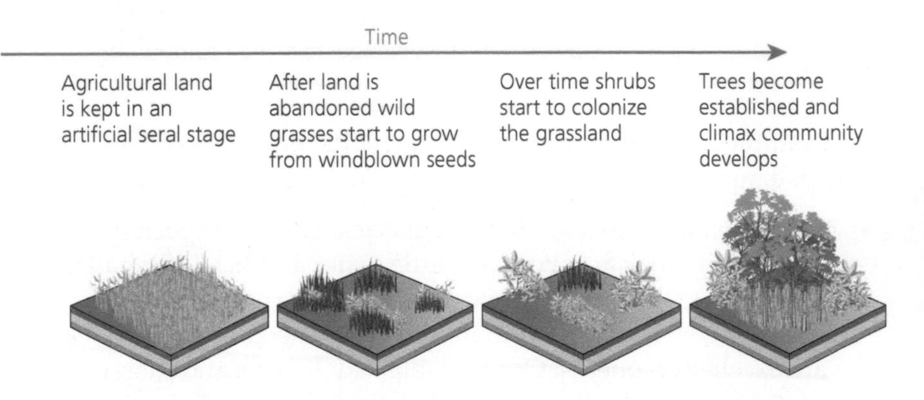

▲ Figure 2.4.4 An example of secondary succession on abandoned farmland

Assessment tip

You need to be able to interpret models or graphs related to both succession and zonation.

Test yourself

2.16 Compare and contrast, using examples, succession and zonation. [4]

2.17 Outline the differences between pioneer and climax communities. [4]

Assessment tip

The concept of succession, occurring over time, should be carefully distinguished from the concept of zonation, which refers to a spatial pattern. You may be asked to distinguish between the two.

▼ Table 2.4.2 Comparing *r*- and *K*-strategist species

r-strategist species	*K*-strategist species
Short life	Long life
Rapid growth	Slower growth
Early maturity	Late maturity
Many small offspring	Fewer, large offspring
Little parental care or protection	High parental care and protection
Little investment in individual offspring	High investment in individual offspring
Adapted to unstable environment	Adapted to stable environment
Pioneers, colonizers	Later stages of succession
Niche generalists	Niche specialists
Prey	Predators
Regulated mainly by external factors	Regulated mainly by internal factors
Lower trophic level	Higher trophic level
Examples are: flour beetles, bacteria, cockroaches	**Examples are: trees, albatrosses, humans**

r- and *K*-strategist species

Species can be classified as *r*-strategist or *K*-strategist (see table 2.4.2).

K-strategists:

- occur in climax communities
- are large organisms that need a developed soil and food web
- are slow growers
- outcompete other organisms for light and nutrients
- have few offspring
- can only grow to maturity when a space appears in the ecosystem.

r-strategists:

- are pioneers and can reach new areas that need colonization
- are involved in the initial colonization of a piece of land or water
- have rapid growth
- can get through the lifecycle quickly
- have many offspring
- are generally small in size
- cannot compete in later stages of a succession.

Effect of human activity on succession

A climax community is more diverse and therefore more stable than a community which has been interrupted by human activity. Climax communities have more resilience due to the following factors.

- A climax community is more productive
  - making more energy to support consumers and decomposers
  - so more niches are available.
- Each stage (sere) of succession helps to create a deeper and more nutrient-rich soil, so allowing larger plants to grow
  - this increases the habitat diversity which leads to greater species and genetic diversity and thus greater stability.
- Climax communities have a more complex system which is more stable.
- They contain more complex food webs and have greater diversity, so they have more stability if one organism goes extinct, creating balanced relationships/feedback mechanisms leading to a steady state.

Human activity can have various effects, for example:

- It can lead to a decrease in productivity due to the removal of primary producers:
  - leading to reduced niches
  - this is a particular threat to more specialized species.
- Human activity can also cause deterioration in abiotic factors (leading to harsh conditions to which few species adapt).
- It can remove certain species, creating shorter food webs.
- It can generate rapid change
  - so relationships have insufficient time to evolve a stable balance (steady state).

Humans often try to recreate pioneer seres in agriculture. These are less stable and so humans have to constantly monitor crops. Monocultures in agriculture are more vulnerable to disease and pests and so less stable.

Assessment tip

You need to be able to discuss the factors that could lead to alternative stable states in an ecosystem, and discuss the link between ecosystem stability, succession, diversity and human activity.

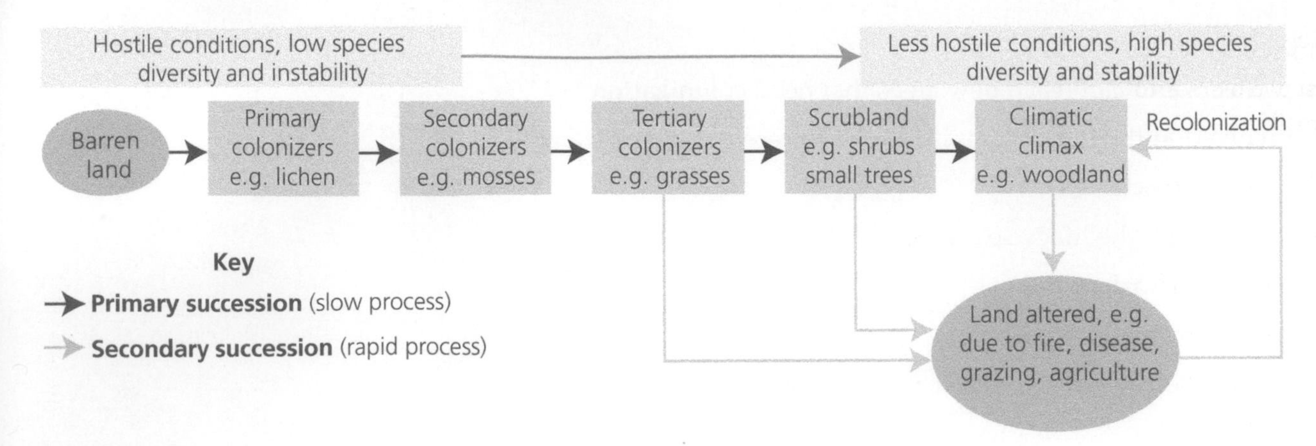

▲ Figure 2.4.5 Comparing primary and secondary succession

▲ Figure 2.4.6 An example of a plagioclimax, Goathland, Yorkshire, UK

An interrupted succession due to human activity is called a plagioclimax. Figure 2.4.6 shows an example of a plagioclimax, where heather has been set on fire to reinvigorate fresh growth and to stop communities proceeding to a climax community (i.e. woodland). This is a strategy employed by moorland management in the North York Moors National Park on a rotational basis.

Test yourself

2.18 Explain the relationship between ecosystem stability, diversity and succession. [4]

Concept link 

BIODIVERSITY: High biodiversity confers resilience in ecosystems.

Content link

The role of resilience and diversity in stabilizing systems has been discussed in section 1.3.

The concepts of species, habitat and genetic diversity are discussed in section 3.1.

Content link

The concept of keystone species is covered in section 3.4. The role of sizes of storages and negative feedback in conferring resilience is discussed in section 1.3.

Concept link 

EQUILIBRIUM: Resilience in ecosystems leads to steady-state equilibrium and avoidance of tipping points.

Resilience of ecosystems

Various factors contribute to the resilience of ecosystems:

- high biodiversity (i.e. genetic, species and habitat diversity)
- the complexity of interactions between components of the ecosystem, i.e. developed food webs, nutrient cycling, and the establishment of keystone species
- establishment of keystone species
- size of storages, e.g. large population sizes and abundant resources such as nutrients, water, sunlight, reproductive rates, and biomass
- larger size of the system
- strong negative feedback systems
- position of tipping points/thresholds of change
- maturity, i.e. later stage of succession (climax community)
- balance of inputs and outputs (i.e. steady-state equilibrium).

Assessment tip

If asked to explain why some ecosystems are more resilient than others, do not simply state that they have "storages" or "tipping points" because all systems have these. It is the size or position of the storage or tipping point respectively that determines its resilience. Responses referring to low human interference are not correct: reduced disturbance may lead to more stable ecosystems, but it does not influence their resilience as such (i.e. their inherent ability to resist disturbance).

QUESTION PRACTICE

1 List two biomes found in Brazil. [1]

2 Outline how four different factors influence the resilience of an ecosystem. [4]

How do I approach these questions?

1 The question requires you to use information from the resource booklet to list *two* different biomes found in Brazil, for one mark. If you list only one biome, you will receive no marks.

2 The command term "outline" means "give a brief account or summary". You do not need to go into detail about why these various factors contribute to resilience. One mark will be awarded for each correct factor identified, up to a maximum of 4 marks. For full marks, you should say how each factor affects resilience. If valid factors are identified but their effect on resilience is not, one mark will be awarded for each of two factors, up to four factors, i.e. a maximum of 2 marks can be awarded.

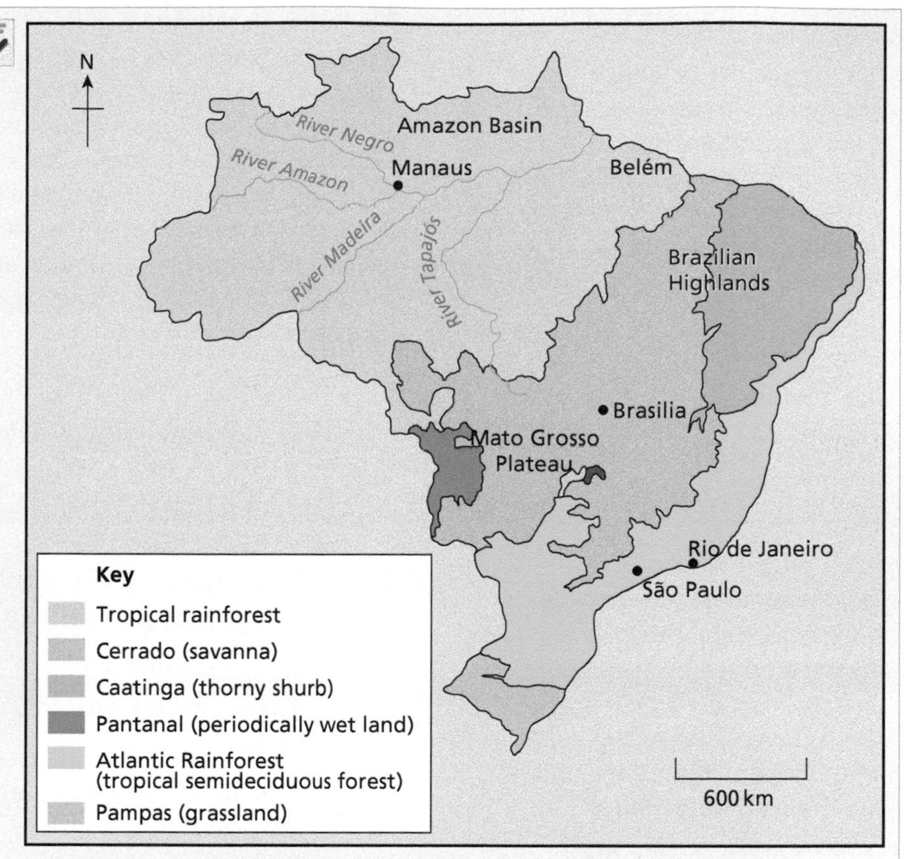

▲ **Figure 1** Map of Brazil's vegetation including location of Cerrado and Atlantic Rainforest

SAMPLE STUDENT ANSWER

1 Tropical rainforest
Grassland

▲ Question asks for two biomes to be listed, using information provided by the Resource booklet. Other answers could have included tropical savanna, thorny shrub, periodically wet land, and tropical semi-deciduous forest

This response could have achieved 1/1 marks.

2 The resilience of an ecosystem can be affected by multiple factors. The species diversity in an ecosystem can affect the resilience of a system. An ecosystem with a large amount of species diversity is more resilient as even if one organism dies out, another one will be able to take its place in the system. The genetic diversity within a species also affects the resilience of an ecosystem. The more genetically diverse a species is, the more resilient the ecosystem becomes. The geographic range of an ecosystem also contributes to its resilience. The larger the area of an ecosystem, the more resilient it is. Finally, the number of organisms in an ecosystem affects its resilience. The more the number of organisms in an ecosystem, the higher the resilience.

▲ Correctly says that greater diversity of components (e.g. species) increases resilience, for 1 mark

▼ There will be only one mark available for mentioning biodiversity or any aspect of it, e.g. genetic, species, or habitat diversity. No additional mark is therefore awarded for "genetic diversity" because 1 mark has already been awarded for discussing species diversity

▲ Larger size of the system increases resilience, and so 1 mark awarded

>> Assessment tip

Strong positive feedback mechanisms, being close to a tipping point, and human impact degrading the structure, diversity or abundance will decrease the resilience of a system.

This response could have achieved 2/4 marks.

Only two distinct points have been made in this answer, and so only 2 marks out of a total of 4 have been awarded. Other answers could have included: strong negative feedback systems increase resilience; strong positive feedback mechanisms may decrease resilience; human impacts degrade structure (e.g. diversity or abundance of resources) which will decrease resilience; a steady-state equilibrium/balanced inputs and outputs (as in climax communities) increases resilience; systems being close to a tipping point decreases resilience.

2.5 INVESTIGATING ECOSYSTEMS

- **Dichotomous key** – a guide where there are two options based on different characteristics at each step. The outcome of each choice leads to another pair of questions. This is done until the organism is identified
- **Percentage cover** – the percentage of the area within the quadrat covered by one particular species. Percentage cover is worked out for each species present
- **Population density** – the number of individuals of each species in a specific area. It is calculated by dividing the number of organisms by the total area of the quadrats
- **Percentage frequency** – the percentage of quadrats in an area in which at least one individual of the species is found
- **Quadrat** – empty square frames of known area (e.g. 1 m^2), used for estimating the abundance of plants and non-mobile animals
- **Environmental gradient** – an area where two ecosystems meet or where an ecosystem ends
- **Transect** – used to measure changes along an environmental gradient, ensuring all parts of the gradient are measured

You should be able to show how ecological investigations can be carried out, using biotic and abiotic sampling methods

✔ Ecosystems need to be named and located, for example the Danum Valley Conservation Area in Sabah, Malaysia – a tropical forest ecosystem.

✔ Organisms in an ecosystem can be sampled and identified using a variety of different techniques.

✔ The biomass and energy of trophic levels can be estimated and extrapolated to whole communities.

✔ The abundance of non-motile and motile species can be estimated using different sampling techniques.

✔ "Species richness" refers to the number of different species in an area whereas "species diversity" takes into account both the number of species and their relative abundances.

✔ The Simpson diversity index can be used to assess communities.

Identifying specimens

In an ecological study it is important to correctly identify the organisms being studied. It is unlikely that you will be an expert in the animals or plants you are looking at and so **dichotomous keys** must be used (figure 2.5.1). Dichotomous means "divided in two parts". The key is written so that identification is done in steps. At each step two options are given, based on different characteristics of

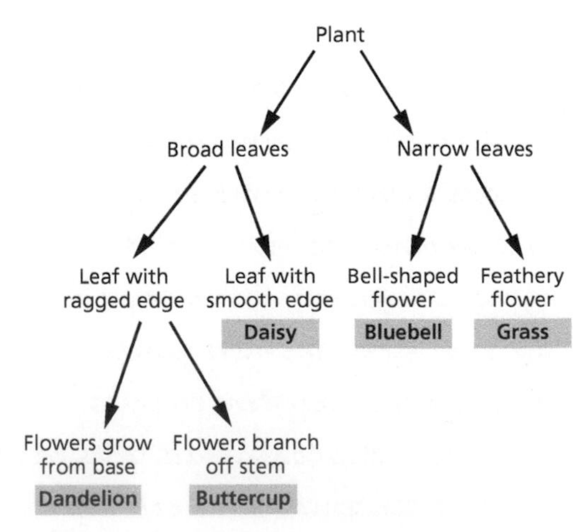

▲ Figure 2.5.1 An example of a dichotomous key to identify a selection of different plants

the organisms being identified. The outcome of each choice leads to another pair of questions, and so on until the organism is identified.

Herbarium or specimen collections can also be used to compare specimens. Scientific expert advice can also be sought. Modern techniques use genetic analysis to separate different species at a molecular level.

Measuring abiotic factors

Ecosystems can be divided into three types: marine, freshwater, and terrestrial. Marine ecosystems include the sea, estuaries, salt marshes, and mangroves, which all have a high concentration of salt in the water. Estuaries are included in the same group as marine ecosystems because they have a high salt content compared to freshwater. Freshwater ecosystems include rivers, lakes, and wetlands. Terrestrial ecosystems include all land-based ecosystems. Each type of ecosystem has its own specific abiotic components as well as factors that they share with other types of ecosystems.

Abiotic components of a marine ecosystem include: salinity, pH, temperature, dissolved oxygen, and wave action.

Abiotic components of a freshwater ecosystem include: turbidity, temperature, flow velocity, dissolved oxygen, and pH.

Abiotic components of a terrestrial ecosystem include: temperature, light intensity, wind speed, soil particle size, amount and angle of slope, soil moisture, drainage, and mineral content.

Methods to measure these abiotic factors have been discussed in 2.1.

Measuring biotic factors

Quadrats

Biotic factors can be measured using **quadrats**. Quadrats can be used to measure **percentage cover, population density** and **percentage frequency**. Quadrats are suitable for measuring vegetation and non-motile animals.

Different types of quadrat can be used. **Frame quadrats** are empty frames of known area such as 1 m^2. **Grid quadrats** are frames divided into 100 small squares with each square representing one percent that helps in calculating percentage cover. **Point quadrats** are made from a frame with 10 holes which is inserted into the ground by a leg. A pin is dropped through each hole in turn and the species touched are recorded. Point quadrats are used for sampling vegetation that grows in layers. In a point quadrat the total number of pins touching each species is converted to percentage frequency data, for example if a species touched 6 out of the 10 pins it has 60% frequency.

Lincoln Index

The Lincoln Index is also known as the capture-mark-release-recapture method.

Animals are captured, marked, and then released. After a specific amount of time the animal population is resampled. Some of the animals initially marked will be caught again, or recaptured. The total population size of the animal is estimated using this equation:

$N = (n_1 \times n_2)/n_m$

N is the total estimated population size.

>> **Assessment tip**

You need to be able to construct simple identification keys for up to eight species.

Test yourself

2.19 State two limitations of using a key to identify organisms. [2]

>> **Assessment tip**

You may be asked to evaluate the use of quadrats in ecological studies. The quadrat method is difficult to use for very large or very small plants. It is difficult to count plants that grow in tufts or colonies. It is possible that plants that appear separate are joined by roots; this will affect calculation of population density. It is also difficult to measure the abundance of plants outside their main growing season when plants are largely invisible.

n_1 is the number caught in the first sample.

n_2 is the number caught in the second sample.

n_m is the number caught in the second sample that were marked.

Assessment tip

You need to be able to evaluate sampling strategies, and to evaluate methods for measuring or estimating populations of motile and non-motile organisms.

For example:

A snail population was sampled. Snails were marked by putting paint on their shell.

Data from the snail population sampled using the capture-mark-release-recapture method was as follows:

- Number of snails captured in first sample and marked, $n_1 = 21$
- Number of snails captured in second sample, $n_2 = 13$
- Number of snails captured in second sample that are marked, $n_m = 5$.

Total population size of snails in the study site, $N = (21 \times 13)/5 = 55$.

The population size is rounded up to the nearest whole number.

Animals may move in and out of the sample area, making the capture-mark-release-recapture method less trustworthy and the data invalid. The density of the population in different habitats might vary: there may be many in one area, few in another. The assumption that they are equally spread all over may not be true. Some individuals may be hidden by vegetation and therefore difficult to find, hence not included in the sample. There may be seasonal variations in animals that affect population size; for example, they may migrate in or out of the study area. A key source of error is when trapping or marking procedures affect the probability of recapture. For example, if yellow paint spots are painted on the shells of snails then the snails will be more visible to predators, and so fewer will be recaptured.

Test yourself

2.20 In a field experiment 60 edible dormice (*Glis glis*) were captured using Longworth mammal traps laid out in a grid within a 500 m × 500 m quadrat. Each individual was marked and released. Two days later a second trapping exercise caught 50 edible dormice, 15 of whom were previously marked.

Use the Lincoln Index to **estimate** the size of the edible dormouse population. [2]

Measuring change along an environmental gradient

Placing quadrats at random is not appropriate for measuring abiotic and biotic factors along an **environmental gradient**. This is because environmental variables change along the gradient and all parts of the gradient need to be sampled. A **transect** is used to ensure that all parts of the gradient are sampled. Using a transect is an example of systematic sampling. The simplest transect is a line transect. A line transect is made by placing a tape measure in the direction of the gradient; for example, on a beach this would be at 90° to the sea. All organisms touching the tape are recorded. Many line transects need to be taken to obtain valid quantitative data.

Larger samples can be taken by using a belt transect. This is a band of chosen width, usually between 0.5 and 1 m, placed along the gradient. Many belt transects need to be taken to obtain valid quantitative data.

If the whole transect is sampled in line and belt transects it is called a continuous transect. If samples are taken at points of equal distance along the gradient in line and belt transects it is called an interrupted transect. Horizontal distances are used if there is no discernible vertical change in an interrupted transect, such as along a shingle ridge succession. If there is a climb or descent in an interrupted transect then vertical distances are normally used, such as on a rocky shore.

Transects should be repeated so that data are reliable and quantitatively valid. A tape measure is laid at right angles to the environmental gradient and transects are located at random intervals along the tape. Locating transects at random avoids bias. The location of transects should be chosen by using a random number generator. The number chosen can relate to the distance along a tape measure placed perpendicularly to the environmental gradient. All transects can be located randomly or they can be systematically located following the random location of the first.

Using transects to sample an environmental gradient can give biased results if the sample is too small. It is also possible that using an interrupted transect results in some parts of the gradient not being recorded. Repeating transects and covering the largest area possible will improve the validity of the data.

It is important that measurements from transects are carried out at the same time of day so that abiotic variables can be compared. Seasonal fluctuations mean that samples should be taken either as close together in time as possible or throughout the whole year. Data-logging equipment allows continuous data to be recorded over long periods of time. Biotic measurements will also vary with time and so must be treated in the same way as the abiotic variables.

Estimating the biomass and energy of trophic levels

The sample of the organism whose biomass is being determined is weighed in a container of known mass. The sample is put in a hot oven at around 80°C and left for a specific length of time. The sample is reweighed and replaced in the oven. This is repeated until the same mass is obtained on two subsequent weigh-ins; no further loss in mass is recorded, as no further water is present.

Biomass is recorded per unit area, such as per metre squared. This is done so that trophic levels can be compared. Not all organisms in an area need to be sampled.

To estimate the biomass of a primary producer within a study area all the vegetation is collected within a series of quadrats. The vegetation will include roots, stems and leaves. Dry biomass measurements of sample areas can be extrapolated to estimate total biomass over the whole area being considered by the study. This means that the mass of one organism, or the average mass of several organisms, is taken. This mass is multiplied by the total number of organisms to estimate total biomass.

Assessment tip

You need to be able to evaluate methods for estimating biomass at different trophic levels in an ecosystem.

Assessment tip

One criticism of the method for estimating biomass is that it involves killing living organisms. It is also difficult to measure the biomass of very large plants, such as trees. There are also problems measuring the biomass of roots and underground biomass, as these are difficult to remove from the soil.

Species richness and diversity

Diversity is not just the number of species in an area (this is the species richness), but the function of two components: the number of different species and the relative numbers of individuals of each species.

Make sure you can calculate the Simpson diversity index using given data, with the formula:

$$D = \frac{N(N-1)}{\Sigma n(n-1)}$$

where D = diversity index, N = total number of organisms of all species found, n = number of individuals of a particular species

You do not need to memorize this formula, but understand what the different symbols mean. Σ indicates that you will sum all of the $n(n-1)$ components.

Content link

The application of species diversity measures is discussed in section 3.1.

A high value of D suggests a stable and ancient site, where all species have similar abundance (or "evenness"), and a low value of D could suggest pollution, recent colonization or agricultural management, where one species may dominate.

>> Assessment tip

You need to be able to calculate and interpret data for species richness and diversity.

Test yourself

2.21 Two streams were sampled for aquatic invertebrates. The following species and numbers of individuals were recorded:

Stream A: mayfly nymph – 4, caddis fly larva – 20, freshwater shrimp – 70, water louse – 34, bloodworm – 10, sludgeworm – 2.

Stream B: mayfly nymph – 0, caddis fly larva – 0, freshwater shrimp – 1, water louse – 4, bloodworm – 45, sludgeworm – 100.

a) Create a table to display these results. **Calculate** the Simpson diversity index for both Stream A and Stream B. You must show your working. [3]

b) **Describe** two differences between streams A and B. [2]

c) **Suggest** a reason for the difference in the Simpson diversity index between the two streams. [2]

Designing and carrying out ecological investigations

>> Assessment tip

You need to be able to design and carry out ecological investigations.

Usual scientific procedures must be applied when carrying out ecological investigations, i.e. an independent variable, dependent variable, and controlled variables identified and applied. When carrying out work in the field, it will not be possible to control certain variables (i.e. they cannot be kept the same), such as temperature and light intensity. If these "uncontrolled" variables may affect the dependent variable (i.e. the variable being measured) then they should be monitored so that any impact on the investigation can be determined.

Content link

The way in which environmental conditions vary according to habitat island size is discussed in section 3.4.

Both abiotic and biotic factors should be measured:

- If investigating the effect of human activity on an ecosystem, measurements must be taken in both undisturbed and disturbed habitats so that comparisons can be made.
- Habitat islands of different sizes will have different environmental conditions, and so a variety of different-sized patches must be measured.
- Samples must be repeated so that data are reliable.
- Abiotic factors can include temperature, humidity and light intensity for example.
- Biotic factors can include the species richness or diversity of plants or animals present, population sizes, or the distribution and abundance of selected indicator species.
- Where environmental gradients are present, factors must be measured along the full extent of the gradient so that valid comparisons can be made.
- Factors must be measured over a long period of time to take into account daily and seasonal variations.

>> Assessment tip

You need to be able to evaluate methods to investigate the change along an environmental gradient and the effect of a human impact in an ecosystem.

See 2.1 and 2.5 for description and evaluation of methods for measuring abiotic and biotic factors.

QUESTION PRACTICE

▼ Table 1 The species richness of Yasuni National Park

Group	Number of species	Unit area (km^2)
Amphibians	139	6.5
Trees	655	0.1

Question

Describe a method that may have been used for collecting the tree data in table 1. [2]

How do I approach this question?

"Describe" means that you need to give a detailed account, i.e. state a collecting method and then say how it would be used to collect the tree species richness data shown in the table. For example, quadrat sampling can be used, where all the different species seen in quadrats can be identified and counted. You only need to state *one* collecting method, not two or more.

SAMPLE STUDENT ANSWER

The tree data could have been collected using a quadrat method. In this method, the total area is divided in equal parts, and randomly selected areas would count the amount of trees. For example, the total area is divided in 100 equal quadrats, 25 randomly selected areas would be looked into and counted. This allows us to estimate an amount of trees in the total areas from where the date is collected.

▲ Correct sampling method identified for 1 mark

▼ Not enough detail. Need to mention the number of trees or number of species

▼ This is too vague. The figure shows number of species and so the sampling method needs to refer to how this would have been recorded, for example by identifying and counting all the different species seen in quadrats

This response could have achieved 1/2 marks.

Test yourself

2.22 Describe the methods that may be used to record changes in a named ecosystem as a result of human activities. [4]

3 BIODIVERSITY AND CONSERVATION

This unit examines the origin and distribution of Earth's biodiversity. It also focuses on the threats faced by animal and plant species, and how they can be best protected into the future. There are links to Unit 2 (Ecosystems and ecology), with references to how biodiversity can be measured, and the role that different environmental value systems have in determining how societies value and protect biodiversity.

You should be able to show:

- ✔ an understanding of how biodiversity can be quantified;
- ✔ the processes through which biodiversity originates, and how mass extinction events have led to biodiversity loss;
- ✔ the threats faced by biodiversity, and how these can be measured;
- ✔ how biodiversity can be conserved through a range of different measures.

3.1 AN INTRODUCTION TO BIODIVERSITY

- **Biodiversity** – a broad concept encompassing the total diversity of living systems, which includes the diversity of species, habitat diversity and genetic diversity
- **Habitat diversity** – the range of different habitats in an ecosystem or biome
- **Genetic diversity** – the range of genetic material present in a population of a species
- **Species diversity** – a function of the number of species and their relative abundance
- **Diversity index** – quantitative measure of species richness, calculated using a combined measure of species richness and evenness
- **Evenness** – a measure of the relative abundance of different species making up the species richness of an area

Concept link

BIODIVERSITY: This is comprised of different elements – species, habitat and genetic.

You should be able to show an understanding of how biodiversity can be quantified

- ✔ The concept of biodiversity encompasses species diversity, habitat diversity, and genetic diversity.
- ✔ Diversity indices can be used to describe and compare communities.
- ✔ The quantification of biodiversity is important for assessing areas that need prioritizing for conservation.
- ✔ The measurement of changes to biodiversity over time is important in assessing the effects of human impacts on environments.

Components of biodiversity

The word "**biodiversity**" is derived from the term "biological diversity". The concept includes habitat, species and genetic diversity. An ecosystem that has high **habitat diversity** (i.e. a broad range of habitats where species can live) will also have high **species diversity** (i.e. a large number of different species each of which occur in similar abundances). Species with high **genetic diversity** have wide genetic variation (i.e. a large number of genes and many different versions of those genes). A species with high genetic diversity will be more stable and show greater resilience to environmental change, because within the population there will be a greater likelihood of some individuals surviving a move away from equilibrium.

>> Assessment tip

You need to be able to distinguish between biodiversity, diversity of species, habitat diversity and genetic diversity.

Various factors can explain why an area has high biodiversity:

- Close to the equator the climate is favourable (i.e. there is high rainfall and insolation) and so there is high primary productivity
- Where there is high habitat diversity there will be high species and genetic diversity
- Active **plate tectonics** create barriers to populations, encouraging speciation
- Altitude variation on mountains creates zonation and a diversity of habitats
- The region may be part of a large protected area or a national park.

Species diversity is a measurement that combines species richness (i.e. the number of species in a particular area) with the relative abundance of each species, or "**evenness**" (i.e. how many individuals of each species there are relative to other species).

Content link

The way in which biodiversity can be quantified is discussed in detail in section 2.5.

Assessment tip

In exam questions, measurements of species richness, abundance, and diversity may often be linked with the practical procedures for collecting field data that can be used to calculate the Simpson diversity index.

Quantification of biodiversity

Diversity indices can be calculated to give a quantitative measure of species diversity (see Unit 2, pages 63–64).

The **diversity index** is not on an absolute scale, in the same way that measurements of temperature are, for example. Two measurements of species diversity can only be interpreted relative to one another, and the values themselves do not signify absolute numbers.

Assessment tip

Different values of the Simpson diversity index have implications for the state of ecological systems.

Assessment tip

You need to know how to calculate the Simpson diversity index. The equation will be given to you, but you need to know how to process data provided.

Test yourself

3.1 Define the term "biodiversity", and explain how species diversity for an area may be calculated. [4]

Assessing changes to biodiversity

Human activities affect biodiversity. Biodiversity indices can be used to assess changes to biodiversity. Disturbance of habitats can lead to a reduction in the realized niches of organisms, and potentially increase the overlap of niches and increase competition. Simplification of habitats can lead to a reduction of biodiversity, with decreased evenness between species due to increased dominance of species adapted to the changed environment. For example, in tropical forests logging can reduce canopy cover. This opens up the forest, increases temperatures on the ground and reduces humidity. Species adapted to the closed undisturbed forest ecosystem can be expected to reduce in number whilst species adapted to these new conditions increase, leading to an overall reduction in species diversity.

▲ **Figure 3.1.1** Deforestation in Borneo, Malaysia, to prepare for oil palm plantation

Assessment tip

Species diversity indices can be used to understand the nature of biological communities and the conservation of biodiversity.

Content link

Examples of using biodiversity indices to assess changes in communities over time are contained in section 2.5.

Test yourself

3.2 Outline the reasons why tropical rainforests should be conserved. [4]

In many parts of the world natural forest ecosystems are being replaced by plantation monocultures (i.e. made up of one species). Such changes will reduce biodiversity even further due to the absence of natural habitats and a significant reduction in available resources and environmental conditions for native species (figure 3.1.2).

▲ Figure 3.1.2 An oil palm plantation in South-East Asia, following clearance of native rainforest

QUESTION PRACTICE

a) **Table 1** shows biodiversity data for the Atlantic Rainforest and the Cerrado region. **Identify two** factors that may have led to these differences in biodiversity. [2]

▼ Table 1 Estimated number of species found within the Atlantic Rainforest and the Cerrado region

	Atlantic Rainforest		Cerrado	
Plants	20 000	(8000 endemic)	10 000	(4400 endemic)
Birds	934	(144 endemic)	607	(17 endemic)
Mammals	264	(72 endemic)	195	(14 endemic)
Amphibians	456	(282 endemic)	186	(28 endemic)
Reptiles	311	(94 endemic)	225	(33 endemic)

◀ Figure 1 Map of Brazil's vegetation including location of Cerrado and Atlantic Rainforest

There have been a number of processes at work in Brazil that have caused significant losses in biodiversity. Of the original Atlantic Rainforest only about 10% remains, and of the original Cerrado vegetation only around 20% remains. These losses have occurred as a result of:

- Land clearance for crop production e.g. sugar cane, coffee, soy beans, and biofuel crops. The amount of food and biofuel crops grown has increased significantly over the past 20 years.
- Land clearance for cattle ranching. Brazil has the largest number of cattle of any country in the world.
- Land clearance for forest plantations e.g. pine and eucalypt plantations.
- Expansion and development of urban areas e.g.:
  - Rio de Janeiro and São Paulo in the Atlantic Rainforest region
  - Brasilia, the capital city in the Cerrado region.
- Infrastructure development e.g. road building schemes to support industrialization.
- Commercial logging.

▲ **Figure 2** Fact file on the human impacts on biodiversity

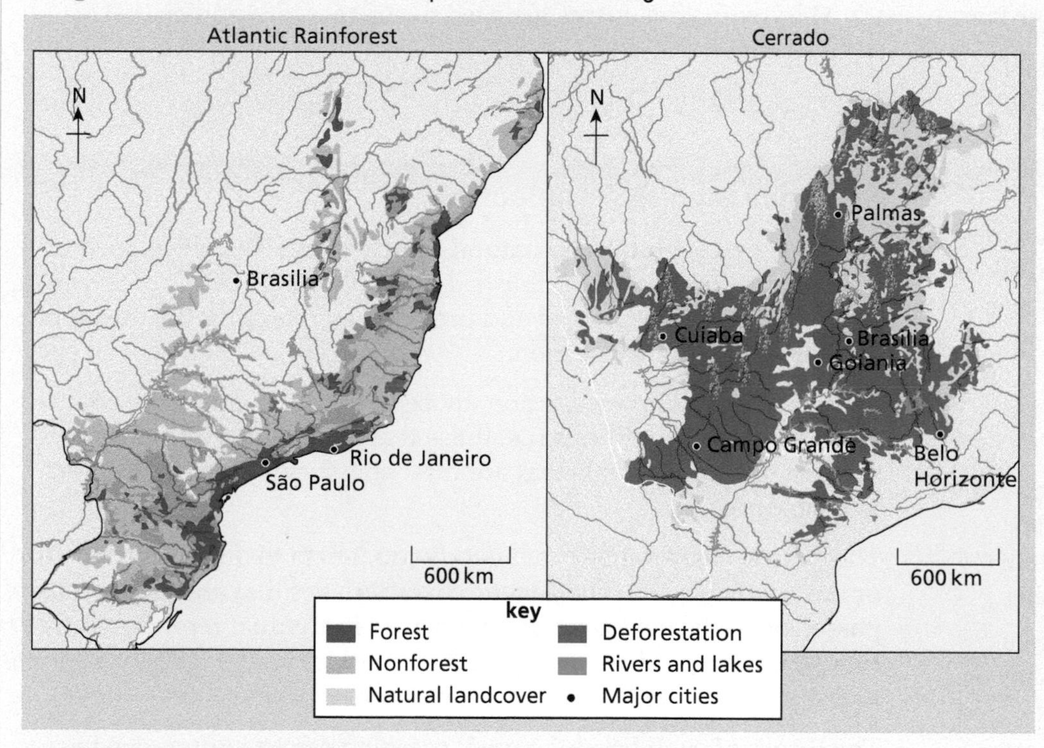

◀ **Figure 3** Deforestation in the Atlantic Rainforest and the Cerrado

How do I approach this question?

To be awarded marks, answers must be clear about which site has higher or lower biodiversity, or where there is greater biodiversity loss. Two different points need to be made to be awarded 2 marks.

SAMPLE STUDENT ANSWER

Tropical rainforests are biological hotspots due to their vast rainfall and temperature as opposed to a savanna which doesn't have the same average rainfall. Rainforests have more complex food webs which leads to more biodiversity and resilience as opposed to the savanna as it has overall less number of species.

▲ Two reasons why biodiversity differs in the two regions are given. Other answers could have included: more complex habitats in the Atlantic Rainforest than the Cerrado and so more niches; the Atlantic Rainforest may have greater variation in elevation resulting in a wider range of habitats; greater loss of biodiversity in the Cerrado due to human disturbance such as deforestation, agriculture expansion, slash and burn clearance, and urban expansion; habitats in the Cerrado are less resilient to external factors such as climate change or pollution

Specific locations (savanna and rainforest) are compared and contrasted, allowing both marks to be awarded.

This response could have achieved 2/2 marks.

3.2 ORIGINS OF BIODIVERSITY

You should be able to show the processes through which biodiversity originates, and how mass extinction events have led to biodiversity loss

✔ The origin of species can be explained by evolution through natural selection.

✔ Speciation is the formation of new species, where populations become isolated and undergo different selection pressures.

✔ Different isolation mechanisms have contributed to biodiversity in various ways.

✔ Mass extinction events have led to changes in biodiversity over time.

- **Evolution** – gradual change in the genetic character of populations over many generations
- **Natural selection** – the process where organisms that are better adapted to their surroundings are more likely to survive and produce more offspring
- **Speciation** – the process through which new species form
- **Isolation** – the process by which one population becomes divided to form sub-populations separated by geographical, behavioural, genetic or reproductive factors. If gene flow between the two subpopulations is prevented, sub-populations become genetically different to the point that they could no longer interbreed and so are different species
- **Plate tectonics** – the movement of the eight major and several minor internally rigid plates of the Earth in relation to each other
- **Tectonic plates** – massive areas of solid rock that make up the Earth's surface, both on land and under the sea. Plate size can vary from a few hundred kilometres to thousands of kilometres across
- **Extinction** – the loss of species from the Earth
- **Fossil record** – the remains of organisms preserved, for example, in rock, with simpler organisms found in older rock and more complex ones in newer rock
- **Geological time** – arrangement of events that have shaped the Earth over long periods of time, such as mountain-building. It is usually presented as a chart, with the earliest event at the bottom and the latest at the top

Evolution by natural selection

The theory of **evolution** by **natural selection** was first developed by Charles Darwin. Darwin published his book *On the Origin of Species* in 1859. The book explained and provided evidence for the theory of evolution by natural selection.

Because species show variation, those individuals that have adapted best to their surroundings (i.e. those that fit their environment) survive. The individuals that are fitter and survive can then go on to reproduce.

The genetic characteristics of an individual help to determine whether or not it will survive. The genetic characteristics that are successful are passed on to the next generation when an individual reproduces. Over time, there is a gradual change in the genetic characteristics of a species and this leads, eventually, to the formation of new species.

The theory of evolution by natural selection can be summarized as follows:

- populations of a species show variation
- all species over-produce
- despite over-production, population levels remain the same
- over-production leads to competition for resources
- the fittest, or best adapted organisms, survive
- the survivors reproduce and pass on their adaptive genes to the next generation
- over time, the population's gene pool changes and new species emerge.

Speciation

With natural selection, the genetic makeup (or "gene pool") of the species changes over time, and this, combined with **isolation** (see below), can lead to **speciation**. If gene flow between the two sub-populations is prevented, sub-populations become genetically different to the point that they could no longer interbreed and so are different species.

- **Mass extinction** – extinction events in which at least 75% of the species on Earth disappear

Concept link 

BIODIVERSITY: Biodiversity has originated through the process of evolution by natural selection.

▲ Figure 3.2.1 Speciation in the Galápagos finches. An ancestral finch arrived from mainland South America (Ecuador) and adapted to local conditions of each of the Galápagos Islands. Different selection pressures led to the evolution of different species, each adapted to their local environment

Test yourself

3.3 State two pieces of evidence which support Charles Darwin's views on the theory of evolution. [2]

Test yourself

3.4 Describe the ways in which natural selection has contributed to biodiversity. [4]

Isolation of populations

Isolation is essential to the process of speciation. Without isolation, populations undergoing separate processes of natural selection, due to environmental differences in the areas they inhabit, would continue to interbreed and genetically distinct species would be unable to develop. Isolating mechanisms separate the populations so that natural selection can act separately on each population, leading to speciation.

There are different types of isolation:

- Geographical isolation: as a result of island formation, loss of land bridges, mountains
- Behavioural differences: different reproductive displays, songs, daily activity
- Anatomical differences: reproductive organs, size.

The role of plate activity on speciation

The outer crust and upper mantle (the lithosphere) of the Earth are divided into many plates that move over the molten part of the mantle (magma, see figure 3.2.2). The plates move apart, slide against each other, or collide. Magma can be released through plate movement, causing new land mass to form.

Concept link 

BIODIVERSITY: The movement of the Earth's plates has caused isolation of populations and climatic variations, and it has contributed to the evolution of new biodiversity.

Assessment tip

Make sure you can explain how plate activity has affected evolution and biodiversity.

Content link

The way in which changes in abiotic and climatic conditions affect habitats and ecosystem development is examined in section 2.4.

Evidence for plate movement comes from rock formation and **fossil records** on different continents, which show a shared history. The shapes of the Earth's main landmasses also show how they were once connected (figure 3.2.3)

Plate activity has led to the generation of new and diverse habitats, thus promoting biodiversity.

- Island populations are separated from each other, allowing speciation.
- Populations have become separated by mountain uplift: the uplift creates new habitats, promoting biodiversity through adaptation to new surroundings.
- Collision of plates allows convergence of land masses and produces a mixing of gene pools, promoting new ecological links and possibly hybridization.
- Land masses are shifted to new climatic conditions, creating new environments with different selection pressures.

Isolation can lead to behavioural differences that cause reproductive isolation, i.e. the isolated populations develop behavioural differences and so are no longer capable of interbreeding. For example, male birds of paradise, found in Papua New Guinea, attract females using elaborate dances (as well as bright plumage) which vary according to species. Over time, populations become so genetically different that they form distinct species.

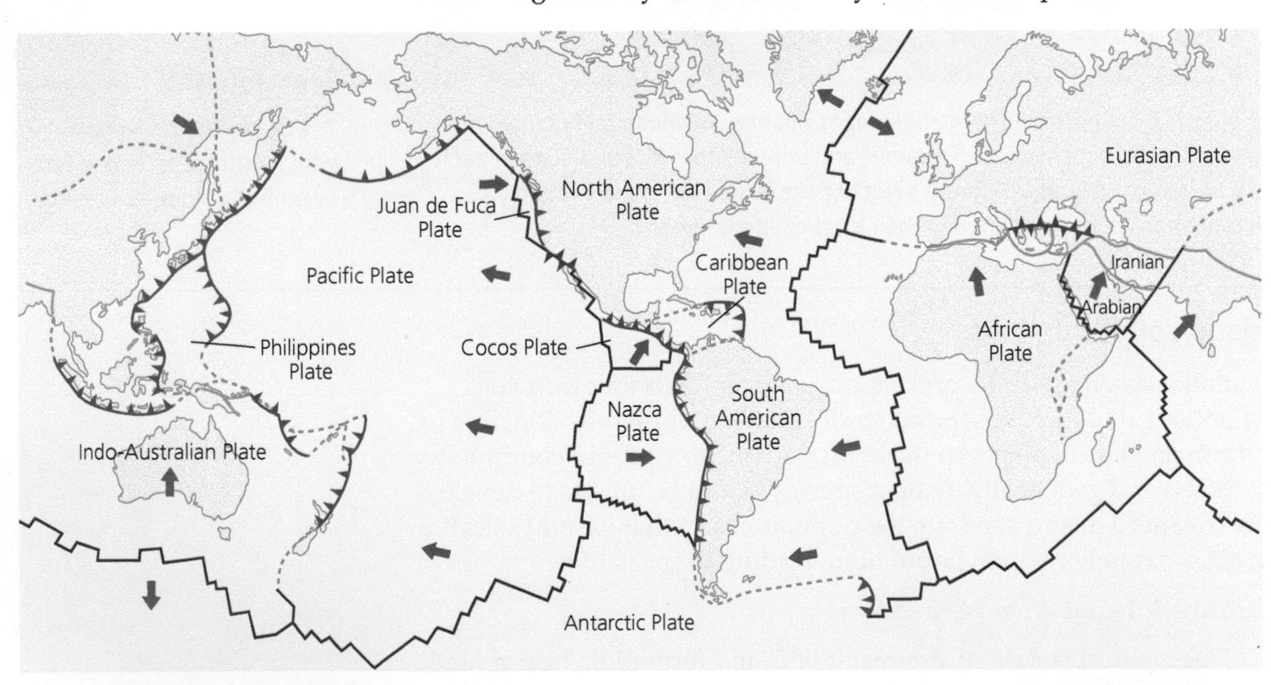

Map key
Direction of plate movement
Destructive margin – one plate sinks under another (subduction)
Destructive (collision) margin – two continental plates move together
Constructive margin – two plates move away from each other
Conservative margin – two plates slide alongside each other
Uncertain plate boundary

▲ Figure 3.2.2 Global plate tectonic movement

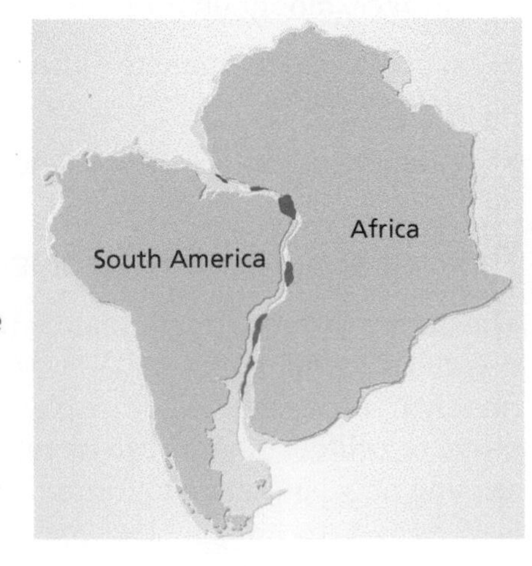

▶ Figure 3.2.3 The eastern coastline of South America matches the western coastline of Africa

Mass extinction

Global rates of **extinction** have been relatively constant over long periods. The levels of extinction have been fairly low as they have been balanced by rates of speciation. However, there have been several major peaks of extinction, known as **mass extinction** events, with unusually high extinction rates.

The fossil record shows that there have been five periods of mass extinction in the past. Mass extinctions occur when species disappear in a geologically short time period, usually between a few hundred thousand to a few million years. Animals and plants died due to both the initial event and the events that followed. Causes of mass extinction have been associated with relatively sudden changes in natural global cycles such as those caused by volcanic activity due to plate tectonic activity, or catastrophic events such as asteroid impact. For example, an asteroid strike about 65 million years ago, on the Gulf of Mexico, is believed to have caused the extinction of the dinosaurs. Changes to the atmosphere either led to climate change or planetary cooling.

The five mass extinctions were caused by natural events, but the current perceived mass extinction event, known as the sixth mass extinction, is the first to be due to biological impacts that are anthropogenic in nature (i.e. due to human causes). Human impacts leading to species extinctions include hunting, urbanization, global warming, and habitat destruction.

Mass extinctions of the past took place over **geological time**, which allowed time for new species to evolve to fill the gaps left by the extinct species. Current changes to the planet are occurring much faster, over the period of human lifetimes. The changes to the planet do not allow time for species to adapt and evolve, which increases extinction rates.

Assessment tip

In an exam answer, refer to a specific example of how plate tectonic movement has led to speciation. For example, when discussing the colonization of new islands, a good example is the Galápagos finches that have undergone speciation to fill many of the niches on these volcanic islands.

Test yourself

3.5 Explain the role of plate activity in speciation. [4]

3.6 Describe the role of isolation in natural selection. [4]

3.7 Distinguish between the mass extinctions evident in the fossil record and extinctions within historic times. [4]

Content link

Not all countries contribute equally to anthropogenic issues – these issues are discussed in section 8.4.

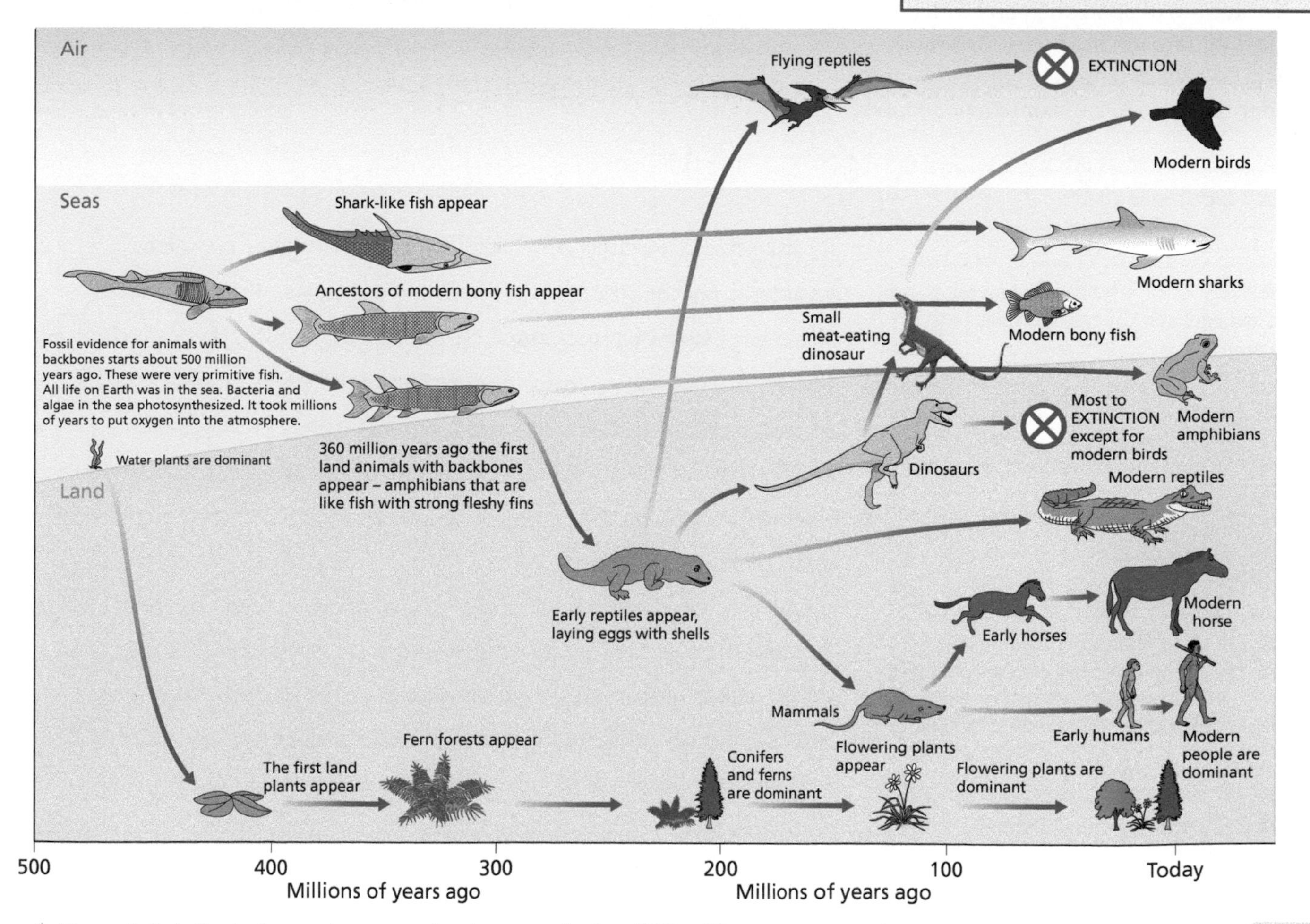

▲ Figure 3.2.4 Evolution and mass extinction over the last 500 million years

QUESTION PRACTICE

a) **Identify four** reasons why the genetic diversity of a population may change over time. [4]

b) **Define** biodiversity. [1]

c) With reference to **Figure 1 identify three** factors that could explain the high biodiversity in Ecuador. [3]

How do I approach these questions?

a) The reasons selected need to be linked to how they affect genetic diversity over time. Responses that identify relevant factors, e.g. "mutation", but do not identify why or how this influences diversity will not be awarded a mark. One mark maximum can be awarded for answers that list three valid factors without identifying how they influence diversity. Two marks maximum can be awarded for answers that identify four such factors. one mark will be awarded for each correct reason identified and saying how it affects diversity.

b) You do not have to revise definitions word for word; as long as you get the meaning of the word across, using your own words, you will be awarded a mark.

c) To answer this question you must use information in the figure as well as your own knowledge of factors that contribute to speciation and which may be illustrated in the figure. One mark will be awarded for each correct factor identified, up to a maximum of 3 marks.

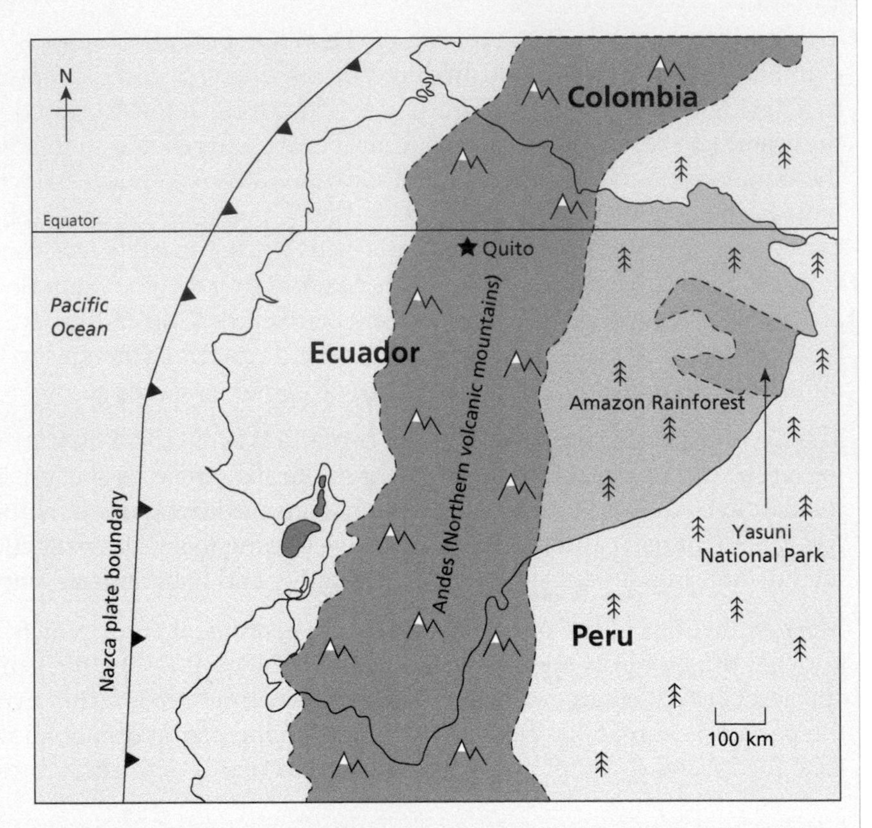

▲ **Figure 1** Map showing the location of Yasuni National Park in Ecuador, a globally significant high biodiversity area

SAMPLE STUDENT ANSWER

a) Genetic diversity of a species is the range of genetic material in the gene pools of its populations. First, genetic diversity may be increased through geographic, behavioural or temporal isolation, which prevents interbreeding and results in speciation as different populations adapt to different environments with different selection pressures, and the fittest individuals survive to reproduce and pass on their favourable genes to the offspring.

Examples of isolation include mountains or oceans, land bridges formed by plate tectonics, differences in mating calls, etc.

Also, genetic diversity of a species may be decreased due to habitat destruction or fragmentation caused by humans, which leads to repeated inbreeding of a population within the fragmented area, resulting in weakened and decreased genetic

▲ 1 factor identified

▲ Reason for why the first factor affects genetic diversity is given for 1 mark

▲ A second factor is identified. 1 mark

▲ A second mark is awarded for giving the reason why natural selection (the second factor given) affects genetic diversity

▲ The candidate says whether the factor mentioned increases or decreases genetic diversity

▲ A third factor is given

▲ A correct reason for the third factor affecting genetic diversity is given for 1 mark

diversity. Additionally, artificial selection by humans through domestication and selective breeding to produce individuals with the most desired traits may decrease the genetic diversity. Further, genetic modification of species by humans to produce pest- or drought-resistant crops can decrease genetic diversity of a species.

▼ This point does not refer specifically to "populations" but to genetically modified species. Escaped individuals may have an impact on natural populations, but this is not mentioned here, so no mark awarded

This response could have achieved 3/4 marks.

b) It is the quantity and diversity of plants and animals living in certain areas. The amount of species living in the same area.

▼ The candidate has confused species diversity with biodiversity. The definition of species diversity needs to refer to species, habitat and genetic diversity within an area, or simply say that it is the amount of biological diversity per unit area.

This response could have achieved 0/1 marks.

c) It is around the equator so limiting factors like sunlight, temperature and rainfall are abundant. The Andes act as a physical barrier, separating populations from the west and the east of this country, probably leading to speciation and thus high biodiversity. The existence of different biomes like tropical rainforest (Amazon) and alpine tundra (the Andes) provides largely different habitats and niches for varied species to coexist in the same country, leading to high biodiversity.

▲ 1 mark awarded. Biomes on or close to the equator have high primary productivity due to a favourable climate with high rainfall and high levels of insolation

▲ 1 mark for saying how barriers to populations have encouraged speciation. A specific example (the Andes) is given to illustrate the point

▲ 1 mark for explaining how the different biomes have led to varied niches for species and therefore high biodiversity

This response could have achieved 3/3 marks.

3.3 THREATS TO BIODIVERSITY

You should be able to show the threats faced by biodiversity, and how these can be measured

- ✔ Estimates of the number of species on Earth vary.
- ✔ The Earth may be entering a sixth period of mass extinction due to humanity's actions.
- ✔ The International Union of Conservation of Nature (IUCN) "Red List of Threatened Species" provides information on the conservation status of species.
- ✔ Threats to tropical rainforest are of special importance.

- **Habitat degradation** – decrease in the quality and complexity of the area where organisms live
- **Habitat fragmentation** – when habitat is divided into smaller areas that are separate from each other

- **International Union for the Conservation of Nature (IUCN)** – the world's oldest and largest global environmental organization. Its aim is to show how biodiversity is essential to solve the problems of climate change, sustainable development, and food security
- **Red List** – information that assesses the conservation status of species on a worldwide basis
- **Conservation status** – a measure of how endangered a species is; a sliding scale operates from being of least concern to extinct

Concept link 

BIODIVERSITY: Global biodiversity is difficult to quantify but it is decreasing rapidly due to human activity.

Content link

The causes and implications of human population growth are discussed in section 8.1.

Content link

Issues concerning water pollution are covered in section 4.4.

Assessment tip

Remember that the rate of loss of biodiversity may vary from country to country depending on the ecosystems present, protection policies and monitoring, environmental viewpoints, and the stage of economic development.

Assessment tip

A useful mnemonic to remember the causes of biodiversity loss is "A HIPPO" (Agricultural practices; Habitat loss; Invasive species; Pollution; Population growth of humans; Overhunting).

Estimates of the total number of species on Earth

The total number of species on Earth today is still not well understood. Estimates of the current number of species on the planet range from 5 million to 100 million. So far, science has identified about 1.8 million species. It is impossible to get an accurate count of the number of species because many species have not been discovered yet. We still know little about many species, such as those in the canopy of rainforest and the deep ocean, where many undiscovered species may live. Without a reliable estimate of the number of species it is difficult to calculate extinction rates.

Species extinctions due to human activities

Factors that lead to the loss of diversity include:

- **Habitat degradation**, **fragmentation** and loss
- Agricultural practices (for example the production of monocultures, the use of pesticides)
- Introduction of non-native species
- Pollution
- Hunting, collecting and harvesting.

Human population growth has led to increased demand on natural resources, exacerbating the issues listed above.

One of the main threats to ecosystem biodiversity is the introduction of non-native (i.e. exotic or invasive) species. Well-known examples of this are the introduction of the grey squirrel from North America into Europe, where it has led to the reduction in population density of the indigenous red squirrel, and the proliferation of the cane toad in Australia following its introduction in 1935 to eradicate the cane beet beetle.

Human population growth has led to conflict with wildlife. Exponential growth has led to greater demand for resources, which has led to clearance of land and reduction of species' habitat. Habitat loss, fragmentation and degradation also lead to a reduction in biodiversity. Pollution such as chemicals, litter, nets, plastic bags and oil spills damage habitats and lead to loss of life and reduction in species' population numbers. Overhunting has led to the reduction in population size of many species. Overfishing of North Atlantic cod in the 1960s and 1970s led to a significant reduction in population numbers. Tigers are hunted for illegal trade in skins and exotic pets. People kill wild animals for food or to sell the meat (bush meat).

Test yourself

3.8 State what is meant by an "invasive species". [1]

3.9 Suggest how invasive species get to an area. [2]

3.10 Describe the impacts of invasive species. [4]

3.11 Give an example of an invasive species and **outline** how it has adversely affected native biodiversity. [2]

3.12 To what extent is it possible to manage invasive species? [4]

3.13 Explain how human actions can reduce species diversity in two named ecosystems you have studied. [4]

The Red List

The **International Union of Conservation of Nature (IUCN)** publishes data in the "**Red List** of Threatened Species" in several categories.

Specific factors are used to determine a species' **conservation status**:

- **Population size.** Species with small populations are more likely to have low genetic diversity and their inability to adapt to changing conditions can be fatal. For example, many of the large cat species have low genetic diversity (e.g. the cheetah, snow leopard and tiger). **Reduction in population** size may indicate that a species is under threat, for example the European eel (*Anguilla anguilla*): European eel numbers are at their lowest levels ever in most of its range and it continues to decline.
- **Degree of specialization.** Specialized species have a narrow niche so, if their surroundings change, they may not be able to adapt and change. For example, a species' food resources may be very specialized, such as the giant panda, which mainly eats bamboo. Some animals can live only on certain tree species, such as the Palila bird (a Hawaiian honeycreeper), which depends on the Mamane tree (*Sophora chrysophylla*) for its food and is therefore losing habitat as the Mamane trees are cut down.
- **Distribution.** Species with small population sizes and limited distribution are more likely to become extinct than common and widespread species. For example, the slenderbilled grackle (*Quiscalus palustris*) was a bird that once occupied a single marsh near Mexico City and became extinct through human activity.
- **Reproductive potential and behaviour.** Species that produce few offspring are vulnerable to extinction. If there is a change in habitat or a predator is introduced, the population drops and there are not enough reproductive adults to support and maintain the population. If animals are slow to reproduce, any loss in numbers will mean a fast decline. The Steller's sea cow was heavily hunted and unable to replace its numbers fast enough.
- **Geographic range and degree of fragmentation.** Species with a limited geographic range may be under greater threat from extinction, for example the peacock parachute tarantula (*Poecilotheria metallica*) which is found in only a single location in the Eastern Ghats of Andhra Pradesh in India. Species that live in a smaller area are under greater threat from extinction than more widespread species, because loss of the area they live in will lead to loss of the species. Species in fragmented habitats may not be able to maintain large enough population sizes, for example the Sumatran rhinoceros (*Dicerorhinus sumatrensis*).
- **Quality of habitat.** Even if a species is not directly under threat, if its habitat is being reduced or degraded this will indirectly affect the species. Species that live in habitats that are poorer in quality are less likely to survive than species in habitats that are better in quality. For example, the fishing cat (*Prionailurus viverrinus*) is found in South-East Asian wetland areas where it is a skilful swimmer, but drainage of wetlands for agriculture has led to a reduction in habitat quality.
- **Trophic level.** Top predators are sensitive to any disturbance in the food chain. Any reduction in numbers of species at lower trophic levels can have dramatic consequences. It is also possible that

Assessment tip

The Red List classification system has several categories. You do not need to know the definitions of these different conservation status categories.

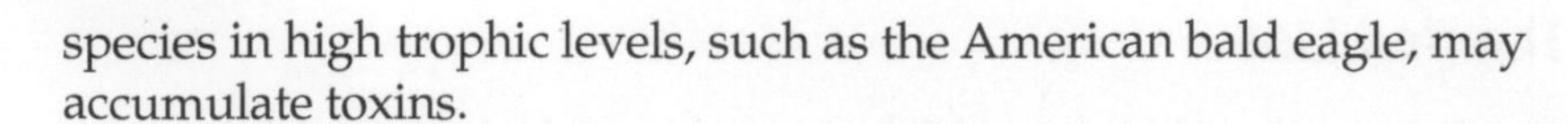

species in high trophic levels, such as the American bald eagle, may accumulate toxins.

- **The probability of extinction.** A species is assessed as being critically endangered if the probability of extinction is calculated to be ≥ 50% in 10 years or 3 generations (up to a maximum of 100 years).

Concept link

EVS: Different societies will view tropical rainforest in different ways, based on their EVS. Some will see value in the land through, for example, sales of timber, whereas others will see the life-giving value of rainforest, through the provision of food, medicine and cultural values.

Impact of human activity on the biodiversity of tropical biomes

Tropical rainforests cover 5.9% of the Earth's land surface but may contain 50% or more of all species – their conservation and protection are therefore extremely important.

Content link

The effect of human activities on soil fertility is covered in detail in section 5.3.

Tropical rainforests are vulnerable because they are under constant threat from logging or the removal of forest for other land use, such as agriculture. An average of 1.5 hectares (the size of a football pitch) of tropical rainforest is lost every four seconds. Deforestation and forest degradation occur as a result of external demands for timber, beef, soya, and biofuels. External demands for beef and soya lead to the destruction of trees to create farmland.

Assessment tip

The regeneration rate of tropical rainforests will depend on the level of impact – selectively logged forests will recover more quickly than areas that have been clear-felled.

Rainforests have thin, nutrient-poor soils. The poor soils have implications when forests are cleared. Because there are not many nutrients in the soil, it is difficult for rainforests to regrow once they have been cleared. Studies in the Brazilian Atlantic Forest have shown that certain aspects of the structural integrity return surprisingly quickly (within 65 years), but for the forest to fully recover, more time is needed. It can take up to 4000 years for cleared tropical rainforest to fully recover. Recovery depends on the level of disturbance. A large area of cleared land will take a lot longer to grow back than small areas which have been used for shifting cultivation (i.e. where the plot has been left fallow after a period of cultivation). A forest which has been selectively logged can grow back if not too much timber has been removed. If too much timber is removed from selectively logged forests, the forest never fully recovers because fast-growing, light-loving species, such as vines and creepers, block out the light for slow growers. As a result, the forest remains at a sub-climax level.

Content link

The use of natural resources by human societies, and whether these are sustainable, is discussed in section 8.2.

Case histories of three different species: extinct, critically endangered, and back from the brink

Extinct

In the early part of the 20th century as many as 100,000 tigers roamed the diverse habitats of Asia, from forests and swamps to open tundra. Now, overall, there may be as few as 3,200 tigers left in the wild, with all subspecies threatened with extinction.

▲ **Figure 3.3.1** A Javan tiger in London Zoo before 1942

The extinctions of the Bali (*Panthera tigris balica*), Caspian (*Panthera tigris virgata*) and Javan (*Panthera tigris sondaica*) tigers have all occurred since 1937, as a result of hunting, competing with humans for food, illegal trade, habitat loss, and fragmentation. Since they are top (or "apex") predators, the extinction of tigers will have led to fundamental changes to the ecosystems in which they live. The population size of prey species will increase due to the absence of natural negative feedback loops involving predator–prey dynamics, and as a result, the ecosystem will change, with increased populations of herbivore species consuming more plants, affecting the natural succession of the forest.

On the brink of extinction

The Amur, or Siberian, tiger (*Panthera tigris altaica*) once lived across a wide geographical area, reaching from northern China, across the Korean Peninsula, to the southern regions of eastern Russia. Expanding human settlements in the early 20th century led to hunting and habitat loss and the near extinction of this largest of the tiger subspecies (and therefore the largest big cat), with a 90% reduction in population numbers from across its range. One of the main threats to tigers in the wild is the trade in body parts, e.g. for medicinal purposes in East Asian countries, and the illegal trade in skins, claws, and teeth. A recent report has suggested that the effective population size of the Amur tiger is now fewer than 14 individuals (source: http://news.bbc.co.uk/earth/hi/earth_news/newsid_9407000/9407744.stm). Very low genetic (i.e. allele) diversity means that the remaining populations are vulnerable to diseases or rare genetic disorders, and a reduced likelihood that the subspecies will survive into the future. The subspecies is therefore on the brink of extinction.

Back from the brink

The Arabian oryx (*Oryx leucoryx*) is a medium-sized antelope that was formerly found throughout the Arabian Peninsula, and north to Kuwait and Iraq. They live in desert habitats, and can survive in areas with low humidity, low rainfall, high, sandy winds, and high temperatures (over 45°C), and they can withstand droughts of up to six months. The species' range had reduced significantly by the early 20th century, and continued to do so at an accelerating rate. The main cause of the reduction in population size was overhunting, by Bedouin for meat and hides, and also sport hunting. Oryx disappeared from the north of their range in the 1950s, and continued to decrease steadily in the south due to hunting. By the 1960s, Oryx populations were restricted to parts of central and southern Oman. The last wild individuals were probably shot in 1972.

Populations of Arabian oryx were maintained in zoos, and breeding programmes proved successful in sustaining population numbers and genetic diversity. Following their extinction in the wild, Arabian oryx were reintroduced to Oman, Saudi Arabia, Israel, the United Arab Emirates, and Jordan. In 2016, the number of individuals in the wild totalled 1,220 (about 850 mature individuals). Overall, reintroduced populations are stable or increasing slowly. An estimated 6,000–7,000 animals are held in captivity worldwide, most of them on the Arabian Peninsula.

Test yourself

3.14 For a named endangered species, **discuss** why it is endangered. [4]

3.15 Outline two reasons for the extinction of a named species and **suggest** how intervention measures can improve the conservation status of a species. [4]

Content link

The concept of genetic diversity is discussed in section 3.1.

▲ Figure 3.3.2 The Amur, or Siberian, tiger

▲ Figure 3.3.3 The Arabian oryx

Assessment tip

You need to be able to discuss the case histories of three different species: one that has become extinct due to human activity, another that is critically endangered, and a third species whose conservation status has been improved by intervention.

Assessment tip

You need to know the ecological, socio-political, and economic pressures that caused or are causing a species' extinction. You should also understand the species' ecological roles and the possible consequences of their disappearance.

Threats to biodiversity from human activity in an area of biological significance

An area will be of biological significance if it:

- has a high number of endemic or unique species
- contains multiple ecosystems
- has high species richness and high biodiversity
- contains a major portion of the world's biodiversity
- is under high threat from human activity such as logging, overhunting or habitat destruction
- provides economic benefits such as tourism
- provides educational benefits
- provides ecological services, such as acting as a carbon sink and flood control
- provides intrinsic value/existence value.

Assessment tip

You need to learn a case study that examines a natural area of biological significance that is threatened by human activities. You should know the ecological, socio-political and economic pressures that caused or are causing the degradation of the chosen area, and the consequent threat to biodiversity.

Content link

The view that the environment can have its own intrinsic value is discussed in section 1.1.

Concept link 

EVS: The range of EVSs, from technocentric to ecocentric, will determine how a person or society views tropical biomes and how to best protect them.

QUESTION PRACTICE

Essay

Discuss the implications of environmental value systems in the protection of tropical biomes. [9]

How do I approach this question?

To obtain high marks the following are needed: relevant ESS concepts and terminology that are used correctly; different EVSs considered in relation to the protection of tropical biomes; examples of topical biomes, how different EVSs relate to their protection, and value systems in specific tropical societies; a balanced analysis of the ways in which different value systems are likely to impact or influence the protection of tropical biomes, acknowledging relevant alternative viewpoints; a valid conclusion.

SAMPLE STUDENT ANSWER

Tropical biomes such as the tropical rainforest (Amazon in Brazil, for example) and the coral reefs (Great Barrier Reef in Australia) are the biomes located on and around the equator. They account for over 40% of the global production of oxygen and are carbon sinks. They also have a very high biodiversity due to their favourable climatic conditions, high rainfall, high temperature, low pressure, high insolation. Nutrients cycle very fast, due to the biome's dynamic nature (high gross productivity, low net productivity). All in all, tropical rainforests have a very high natural capital and play an important role in regulating the environment.

▲ Examples of tropical rainforest and coral reef given – both tropical biomes. A greater range of examples would have allowed greater illustration of points made. Other tropical biomes include grasslands, savannas, lakes and rivers. Most of the points in this essay relate to tropical rainforests

▲ The importance of tropical rainforests is stated – this puts subsequent points about how and why they should be managed into perspective

However, the tropical rainforest faces many threats such as deforestation (for palm oil, cattle farming etc), leading to habitat destruction, loss of biodiversity, less photosynthesis, erosion, water sedimentation etc.
One's environmental value system is a world-view that determines the way in which one regards environmental issues. It is affected by social, political, and cultural factors.
Ecocentrists emphasize tropical biome's intrinsic value as it plays an essential role in stabilizing and regulating the Earth's environment. They also consider the bio-rights of organisms in the tropical biomes. Hence, ecocentrists aim to change the attitudes people have towards the environment to mitigate the destruction of tropical coral reefs and rainforests. E.g. eating less meat, using less timber etc. Ecocentrists thus emphasize that the tropical biomes are more valuable and important than humans.
Anthropocentrics believe that humans can protect the environment by introducing laws, policies, educational programmes etc. Projects to plant trees in rainforests and put stronger regulations in place to decrease the rate at which the biomes are destroyed. Moreover, international agreements such as the Kyoto Protocol are another way through which anthropocentrics aim to protect tropical biomes.
Technocentrics believe that technological development can regulate the environment, while still gaining natural income (economic development). For instance, they would use fertilizers to increase the yield obtained from timber (more tree growth). Technocentrics deem the tropical rainforest as important because it provides high revenues, yet put less emphasis on its protection because they believe that – if it is exhausted – technologies will regulate the Earth's environment (for example extracting CO_2 from the atmosphere with pumps).
Depending on one's EVS, the extent to which and the way in which they protect tropical biomes varies.

- The answer shows substantial understanding of concepts and precise terminology.
- It is effective at linking the question context and EVS to protection measures.
- Relevant examples are given to illustrate points being made, but not in all cases.

▲ Threats to tropical rainforest are introduced

▲ Three valid arguments about how an ecocentric approach can lead to the protection of tropical rainforest

▲ The candidate clearly identifies the different EVSs, allowing each to be discussed in turn. This shows good planning, and allows the examiner to clearly see that the answer has breadth

▲ An example of an anthropocentric approach is given to illustrate the answer

▲ A second example is given

▲ A valid example of the technocentric approach is given

▼ It is unlikely that fertilizers would be used to increase the growth of rainforest. Most rainforests are directly harvested for their timber. Fertilizer may be used to help subsequent crops of plantations grow should the forest be cleared, but this is not relevant to the question

▲ Valid point about the technocentrist approach putting more value on the timber rather than protection of the forest

▲ A valid point about how a technocentrist would approach the issue

▼ A conclusion is given, but this is very brief, leading to a lower mark overall

- The analysis clearly states the implication of each EVS for protection of tropical biomes, and is therefore balanced.
- The analysis also identifies threats and is therefore insightful.
- The answer summarizes the tropical rainforest characteristics needed and establishes their relevance, and is very thorough.
- The conclusion is brief and not supported by analysis or examples contained within the answer.
- A better conclusion would have been: "Because ecocentric values embrace the biorights of all living species and habitats their implications are bound to be the most fundamentally protective. However, to be practical in current society, more is likely to be achieved in protecting tropical biomes through some compromise with other value systems."

This response could have achieved 7/9 marks.

3.4 CONSERVATION OF BIODIVERSITY

- **Intergovernmental organization (IGO)** – an organization that is established through international agreements to protect the Earth's natural resources
- **Non-governmental organization (NGO)** – an organization that is not run by the government of any country; NGOs are not funded or influenced by governments in any way
- **Convention on Biological Diversity (CBD)** – an international legally binding agreement created at the Rio Earth Summit. It has three main goals: the conservation of biodiversity, the sustainable use of its components, and the fair and equitable sharing of the benefits arising from genetic resources
- **Rio Declaration** – a document produced at the Earth Summit in 1992 that outlined future sustainable development around the world

You should be able to show how biodiversity can be conserved through a range of different measures

✔ There are economic, ethical, aesthetic, and ecological reasons for conserving species and habitats.

✔ International, governmental, and non-governmental organizations (NGOs) work to conserve and restore ecosystems and biodiversity.

✔ International conventions on biodiversity work towards conservation.

✔ Protected areas should be designed to maximize their conservation value.

✔ A species-based approach to conservation includes CITES, captive breeding and reintroduction programmes, zoos, and the selection of flagship and keystone species.

✔ The location and organization of a conservation area are essential for its success.

Arguments about species and habitat preservation

There are many arguments for preserving species and habitats. Arguments are based on ethical, aesthetic, ecological, and economic reasons. Habitats also have life-support functions (e.g. forests providing oxygen) and ecosystem-support functions (e.g. water cycle and impacts on weather).

Concept link 

EVS: A person's or society's EVS will determine how they perceive the natural world, and the measures they take to protect it.

For example, when considering why a tiger species should be conserved, the following arguments could be applied.

Ethical reasons:

- the tiger has intrinsic value
- the species has its own biorights and humans have no rights to destroy species
- many religions and local cultures attach particular spiritual or cultural significance to the tiger
- we owe it to future generations to pass on the same environmental heritage we received.

Economic reasons:

- each species in the ecosystem contributes to the overall natural capital, i.e. the sustainable goods and services, that the system provides
- ecotourism provides a significant source of income, e.g. whereby tourists pay to see tigers in the wild.

Aesthetic reasons:

- tigers are beautiful, unique, and make the world more pleasurable
- all species add to the diversity and quality of human experience.

Ecological reasons:

- as top carnivores, they have an important role to play in maintaining the balance within their ecosystem
- the tiger may be a keystone or flagship species (see below)
- all species add to the diversity of a system, making it more stable and resistant to change.

Assessment tip

Other arguments why a species or habitat should be conserved include: genetic variation in plant and animal species provides the means for sustaining and improving production in farms, forestry, animal husbandry and fisheries; many organisms contain chemicals which may be of medical or commercial value.

Test yourself

3.16 Explain whether or not you believe it is justified to place an economic value on natural systems.

The role of governmental and non-governmental organizations (NGOs)

Governmental organizations (GOs) and **non-governmental organizations** (NGOs) have different strengths and weaknesses. The United Nations Environment Programme (UNEP) is an **intergovernmental organization** (IGO), and the World Wide Fund for Nature (WWF) and Greenpeace are non-governmental organizations. GOs and NGOs can be compared and contrasted in terms of their use of the media, speed of response to environmental issues, diplomatic constraints, and political influence.

- **Agenda 21** – a plan of action to achieve sustainable development worldwide in the 21st century, to be carried out at the local level
- **Rio +20** – a meeting that took place in 2012, 20 years after the first Earth Summit
- **World Conservation Strategy** – an international agreement that set the priorities of the maintenance of essential life support systems, the preservation of genetic diversity, and the need to use species and ecosystems in a sustainable way
- **Island biogeography** – a theory that predicts that smaller islands of habitat will contain fewer species than larger islands
- **Edge effect** – changed environmental conditions at the edge of habitats
- **Corridor** – a piece of land containing habitats that joins two other areas
- **Species-based approach to conservation** – a method which focuses on specific individual species that are vulnerable, with the aim of attracting interest in their conservation.
- **CITES** – the Convention on International Trade in Endangered Species of wild fauna and flora. This is an international agreement between governments that aims to ensure that international trade in wild animals and plants does not threaten their survival
- **Captive breeding** – the process of raising animals outside their natural surroundings in controlled environments such as zoos, with the aim of increasing population numbers for reintroduction to the wild

- **Reintroduction programme** – a scheme in which animals raised or looked after in zoos are released into their natural habitat
- **Flagship species** – charismatic species that are used to publicize conservation campaigns, stimulate public action, and raise economic support
- **Keystone species** – species that have a disproportionately large effect on their environment, determining the structure of an ecosystem and having many other species dependent on them

Similarities between them are that both types of organization are trying to promote conservation of habitats/ecosystems and biodiversity.

Differences between GOs and NGOs include:

- intergovernmental organizations work within the law, and often NGOs can be more confrontational
- NGOs use the media more to get specific messages about conservation across
- NGOs often run campaigns focused on large charismatic species such as whales, seals, or pandas
- UNEP (which is a GO) works more slowly and is concerned about government-level changes to protect the environment
- NGOs often lobby at UNEP-organized conventions to encourage countries to sign treaties
- NGOs often protest at the UNEP conventions to highlight single issues they are concerned with
- NGOs often carry out publicity stunts that aim to draw attention to the conservation issue
- NGOs work at a local scale, at the grassroots level
- NGOs tend to have local groups to encourage community involvement, more actively including communities
- NGOs provide education/information on issues.

Concept link 

STRATEGY: International conventions provide a framework through which countries can plan and implement conservation strategies.

Concept link 

SUSTAINABILITY: International conventions have allowed governments to discuss issues concerning sustainability, and to plan strategies for sustainable development.

International conventions on biodiversity

International conventions have been influential in developing attitudes towards sustainable development. They include the following.

UN Conference on the Human Environment

The UN Conference on the Human Environment took place in Stockholm in 1972. The conference was the first time that the international community had met to discuss the global environment and development needs. The conference led to the Stockholm Declaration which played an essential role in setting targets and forming action concerning sustainable development at both local and international levels.

UN Rio Earth Summit

The Earth Summit was the first ever UN conference to focus on sustainable development, and took place in Rio de Janeiro, Brazil, in 1992. The Earth Summit was attended by 172 governments. It set the agenda for the sustainable development of the Earth's resources. The conference resulted in the **Rio Declaration** and **Agenda 21**. The Earth Summit led to agreement on legally binding conventions. One of the legally binding conventions was the UN **Convention on Biological Diversity** (CBD). The CBD is governed by the Conference of the Parties (CoP) which meet either annually or biennially in order to assess the success and future directions of the Convention. For example, CoP 14 of the CBD took place in 2018 at Sharm El-Sheikh, Egypt.

Rio +20

A conference took place in Rio in 2012, 20 years after the first Earth Summit. This meeting was known as **Rio +20**. One of the conference aims was to obtain political commitment from nations to sustainable development. Another aim was to assess how much progress there had been on internationally agreed commitments, such as CO_2 reductions. The conference also discussed new and emerging challenges.

The World Conservation Strategy

The **World Conservation Strategy (WCS)** was established in 1980 by the **International Union for the Conservation of Nature (IUCN)**. The WCS focused on three factors:

- maintaining essential life support systems (climate, water cycle, soils) and ecological processes
- preserving genetic diversity
- using species and ecosystems in a sustainable way.

The WCS recommended that each country should prepare its own national strategy for conserving natural resources for the long-term good of humanity. The WCS emphasized how important it was that the users of natural resources became their protectors. It recognized that conservation plans can only succeed if they are supported and understood by local communities.

Concept link

SUSTAINABILITY: The WCS has encouraged countries to focus on the sustainable use of natural resources, and implementing conservation plans to maintain biodiversity.

Assessment tip

You need to know about recent international conventions on biodiversity. These include the Stockholm Declaration and international conventions set up by UNEP, e.g. Rio Earth summit of 1992 and the Johannesburg Sustainability summit of 2002. Recent international conventions on biodiversity include conventions signed at the Rio Earth Summit (1992) and subsequent updates such as the 2012 Rio +20 conference.

Habitat conservation

Criteria used to design protected areas follow the principles of **island biogeography**, with appropriate criteria including size, shape, **edge effects**, **corridors**, and proximity. The best-protected areas are large, they are connected to other areas through corridors, and there are minimum edge effects (see below).

Test yourself

3.17 Suggest why international agreements are needed. [3]

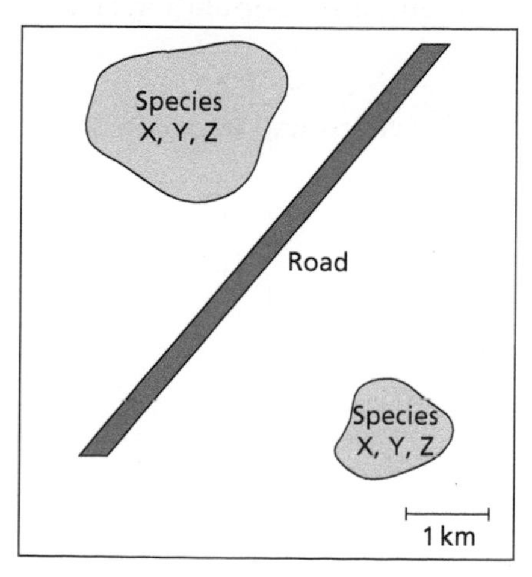

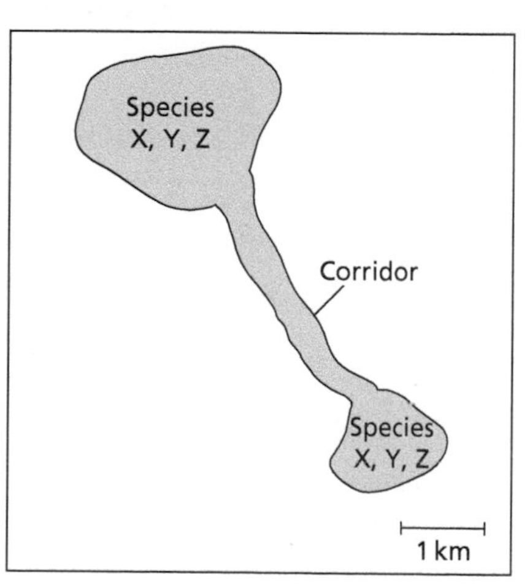

▲ Figure 3.4.1 Two wildlife reserves – one has two isolated reserves whereas the other has a corridor connecting two habitat areas. Species X, Y and Z are found in all three reserves

In figure 3.4.1, the reserve with a corridor connecting two different habitat areas is better for conserving the genetic diversity of species X, Y and Z because the corridor allows organisms to migrate between reserves, leading to a greater number of opportunities for mating with a wider population. There is greater genetic diversity because more individuals can mix.

At the edge of a protected area there may be a change in abiotic components. This change includes more wind, more warmth, and less humid conditions compared to the interior of the reserve. These are called edge effects. Edge effects will attract species that are not found deeper in the reserve but that survive successfully in the edge conditions. Edge effects may also attract exotic species from outside the reserve. Larger reserves have fewer edge effects as they have a low perimeter to area ratio. Fewer edge effects will mean that less of the area is disturbed. The shape of a protected area should minimize edge effects.

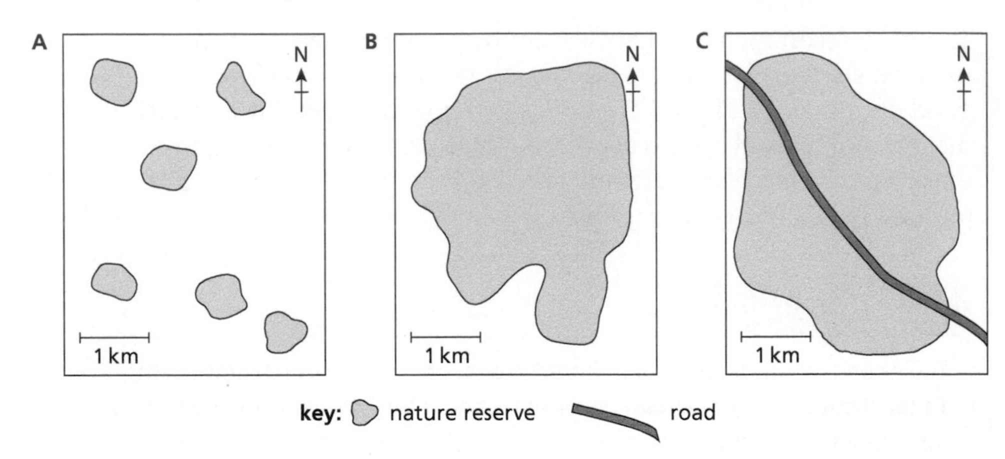

▲ Figure 3.4.2 Three different reserves: A has several isolated habitat areas, B is a large single reserve, and C is a large reserve with a road running through the middle

>> Assessment tip

You need to be able to explain the criteria used to design and manage protected areas.

The strengths or weaknesses of the shapes of the nature reserves in figure 3.4.2 are as follows.

Area A:

- fragmented and small with a large perimeter to area ratio, so there are large edge effects resulting in lots of disturbance
- fragmented so migration between fragments is difficult
- small size may limit species contained and limit population sizes.

Area B:

- large perimeter to area ratio and so relatively small edge effect, meaning less disturbance
- large size promotes high biodiversity
- large size, so good for large vertebrates such as top carnivores
- large size supports large species populations.

Area C:

- as large as B but dissected by a road which acts as a barrier to species migration
- road increases edge effect so more disturbance
- road allows easier access to the interior of the reserve for monitoring
- road gives easier access for poachers.

Species-based conservation

Test yourself

3.18 Suggest two possible reasons why the snow leopard has received special attention from conservationists. [2]

3.19 There are different approaches to snow leopard conservation, based on different value systems (EVSs). **Suggest** one approach to conservation that would be typical of an anthropocentric approach and one that would represent a technocentric value system. [2]

CITES

CITES is the Convention on the International Trade in Endangered Species of Wild Fauna and Flora. It was established in 1973 and is an international agreement aimed at preventing trade in endangered species of plants and animals. Trade in animal and plant specimens (whole organisms, whether alive or dead, or their parts and derivatives), as well as factors such as habitat loss, can seriously reduce their wild populations and bring some species close to extinction. The aim of CITES is to ensure that international trade in specimens of wild animals and plants does not threaten the survival of the species in the wild.

Countries who sign up are agreeing to monitor trade in threatened species and their products that are exported and imported. Illegal imports and exports can result in seizures, fines, and imprisonment, which discourages illegal trade.

The scheme has its limitations: it is voluntary and countries can disagree about the listing of specific species when they join and penalties may not necessarily match the gravity of the crime or be a sufficient deterrent to wildlife smugglers.

In addition, unlike other international agreements such as the Montreal Protocol, CITES lacks its own financial mechanism for implementation at the national level and member states must contribute their own resources. However, taken overall, CITES has been responsible for ensuring that the international trade in wild animals and plants remains sustainable.

Test yourself

3.20 Evaluate CITES as a means of preserving biodiversity. [4]

Zoos, captive breeding and reintroduction programmes

International zoos and wildlife parks play an important role in conservation, although there are some limitations. They are important for promoting conservation during visits by members of the public and education. For many years, international zoos have focused on keeping endangered or critically endangered animals so that they can protect them from the dangers they face in the wild. **Captive breeding** programmes maintain population numbers and genetic diversity, with a view to returning animals to the wild once their natural habitats have been protected.

Advantages of zoos include:

- they provide a safe haven for critically endangered animals
- they provide an opportunity to research the biology and behaviour of endangered species thereby increasing scientific knowledge about the species they protect

- they can be used to raise awareness and educate the public about the threats faced by the species in the wild
- they can be used to obtain funds to help conservation efforts
- breeding pairs can be used to increase the number of endangered animals, which can then be used to reintroduce the animal into the wild.

Disadvantages of zoos include:

- it can be difficult to recreate suitable, natural habitats for animals in captivity
- it can be argued that it is ethically wrong to keep animals in captivity
- captive animals can develop health problems, and species can become stressed in captivity, thereby experiencing behavioural problems
- international zoos and wildlife parks are expensive to create and maintain, and funds could instead be spent on habitat conservation efforts
- they do not address the causes of reduction in wild populations, e.g. deforestation and hunting.

Although international zoos provide an opportunity to increase species numbers through breeding programmes, without tackling the issue of habitat protection the species will remain under threat. Ideally, a mixed approach is needed (see page 89).

>> **Assessment tip**

Flagship species are selected because they appeal to humans by having ideological, cultural or religious significance. Their value is primarily subjective, relative to a society, rather than ecological.

Flagship species

Flagship species are "charismatic" species that help protect others in an area. They have a wide appeal to the public, and conserving these species will result in habitat protection that will also protect other species. For example, the golden lion tamarin is a monkey that lives in southern Brazil; as a primate it is closely related to humans and so it is easy for humans to empathize with, is aesthetically attractive, and so is a potential attraction for ecotourism. The monkey is not found anywhere else and is unique to Brazil. So it is highly suitable for publicity and fund-raising (e.g. posters and soft toys).

▲ Figure 3.4.3 A golden lion tamarin monkey

Keystone species

A **keystone species** is a plant or animal that plays a central role in the way an ecosystem functions. Loss of a keystone species can lead to a significantly different ecosystem or the collapse of the ecosystem altogether. Conservation of keystone species helps to protect the integrity of food webs. For example, sea otters (*Enhydra lutris*; figure 3.4.4) of the North American west coast help to preserve kelp forest by eating invasive sea urchins and so their loss would have an impact on many other species.

▲ Figure 3.4.4 A sea otter (*Enhydra lutris*)

Keystone species are identified through ecological, objective study of their relationships with the entire ecosystem. Although being of ecological value, they may be publicly unpopular (unlike flagship species) and in some cases they may even be considered pests, for example elephants entering plantation forests and destroying trees.

Content link

The structure and function of ecosystems is covered in section 2.2.

A mixed approach

Chengdu Zoo began breeding giant pandas in 1953, and Beijing Zoo in 1963. From 1963 to the present time, the giant panda has been bred in 53 zoos and nature reserves within China and internationally.

Reserves for giant pandas in China were established in 1963, of which four are in Sichuan province. The giant panda nature reserves had expanded from the initial 5 to 67 by 2017. Beijing Zoo has an impressive giant panda house, and has established a successful breeding programme. By combining captive breeding with the preservation of native habitat, conservation goals are more likely to be attained. This "mixed approach" to conservation combines *in situ* and *ex situ* conservation as a means of successfully protecting and prolonging the lives of endangered and critically endangered animals.

Test yourself

3.21 Ecocentrics recognize that all species have an intrinsic value. **Explain** the strengths and weaknesses of using intrinsic value when making decisions about development and conservation. [4]

Concept link 

STRATEGY: There are different strategies for conservation. *In situ* conservation protects habitats (e.g. protected areas); *ex situ* conservation preserves biodiversity in zoos, wildlife parks and herbaria. A mixed approach combines both *in situ* and *ex situ* methods.

Test yourself

3.22 What methods exist for protecting and conserving endangered animals? **List** three. [3]

3.23 Discuss the strengths and weaknesses of the species-based approach to conservation. [3]

3.24 List the threats faced by endangered species in the wild. [3]

3.25 List the advantages and disadvantages of different approaches to conservation provided by protected areas and zoos. [4]

Evaluating the success of a protected area

The strengths of a protected area may include:

- funding from the local government
- involvement of government agencies (such as forestry and wildlife services)
- the presence of high-profile (i.e. flagship) species
- a scientific research programme to assess the local flora and fauna
- ecotourism to raise money for conservation
- the support of local people.

Assessment tip

When evaluating the success of a protected area, both strengths and limitations must be taken into account.

Limitations may include:

- access to the area by poachers due to proximity to urban areas or roads
- lack of protected status in law (i.e. the area may still be subject to disturbance or development)
- lack of long-term revenue to support the protected area.

Assessment tip

When evaluating the success of a named protected area, you need to bear in mind that the granting of protected status to a species or ecosystem is no guarantee of protection without community support, adequate funding, and proper research (i.e. a holistic approach is needed). Choose a specific local example to study and write about.

Test yourself

3.26 Evaluate the importance of species-based conservation and protected areas in the preservation of biodiversity for future generations. [4]

QUESTION PRACTICE

Paraná pine trees (*Araucaria angustifolia*) IUCN Red List status – critically endangered

Pau Brasil (*Caesalpina echinata*) IUCN Red List status – endangered

Golden lion tamarin (*Leontopithecus rosalia*) IUCN Red List status – endangered

Brasilian merganser (*Mergus octosetaceus*) IUCN Red List status – critically endangered

Broad snouted caiman (*Caiman latirostris*) IUCN Red List status – least concern

Giant metallic ceiba borer (*Euchroma gigantea*) IUCN Red List status – none; has not yet been assessed by the IUCN

▲ **Figure 1** Photographs showing examples of species found in Brazil

Questions

a) With reference to the images in Figure 1, **justify** your choice of one animal as the most suitable to promote conservation. [2]

b) Distinguish between the concept of a "charismatic" (flagship) species and a keystone species, using named examples. [4]

c) Outline one advantage of increased tourism on wildlife conservation. [1]

d) Outline one disadvantage of increased tourism on wildlife conservation. [1]

How do I approach these questions?

a) Marks are available for stating that such species are flagship species, and you should say what that means. You need to select a flagship species from those species shown in Figure 1. The command term "justify" means that you need to provide evidence to support your choice.

b) "Distinguish" means "make clear the differences between two or more concepts or items". One mark is available for giving two valid examples (one of each kind). A maximum of three marks can be awarded for valid points of distinction. While a mark can be awarded for stating a discriminatory feature of just one kind of species (without referring to contrasting feature of the other), converse statements will not be credited twice, e.g. "flagship species are X ... keystone are not X" would gain only 1 mark if "X" was valid.

c) A detailed answer is not needed here. However, it is not sufficient to simply say "tourism raises awareness" or "allows more support" – the positive effect on wildlife conservation needs to be mentioned.

d) Answers must refer to a specific impact of tourism and outline the effects on wildlife conservation. It is not enough simply to say that "tourism produces pollution": the type of pollution, with the associated impact on wildlife, should also be stated.

SAMPLE STUDENT ANSWERS

a) The Golden lion tamarin as firstly it is endangered and it is also aesthetically good looking. It could also be a keystone species.

▲ One mark awarded for making a valid point about why the animal would attract interest for conservation

▼ Confusion between the terms "keystone" and "flagship" species. There is no evidence from the image that the animal is a keystone species, but it could be a flagship species (i.e. a charismatic species that is used to publicize conservation campaigns, stimulate public action and raise economic support)

This response could have achieved 1/2 marks.

b) Keystone species like the sea otter are those which have an important role in maintaining the functioning of their ecosystem and habitat. Flagship species are those that the public like because they find them cute, like giant pandas. Hence, they attract donations from people for their conservation.

▲ Correct description of the role of a keystone species, for 1 mark, i.e. that they have a disproportionately large effect on their environment, may determine structure of an ecosystem, and have many other species dependent on them

▲ Correct statement referring to how charismatic species are selected because they appeal to humans , for 1 mark

▲ A second valid point about flagship species and how they raise economic support for conservation . 1 mark awarded

▲ An example of each type of species is given. This is good practice, and led to 1 mark being awarded

This response could have achieved 4/4 marks.

c) Tourism might raise money to fund conservation of species through establishment of National Parks and reserves.

▲ One valid example given, which directly links increased tourism with an advantage to wildlife conservation

This response could have achieved 1/1 marks.

d) The use of land for building hotels, roads, facilities will result in deforestation and habitat loss.

▲ A good, detailed answer that directly links increased tourism to a specific impact on wildlife conservation. 1 mark awarded

This response could have achieved 1/1 marks.

4 WATER AND AQUATIC FOOD PRODUCTION SYSTEMS AND SOCIETIES

This unit examines the importance of water systems to humans, and the ways in which humans impact water systems. It looks at the natural water (hydrological) cycle, variations in access to water, the importance of water in food production systems, and finally, ways in which water is polluted as a result of human activities.

You should be able to show:

✔ the natural hydrological cycle and the impact of human activities on it;

✔ variations in access to freshwater;

✔ the characteristics of aquatic food production systems – capture fisheries and aquaculture;

✔ the causes and consequences of water pollution.

4.1 INTRODUCTION TO WATER SYSTEMS

- **Hydrological cycle** – the continuous movement of water between atmosphere, land, and sea
- **System** – any set of interrelated components which are connected to form a working unit
- **Evapotranspiration** – the combined losses of evaporation and transpiration
- **Irrigation** – extra water added to a soil/plant to encourage growth

You should be able to show the natural hydrological cycle and the impact of human activities on it

✔ The hydrological cycle has a single input and two major losses (outputs).

✔ There are several different storages and flows in the hydrological cycle.

✔ Human activities have a significant effect on the hydrological cycle, e.g. agriculture, deforestation, and urbanization.

✔ The ocean circulation systems, including the oceanic conveyor belt and ocean currents, are driven by differences in temperature and salinity.

Freshwater

The Earth's freshwater resources (figure 4.1.1) are unevenly distributed, both spatially (by area/place) and temporally (by time). Not all freshwater is readily available all of the time, e.g. some glaciers provide seasonal meltwater whereas some ice caps remain frozen year round. In addition, some sources, such as groundwater, may be polluted

Test yourself

4.1 State the proportion of the world's water that is made up of freshwater. [1]

4.2 State the amount of freshwater stored in the two largest storages of global freshwater. [1]

The main inputs and outputs from the hydrological cycle

The **hydrological cycle** refers to the cycle of water between the biosphere, atmosphere, lithosphere, and hydrosphere. At a local scale—the drainage basin—the cycle has a single input (precipitation (PPT)), and two major losses (outputs): **evapotranspiration** (EVT) and runoff. A third output (leakage) may also occur from the deeper subsurface to other basins. The drainage basin **system** is an open system as it allows the movement of energy and matter across its boundaries. Solar radiation drives the hydrological cycle. Water can be stored at a number of stages or levels within the cycle. These storages include vegetation, surface, soil moisture, groundwater, and water channels. These storages are linked by a number of flows (figure 4.1.2).

Concept link

EQUILIBRIUM: The hydrological cycle is in balance between the inputs and the outputs. As either changes, the whole system responds and reaches a new equilibrium.

Assessment tip

The hydrological cycle can be studied at any scale, from the global to the local.

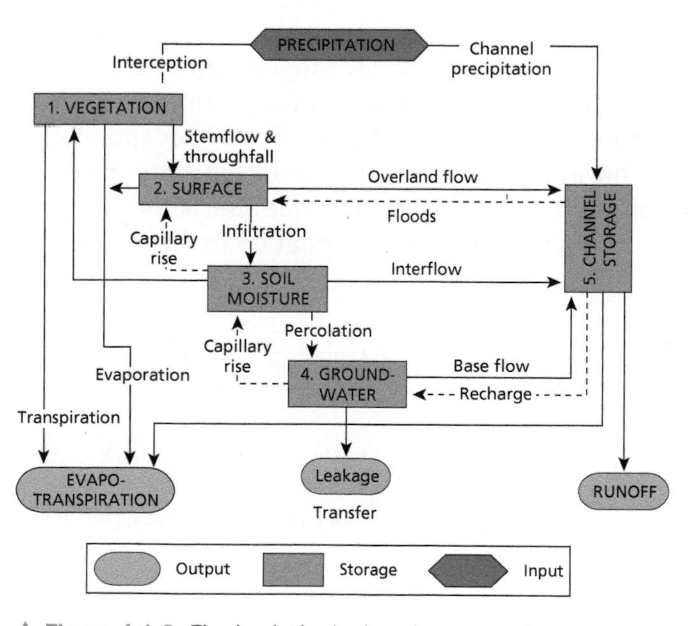

▲ Figure 4.1.2 The hydrological cycle as a system

Assessment tip

When drawing diagrams of cycles, e.g. water, carbon and/or nitrogen cycles, use a key to distinguish between storages and flows.

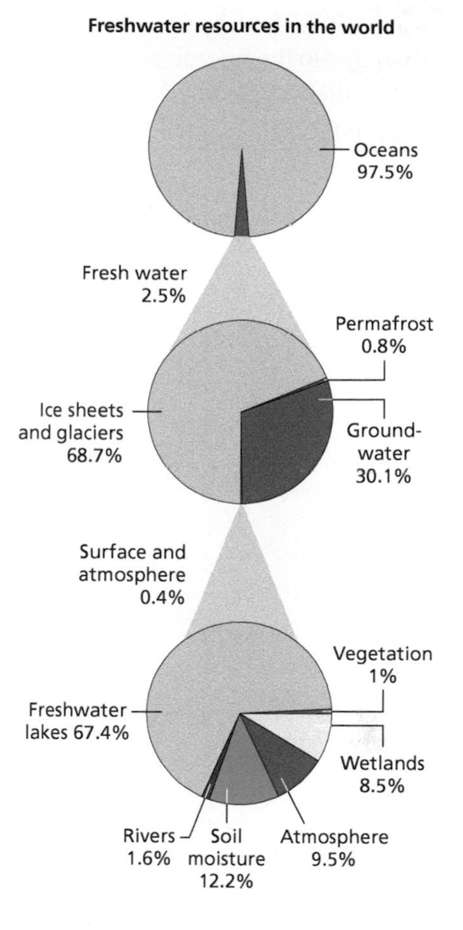

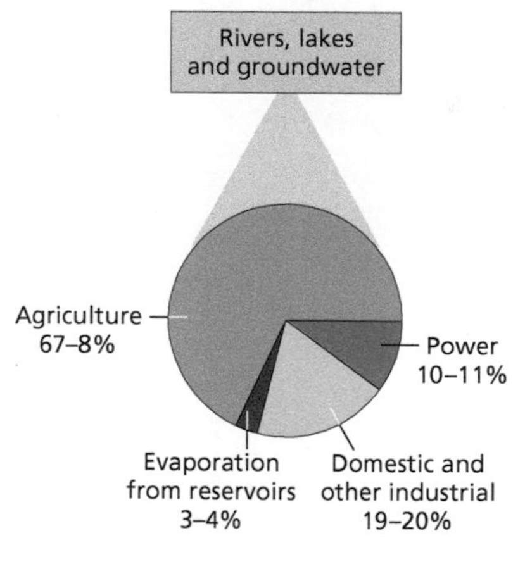

▲ Figure 4.1.1 Availability of freshwater supplies

Content link

Changes in the hydrological cycle, e.g. rainfall and evaporation, are explored further in section 7.2.

Inputs

Precipitation

The main input into the drainage basin is precipitation. This includes all forms of rainfall, snow, frost, hail, and dew. It is the conversion and transfer of moisture in the atmosphere to the land. The main characteristics of precipitation that affect local hydrology are:

- the total amount of precipitation
- intensity
- type (snow, rain and so on)
- geographic distribution and variability.

Other inputs could include **irrigation** water, water transfer schemes and the use of desalinated water.

Outputs

Runoff

Runoff (also known as overland flow) is a major output from drainage basins. Runoff refers to water flowing over the Earth's surface. It increases with gradient and when there is a greater proportion of impermeable surface (e.g. rocks and soil) and reduced vegetation cover. Runoff also increases in intense rainfall events and when the soil is saturated.

Evapotranspiration

Transpiration is the process by which water vapour escapes from living plants, mainly from the leaves, and enters the atmosphere. Evaporation is the process by which a liquid or a solid is changed into a gas. It is the conversion of solid and liquid precipitation (snow, ice, and water) to water vapour in the atmosphere. The combined effects of evaporation and transpiration are normally referred to as evapotranspiration (EVT). EVT represents the most important aspect of water loss, accounting for the loss of nearly 100% of the annual precipitation in arid areas and 75% in humid areas.

The main storages and flows in the hydrological cycle

Flows

Infiltration

Infiltration is the process by which water soaks into or is absorbed by the soil.

Overland flow

Overland flow (surface runoff) is water that flows over the land's surface. It occurs in one of the following two situations:

- when precipitation exceeds the infiltration rate
- when the soil is saturated (all the pore spaces are filled with water).

In areas of high precipitation intensity and low infiltration capacity, overland runoff is common. This is seen clearly in semi-arid areas and in cultivated fields. By contrast, where precipitation intensity is low and infiltration is high, most overland flow occurs close to streams and river channels.

Base flow relates to the part of a river's discharge that is provided by groundwater seeping into the bed of a river. It is a relatively constant flow although it increases slightly following a wet period.

Storages

Vegetation

Interception refers to water that is caught and stored by vegetation.

Soil

Soil moisture refers to the subsurface water in the soil.

Aquifers

Groundwater refers to subsurface water. Aquifers (rocks that contain significant amounts of water) provide a great reservoir of water. Aquifers are permeable rocks such as sandstone and limestone. Water moves slowly downwards from the soil into the bedrock – this is known as percolation. Depending on the permeability of the rock this may be very slow, or in some rocks, such as carboniferous limestone and chalk, it may be quite fast locally. The permanently saturated zone within solid rocks and sediments is known as the phreatic zone. The upper layer of this is known as the water table. The water table varies seasonally. It is higher in the wet season following increased levels of precipitation. The zone that is seasonally wetted and seasonally dries out is known as the aeration zone.

Recharge refers to the refilling of water in pores where the water has dried up or been extracted by human activity. Hence, in some places where recharge is not taking place, groundwater is considered a non-renewable resource.

Assessment tip

Not all aquifers are recharged. Those under the Saharan and Arabian deserts are relicts of fossil aquifers, as their source of water fell 20,000 years ago.

Cryosphere

The cryosphere is the snow and ice environment. Up to 66% of the world's freshwater is in the form of snow and ice. High latitude regions and high altitude areas may have important storages of snow and ice. Some of this may melt seasonally to produce major changes in the basin hydrological cycle.

Channel flow

Channel flow refers to the flow of water in channels such as rivers and streams. Most rivers will eventually transfer water to the ocean, although there are some that drain into inland lakes and seas, e.g. the River Jordan.

Test yourself

4.3 Explain why the hydrological cycle can be considered to be an open system. [2]

4.4 Identify two storages and two flows in the hydrological cycle. [2]

4.5 Distinguish between infiltration and throughflow. [2]

4.6 Outline the main characteristics of aquifers. [3]

The effects of human activity on the hydrological cycle

Agriculture

Agriculture can lead to a decline in infiltration and an increase in overland flow. However, it depends on the type of farming, its scale, the size of fields, use of machinery, and any soil conservation methods that are used, for example. The decrease in infiltration and increase in overland flow and soil erosion are due to a number of factors:

- intensification of agriculture
- heavier and more powerful machinery
- compaction by machines and/or animals, especially near waterholes

Concept link 

SUSTAINABILITY: Many agricultural systems are unsustainable and may lead to a decline in biodiversity, decreased soil quality and reduced groundwater. However, some forms of agriculture, such as agro-forestry, may be sustainable.

- cultivation of steeper slopes
- field enlargement allowing wind speeds to increase
- hedgerow removal, which also allows winds to be faster
- planting of winter cereals, which leaves the soil relatively bare for some months.

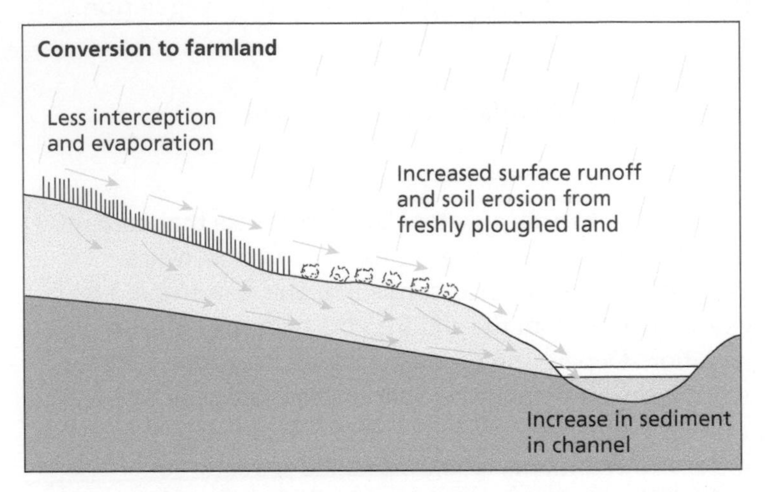

▲ Figure 4.1.3 The effect of agriculture on overland flow and infiltration

Test yourself

4.7 State the main human use of freshwater. [1]

4.8 Outline how agriculture can lead to reduced infiltration and increased overland flow. [4]

Deforestation

There are many types of deforestation—from total clear felling to selective removal—and the impacts of each type of deforestation will vary. For example, after deforestation flood levels in rivers increase. Following forest regeneration, flood levels and water quality return to pre-removal levels. However, this may take decades to occur. The reasons for the changes include:

- higher interception rates of mature forests
- decreased overland runoff beneath a mature forest
- higher infiltration rates beneath forests
- deeper soils beneath a cover of trees.

Hence replacement of natural vegetation by crops needs to be carefully managed. The use of shade trees and cover crops is a useful way of reducing soil erosion following deforestation. Grazing tends to increase overland runoff because of surface compaction and vegetation removal.

▲ Figure 4.1.4 Pollarding

Deforestation is also linked with increases in the sediment load and chemical load of streams (figure 4.1.5). In an extreme case of deforestation in the north-east of the USA, sediment loads increased fifteen-fold and nitrate loads increased by almost fifty times! Other examples are not as extreme—much depends upon how the forest is managed. If deforestation is only partial there is less sediment load. Similarly, if replanting takes place quickly the effects of deforestation are reduced.

Concept link 

SUSTAINABILITY: Deforestation can be unsustainable as it leads to a decrease in biodiversity. However, some forms, such as pollarding (figure 4.1.4) and coppicing, are less destructive and can be sustainable.

Young plants that are growing rapidly take up large amounts of water and nutrients from the soil, thereby reducing the rate of overland runoff and the chemical load of streams. Much depends upon the type of vegetation and its relative density, size, and rate of growth.

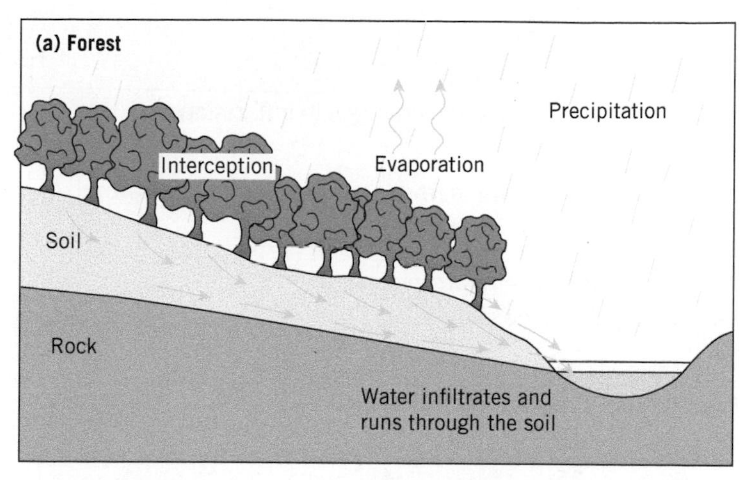

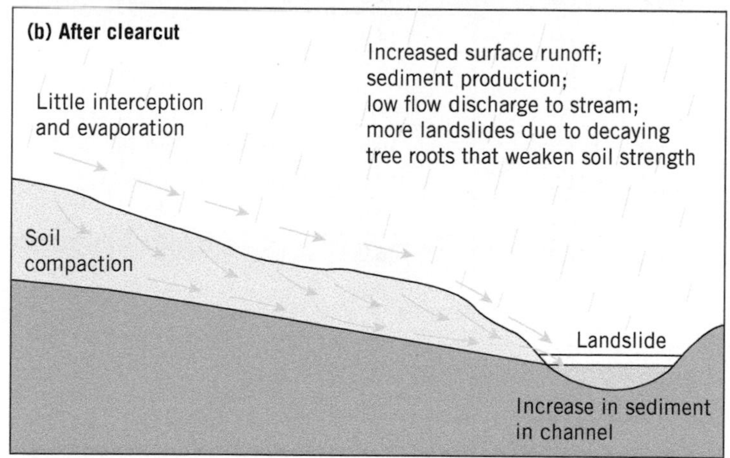

◀ Figure 4.1.5 The effect of deforestation on overland flow and infiltration

Urbanization

Storm water sewers:

- reduce the distance that storm water must travel before reaching a channel
- increase the velocity of flow because sewers are smoother than natural channels
- reduce storage because sewers are designed to drain quickly away.

Encroachment on the river channel:

- includes embankments, reclamation, and riverside roads
- usually reduces channel width, leading to higher floods
- by bridges can restrict free discharge of floods and increase levels upstream.

Replacement of vegetated soils with impermeable surfaces:

- reduces storage and so increases runoff
- decreases evapotranspiration because urban surfaces are usually dry
- increases velocity of overland flow
- reduces infiltration and percolation.

Building activity:

- clears vegetation which exposes soil and increases overland flow
- disturbs and dumps the soil, increasing erodability
- eventually protects the soil with an armour of concrete or tarmac.

>> **Assessment tip**

The impact of urbanization varies from settlement to settlement, depending on the amount of impermeable surfaces and the amount of channels, sewers and drains.

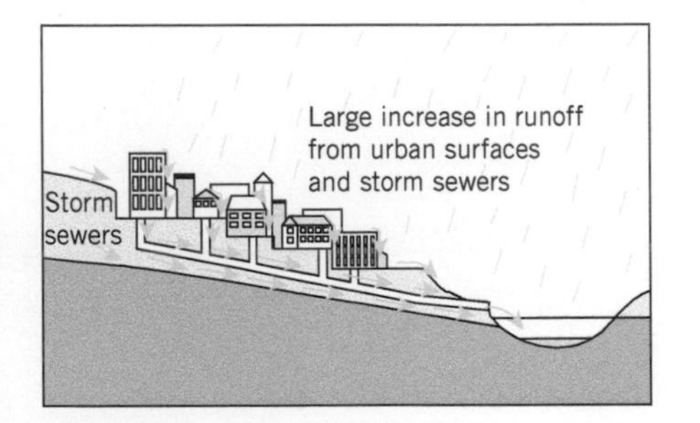

▲ Figure 4.1.6 The effect of urbanization on overland flow and infiltration

Test yourself

4.9 Outline how deforestation leads to changes in infiltration and overland runoff. [4]

The following data shows discharges from a rural stream and an urban stream during the same storm.

▼ Table 4.1.1 Discharges from a rural stream and an urban stream during the same storm

Discharge (litres per second)		
Time (min)	**Rural stream**	**Urban stream**
0	2.0	2.0
30	2.5	4.0
60	3.0	9.0
90	5.0	16.0
120	7.0	21.0
150	9.0	16.0
180	11.0	10.0
210	13.0	7.0
240	12.0	6.0
270	10.0	5.0
300	8.0	4.0
330	6.0	3.0
360	4.0	2.0
390	3.0	2.0
420	2.0	2.0

4.10 Plot the data for the rural and urban hydrograph. [4]

4.11 State the peak flow and the time lag for (a) the rural stream and (b) the urban stream. [2]

4.12 Compare and contrast the differences in the rising and recessional limbs of the rural and urban hydrographs. [4]

4.13 Explain why rural and urban hydrographs differ. [2]

The importance of ocean circulation systems

Surface ocean currents are caused by the influence of prevailing winds blowing steadily across the sea. The dominant pattern of surface ocean currents (known as **gyres**) is a roughly circular flow. The pattern of these currents is clockwise in the northern hemisphere and anticlockwise in the southern hemisphere. The main exception is the circumpolar current that flows around Antarctica from west to east. There is no equivalent current in the northern hemisphere because of the distribution of land and sea.

Warm ocean currents move water away from the equator. In contrast, cold ocean currents move water away from cold regions towards the equator. The major currents move huge masses of water over long distances. The warm Gulf Stream, for instance, transports 55 million cubic metres per second. Without it, the temperate lands of north-western Europe would be more like the sub-Arctic. The cold Peru Current and the Benguela Current of south-west Africa bring in nutrient-rich waters dragged to the surface by offshore winds. Ocean currents are important for the transfer of energy and nutrients.

Content link

In areas where there are upwelling ocean currents, there is increased nutrient availability which increases productivity and fish growth. This is examined in detail in section 4.3.

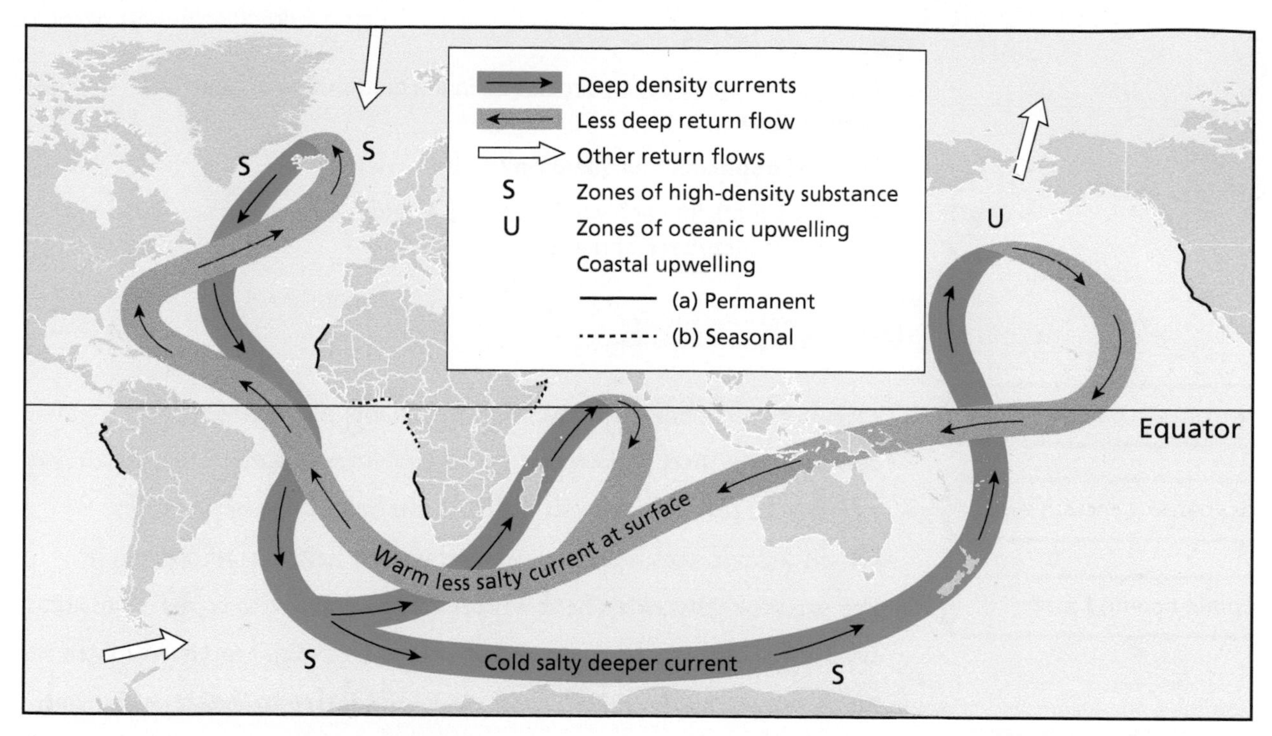

▲ **Figure 4.1.7** The oceanic conveyor belt

Oceanic convection occurs from polar regions where cold salty water sinks into the depths and makes its way towards the equator. The densest water is found in the Antarctic area. Here seawater freezes to form ice at a temperature of around −2°C. The ice is freshwater, hence the seawater left behind is much saltier and therefore denser. This cold dense water sweeps round Antarctica at a depth of about 4 km. It then spreads into the deep basins of the Atlantic, the Pacific and the Indian Ocean. Surface currents bring warm water to the North Atlantic from the Indian and Pacific Oceans. These waters give up their heat to cold winds which blow from Canada across the North Atlantic. This water then sinks and starts the reverse convection of the deep ocean current (figure 4.1.7).

Because the ocean conveyor belt operates in this way, the North Atlantic is warmer than the North Pacific, so there is proportionally more evaporation in the North Atlantic. The water left behind has more salt in it due to the evaporation, and therefore it is much denser, which causes it to sink. Eventually, the water is transported into the Pacific where it mixes with more water and its density is reduced.

Temperature, salinity, and pressure affect the density of seawater. Large water masses of different densities are important in the layering of the ocean water (denser water sinks). As temperature increases, water becomes less dense. As salinity increases, water becomes denser. As pressure increases, water becomes denser. A cold, highly saline deep mass of water is very dense, whereas a warm, less saline surface water mass is less dense. When large water masses with different densities meet, the denser water mass slips under the less dense mass. These responses to density are the reason for some of the deep ocean circulation models.

Test yourself

4.14 Identify one warm current and one cold current. [2]

4.15 Outline the importance of ocean currents. [2]

QUESTION PRACTICE

Outline the role of ocean circulation in the distribution of heat around the world. [4]

How do I approach this question?

There is 1 mark for each valid point and an extra 1 mark for exemplification/development.

SAMPLE STUDENT ANSWER

▲ Identifies causes; 1 mark

▲ Identifies consequences; 1 mark

▲ Development point; 1 mark

▲ Exemplification – two examples – only one would be needed for the final mark

Ocean circulation systems are driven by differences in temperature and salinity that affect water density; the resulting difference in water density drives the ocean conveyor belt; which distributes heat around the world; heating of oceans at the equator generates heat which is transferred to cooler climates in the northern and southern hemispheres; e.g. North Atlantic Drift moves warm water from Gulf of Mexico to Western Europe, ameliorating the climate; in contrast, cold water may also move to warmer areas, cooling the climate; e.g. Labrador Current cools NE USA.

This response could have achieved 4/4 marks.

4.2 ACCESS TO FRESH WATER

- **Aquifer** – a water-bearing rock, i.e. a permeable rock with pores that contain water
- **Grey water** – waste water that has been produced in homes and offices. It may come from sinks, showers, baths, dishwashers, washing machines, etc, but it does not contain faecal material
- **Water stress** – the total annual extraction of water as a proportion of the renewable supply in a given area
- **Water scarcity** – a lack of water due to either physical or economic reasons
- **Physical water scarcity** – where water consumption exceeds 60% of the usable supply

You should be able to show variations in access to freshwater

✔ Access to adequate supplies of freshwater varies.

✔ Climate change is having a major impact on access to freshwater.

✔ There are different types of water scarcity, e.g. physical water scarcity and economic water scarcity.

✔ The demand for fresh water is growing as populations, irrigation, and industrialization increase.

✔ Contamination and unsustainable abstraction may cause freshwater supplies to become limited.

✔ Water supplies can be enhanced through large dams, water redistribution schemes, desalination, water harvesting, groundwater recharging, etc.

✔ Conflict between human populations can occur due to the scarcity of water resources, particularly where sources are shared.

▲ Figure 4.2.1 Access to freshwater in Chalumna, Eastern Cape, South Africa

- **Economic water scarcity** – where a country physically has sufficient water to meet its needs, but requires additional storage and transport facilities
- **Water insecurity** – a lack of access to sufficient amounts of safe drinking water

Access to adequate supplies of freshwater varies

Less than 1% of all freshwater is available for people to use (the remainder is locked up in ice sheets and glaciers). Globally, around 12,500 cubic kilometres of water are considered available for human use on an annual basis. This amounts to about 6,600 cubic metres per person per year. If current trends continue, only 4,800 cubic metres will be available in 2025.

The world's available freshwater supply is not distributed evenly around the globe, either seasonally or from year to year.

- About three-quarters of annual rainfall occurs in areas containing less than a third of the world's population.
- Two-thirds of the world's population live in the areas receiving only a quarter of the world's annual rainfall.

In 2015, some 844 million people lacked a basic level of drinking water service (having an improved source of water within 30 minutes of their home). The average North American and Western European adult consumes 3 cubic metres a day, compared with around 1.4 cubic metres per day in Asia and 1.1 cubic metres per day in Africa.

Concept link 

SUSTAINABILITY: The distribution of water resources around the world is uneven. This means that in some places there is not enough water to meet demand. Therefore, water use in those places is unsustainable.

Test yourself

4.16 State by how much water availability per person is predicted to decline by 2025. [1]

4.17 State the number of people that lacked access to improved water sources in 2015. [1]

4.18 Calculate the average daily water intake of (i) an Asian adult and (ii) an African adult compared with a North American or European adult. [2]

Climate change and access to water

Climate change is likely to lead to increased rain in some areas, longer droughts in others and melting glaciers which may threaten the long-term security of water supplies for millions of people.

Water availability is likely to decrease in many regions. For example, 300 million people in sub-Saharan Africa live in a water-scarce

Concept link 

SUSTAINABILITY: In some areas global climate change is resulting in less water being available to match demand—water use is becoming unsustainable.

Assessment tip

Probable impacts of climate change include:

- 200 million people at risk of being driven from their homes by flood or drought by 2050
- an increase in storm activity such as more frequent and intense hurricanes (owing to more atmospheric energy)
- reduced rainfall over the USA, southern Europe and the Commonwealth of Independent States (CIS), leading to widespread drought
- up to 4 billion people suffering from water shortages if temperatures rise by 2°C.

Test yourself

4.19 Distinguish between physical and economic water scarcity. [1]

4.20 Suggest reasons why water scarcity is predicted to increase by 2040 in some areas. [3]

Assessment tip

When describing a map, use the key and try to identify the areas with the maximum readings, minimum readings and any exceptions.

environment, and climate change increases **water stress** in many areas. Central and Southern Europe are predicted to get drier as a result of climate change.

Climate change can lead to changes in weather patterns and rainfall (in both quantity and distribution). Climates may become more extreme, and more frequent and unpredictable extreme weather events can be expected as atmospheric patterns are disturbed.

According to the 7th World Water Forum (2015), water stress and **water scarcity** are global challenges with far-reaching economic and social implications. Driven by increasing population, growing urbanization, changing lifestyles and economic development, the total demand for water is rising from urban centres, agriculture and industry.

Physical water scarcity and economic water scarcity

The level of water scarcity in a country depends on precipitation and water availability, population growth, demand for water, affordability of supplies, and infrastructure. Where water supplies are inadequate, two types of water scarcity exist:

- **Physical water scarcity**, where water consumption exceeds 60% of the usable supply. To help meet water needs some countries, such as Saudi Arabia and Kuwait, have to import much of their food and invest in desalinization plants.
- **Economic water scarcity**, where a country physically has sufficient water to meet its needs, but requires additional storage and transport facilities. This means having to embark on large and expensive water-development projects, as in many sub-Saharan countries.

The growing demand for fresh water as populations, irrigation and industrialization increase

More water will be required to produce food for the world's growing population, partly because of changes in diet. Many industries, in particular the food, drinks, textiles, and pharmaceuticals industries, need large quantities of water for their products, which will increase demand for water over the coming decades. Much of the growth will be in LICs, many of which are already experiencing water stress.

In LEDCs, access to adequate water supplies is most affected by the exhaustion of traditional sources, such as wells and seasonal rivers. Access may be worsened by cyclical shortages in times of drought, inefficient irrigation practices, and lack of resources to invest to meet demand and to increase the efficiency of irrigation system. The International Water Management Institute estimates that 26 countries, including 11 in Africa, can be described as water-scarce, and that over 230 million people are affected.

As world population and industrial output have increased, the use of water has accelerated, and this is projected to continue. By 2025 global availability of fresh water may drop to an estimated 5,100 cubic metres per person per year, a decrease of 25% on the 2000 figure. Rapid urbanization results in increasing numbers of people living in urban shanty towns where it is extremely difficult to provide an adequate supply of clean water or sanitation.

As well as the need for an adequate quantity of water for consumption, the quality of the water also needs to be adequate. However, WHO estimates that around four million deaths each year can be attributed to water-related disease, particularly cholera, hepatitis, dengue fever, malaria, and other parasitic diseases. The incidence and effects of these diseases is most pronounced in developing countries, where 66% of people have no toilets or latrines.

Irrigation

The use of water for irrigation accounts for about 70% of all water used for human purposes. Withdrawals for industry are about 20%, and those for municipal use are about 10%. Although we are withdrawing only 10% of renewable water resources, and consuming only about 5%, there are still problems for human use. Water is unevenly distributed in space and in time—and we are degrading the quality of much more water than we withdraw and consume.

People have irrigated crops since ancient times: there is evidence of irrigation in Egypt going back nearly 6,000 years. Water for irrigation can be taken from surface storages, such as lakes, dams, reservoirs and rivers, or from groundwater. Types of irrigation range from total flooding, as in the case of paddy fields, to drip irrigation, where precise amounts are measured out to each individual plant (figure 4.2.2). Irrigation may lead to unwanted consequences, for example in relation to groundwater quality and salinization.

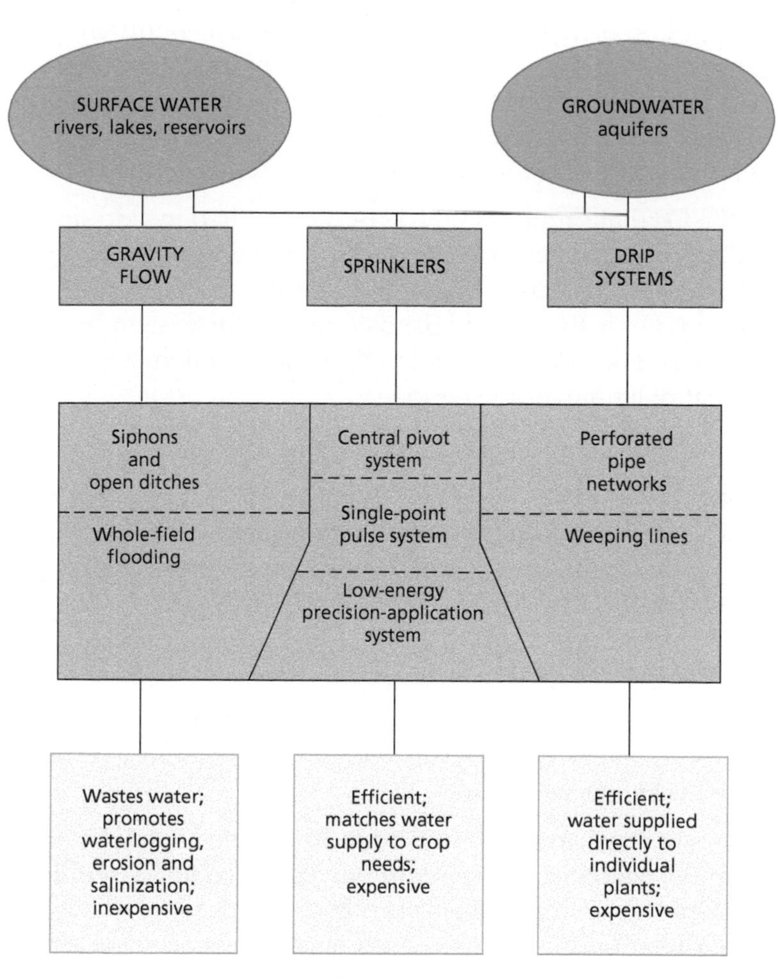

▲ Figure 4.2.2 Types of irrigation

Assessment tip

A number of trends are increasing the pressure to manage water more efficiently. These include:

- population growth – set to reach 9 billion by 2050; some estimates say it may eventually peak at 11 billion
- the growing middle class – increasing affluence leads to greater water consumption, for example showers, baths, gardening
- the growth of tourism and recreation, for example golf courses, water parks, swimming pools
- urbanization – urban areas require significant investment in water and sanitation facilities to get water to people and to remove waste products hygienically
- climate change – no one is sure exactly how this will influence the water supply, but there will be winners and losers in the supply of fresh water.

Content link

Global climate change will reduce water availability in some areas. This is explored in section 7.2.

Concept link

STRATEGY: Different environmental value systems (EVSs) suggest different methods for irrigation. An ecocentrist approach may suggest natural flooding, while a technocentrist approach may suggest drip irrigation, an expensive but efficient method of irrigation.

▲ Figure 4.2.3 Drip irrigation in the UAE

Impacts of irrigation

Irrigation occurs in developed as well as developing countries. For example, large parts of the USA and Australia are irrigated. The advent of diesel and electric motors in the mid-20th century led for the first time to systems that could pump groundwater out of aquifers faster than it was recharged. In some regions this has led to loss of **aquifer** capacity, decreased water quality and other problems.

Irrigation can reduce the Earth's albedo (reflectivity) by as much as 10%. This is because a reflective sandy surface may be replaced by one with dark green crops. Irrigation can also cause changes in precipitation. Large-scale irrigation in semi-arid areas, such as the High Plains of Texas, has been linked to increased rainfall, hailstorms and tornadoes.

Contamination and unsustainable abstraction may cause freshwater supplies to become limited

Growing pressure on rivers has meant that a number of them no longer reach the sea, for example the Colorado river. In addition, of those that do, they may be heavily polluted, such as the Ganges, Yangtze, and Hwang He rivers, making them unsafe for domestic consumption. The Rio Nuevo (New River) that flows from Baja California into California is now contaminated with agricultural, industrial, and municipal waste. It is the most polluted river of its size in the USA. The pollutants cause regular algal blooms. The river feeds the inland Salton Sea in California, which has become increasingly saline over time. It is now saltier than seawater and there have been major fish deaths due to increasing temperatures, salinity, and bacteria present in the river.

Overuse of the Ogallala aquifer in the USA has resulted in a decline of 9% since 1950, and 2% between 2000 and 2009. It has been estimated that it would take 6,000 years to recharge at current rainfall levels. Another vast aquifer is the Nubian Sandstone Aquifer System under the Sahara Desert. This is more than 2 million square kilometres in area and offers great potential for the region.

>> **Assessment tip**

The impacts of irrigation vary. In Texas, USA, irrigation has reduced the water table by as much as 50 metres. By contrast, in the fertile Indus Plain in Pakistan, irrigation has raised the water table by as much as six metres since 1922 and caused widespread salinization.

Enhancing water supplies

Water supplies can be enhanced through reservoirs, redistribution, desalination, artificial recharge of groundwater, and rainwater harvesting schemes. Water conservation (including grey-water recycling) also helps.

The construction of large dams

The number of large dams (more than 15 metres high) being built around the world is increasing rapidly and has reached the level of almost two completions every day.

The advantages of dams are numerous. They include flood and drought control, irrigation, hydroelectric power, improved navigation, recreation and tourism. On the other hand, there are numerous costs. For example, these include water losses through evaporation, salinization, displacement of population, drowning of archaeological sites, seismic stress, channel erosion (clear water erosion) below the dam, silting upstream of the dam, and reduced fertility (sediment decline) downstream from the dam.

Concept link

STRATEGIES: Large dams are an example of a technocentric (environmental value system) strategy to enhance water supplies. They rely on capital and technology to increase water availability.

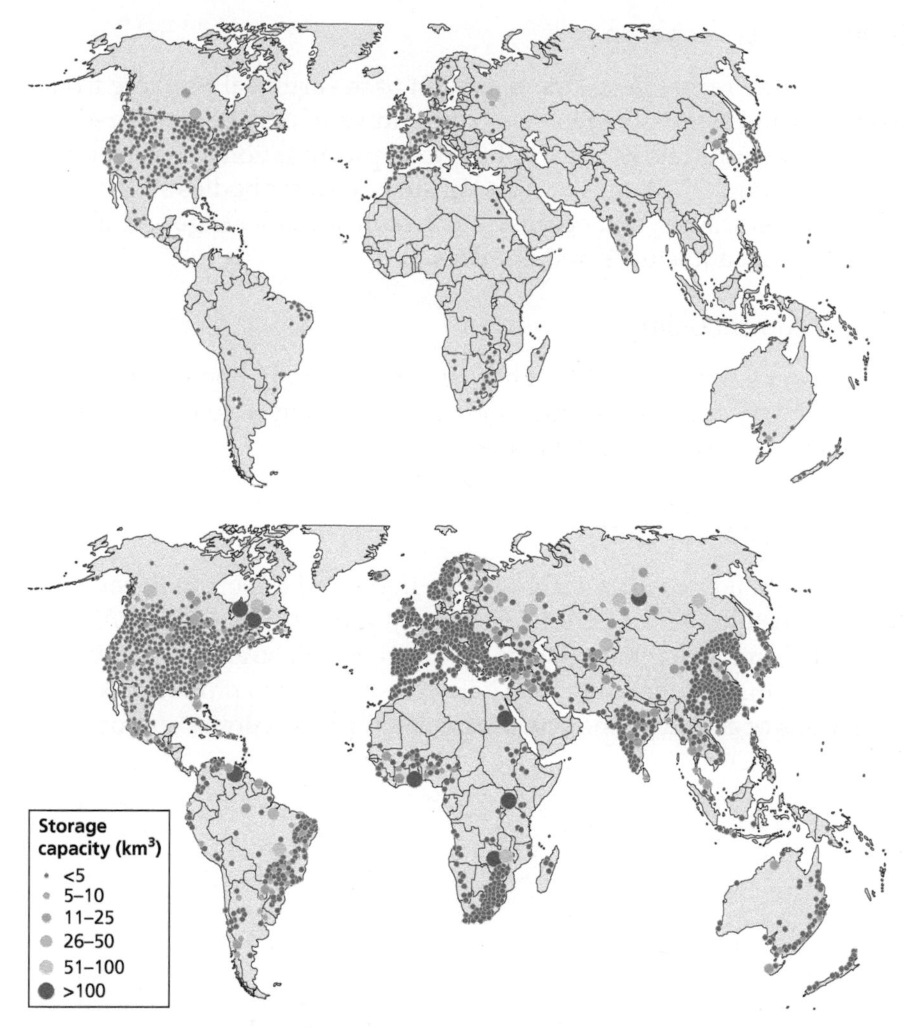

▲ Figure 4.2.4 The number of large dams worldwide, 1945 and 2005

▲ Figure 4.2.5 Visitors at the Three Gorges Dam

Redistribution schemes

China's controversial South–North Water Diversion Project channels water from the south of Hebei province to Beijing in the north. The South–North Diversion Project cost £48 billion and consists of 2,400 kilometres of canals and tunnels designed to divert 44.8 billion cubic metres of water from the wetter south to the north. Some 350,000 people were relocated due to the building of reservoirs and canals.

Desalination

Desalination or desalinization removes salt from seawater. More generally, desalination may also refer to the removal of salts and minerals from seawater and from soil. Seawater is desalinated to produce fresh water fit for human consumption (potable water) and for irrigation.

Water harvesting schemes

Water harvesting aims to capture and channel a greater share of rainfall into the soil, and conserve moisture in the root zone where crops can use it. In many drainage basins, constructing dams across stream headwaters can trap large amounts of runoff, which can either be channelled directly to a field, stored in a tank or small reservoir for later use, or allowed to percolate through the soil to recharge the groundwater.

Concept link 

STRATEGY: Redistribution is another example of a technocratic environmental value system – the use of engineering and capital to move water from one place to another.

Concept link 

STRATEGY: Desalinization is an example of a technocratic environmental value system—the use of engineering and capital to create fresh water from seawater

Concept link 

STRATEGY: Water harvesting could be considered as an ecocentric environmental value system. It captures rainfall using simple techniques such as water butts to catch and store water.

Groundwater recharge

Groundwater recharge occurs as a result of the artificial recharge from irrigation and reservoirs. However, groundwater may be recharged naturally by infiltration of part of the total precipitation at the ground surface; through the banks and bed of surface water bodies such as ditches, rivers, lakes, and oceans, and groundwater leakage and inflow from adjacent aquicludes and aquifers.

Test yourself

4.21 Outline the changes in the number of dams, and their location, between 1945 and 2005. [4]

4.22 Distinguish between water harvesting and desalination. [2]

Grey-water recycling

Grey water is waste water that has been produced in homes and offices. It may come from sinks, showers, baths and washing machines, but it does not contain faecal material.

Scarcity and conflict

Ethiopia is building Africa's largest dam, the Grand Ethiopian Renaissance Dam, on the Blue Nile (figure 4.2.6). When it is finished, it will be 170 metres tall and 1.8 kilometres wide. Its reservoir will be able to hold more than the volume of the entire Blue Nile. It is designed to produce 6,000 megawatts of electricity, more than double Ethiopia's current output.

Concept link 

STRATEGY: The Grand Ethiopian Renaissance Dam is an example of a technocratic strategy (that uses technology and capital) to capture and store water.

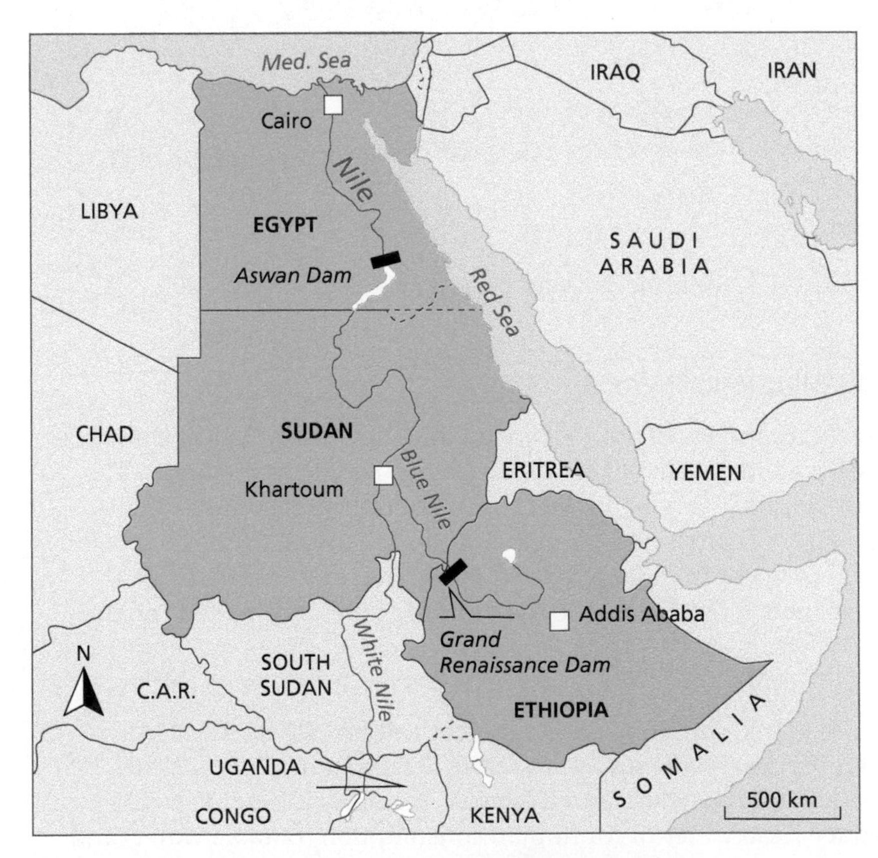

▲ Figure 4.2.6 Scarcity and conflict over the Nile's resources

This opportunity for Ethiopia could spell disaster for Egypt.

- The Nile provides nearly all of Egypt's water.
- Egypt claims two-thirds of that flow based on a treaty it signed with Sudan in 1959.
- This is no longer enough water to satisfy the growing population (1.8% growth in 2015) and agricultural sector.

- Annual water supply per person has fallen by well over half since 1970.
- The United Nations (UN) has warned of a looming crisis.
- The stakeholders include the governments of Egypt, Ethiopia and Sudan, as well as the people who will make use of the water.
- Egyptian leaders have been very forceful in the protection of their water supply. This has soured relations with the other countries that share the Nile Basin.

In March 2015 the leaders of Egypt, Ethiopia, and Sudan signed a declaration that approved construction of the dam as long as there is no "significant harm" to downstream countries.

There is uncertainty over the dam's ultimate use. Ethiopia insists that it will produce only power and that the water pushing its turbines will ultimately flow downstream. Egyptians fear it will also be used for irrigation, reducing downstream supply.

A more reasonable concern is over the dam's large reservoir. If filled too quickly, it would for a time significantly reduce Egypt's water supply and affect the electricity-generating capacity of its own Aswan Dam. But the Ethiopian government faces pressure to see a quick return on its investment. The project, which is mostly self-funded, will cost around US$4.8 billion.

Test yourself

4.23 Outline the benefits of the Great Ethiopian Renaissance Dam for Sudan. [3]

4.24 Suggest reasons why Egypt is against the construction of the Great Ethiopian Renaissance Dam. [3]

QUESTION PRACTICE

Essay

To what extent have human societies contributed both to the problem and the solution of water scarcity around the world? Justify your response with the help of named examples. [9]

How do I approach this question?

There are two parts to the question: what are the causes of, and potential solutions to, water scarcity; and to what extent have human societies contributed to the causes and potential solutions. Both parts need to be addressed.

The causes of water scarcity may include global climate change, over-extraction, wastage, broken pipes, etc. Potential solutions include construction of large dams, irrigation, and use of groundwater. Both parts should be illustrated with the use of named examples, of places as well as technologies.

Answers may include:

- Understanding concepts and terminology, e.g. water scarcity, water stress, water shortages, climate change, irrigation, desalination, groundwater, over-extraction, water harvesting, conservation, regulation, adaptation, mitigation.
- Breadth in addressing and linking (concepts and terminology), e.g. ecocentrism and self-restraint, anthropocentrism and environmental regulation, the use of taxes and legislation, technocentrism and desalination.
- Examples of over-extraction, e.g. Ogallala aquifer, USA; climate change, melting glaciers, e.g. Himalayas; rapid population growth, e.g. Niger and Mali; rapid urbanization, e.g. megacity growth in Asia; drought and water shortages in Cape Town, South Africa.
- Balanced analysis of both the causes and the potential solutions.
- A conclusion that is consistent with, and supported by, examples. For example, "some causes of water scarcity are natural, e.g. cyclical changes such as El Nino events, but human activities are increasingly causing water shortages, e.g. those linked with global climate change. In many places technical developments, such as desalination, are helping to solve water scarcity, but for many poor countries, this is not an option that they can afford, e.g. Yemen."

SAMPLE STUDENT ANSWER

▲ Good definition

▲ Good support

▲ Another definition with some exemplification

▲ Human cause identified and supported with example

▼ Second cause but no support

▼ Third cause but no support

▲ A fourth cause with supporting example

▲ Generic cause

▲ Potential solution – technocentric – with named support

▼ Could name the specific project

▲ Good range of technocentric approaches

▼ All generic

▲ Brief conclusion derived from evidence provided

▼ Mainly limited to human causes/potential solutions rather than a more balanced appraisal

Physical water scarcity occurs where water consumption exceeds 60 per cent of the usable supply. To help meet water needs some countries such as Saudi Arabia and Kuwait have to import much of their food and invest in desalinization plants. In contrast, economic water scarcity occurs where a country physically has sufficient water to meet its needs, but requires additional storage and transport facilities. This means having to embark on large and expensive water-development projects, as in many sub-Saharan countries.

Over-extraction of water for irrigation in agriculture can increase water scarcity, e.g. Ogallala aquifer in Great Plains of USA. Deforestation or changes in land-use can result in water not being able to recharge aquifers as water runs off surfaces too quickly. Urbanization changes water flow and can result in less water infiltration into the ground to recharge aquifers. Rapid population growth may result in a trend to physical or economic water scarcity, e.g. sub-Saharan Africa or South Asia. Climate change will alter the availability of water around the world, e.g. warmer temperatures lead to increased evaporation.

Some large countries may try to solve this challenge of regional physical scarcity with large infrastructure projects, e.g. water transfer projects in China. Countries can enhance water availability through the use of technology, e.g. reservoirs, desalination, artificial recharge of aquifers. Some societies may choose to increase water efficiency, e.g. variable flush toilets, aerated taps and shower heads. Regulation and laws can enforce water metering and cap water use, e.g. hosepipe bans. Low-technology solutions such as rainwater harvesting schemes.

In conclusion, there are many potential causes of water shortages around the world, and human activity appears to be among the main causes. Similarly, there are many potential solutions to the causes of water scarcity, and technocratic solutions appear to hold the key.

- Understanding of concepts and terminology present, e.g. water scarcity, climate change, irrigation, desalination, groundwater, over-extraction.
- Some breadth in addressing and linking the causes of water scarcity (mainly limited to human activities) and only deals with technocentric solutions such as large dams and desalination.
- Examples of over-extraction, e.g. Ogallala aquifer, USA, but most of the examples are very generic.
- Unbalanced analysis – better on the causes than the potential solutions.
- A conclusion that fits with the unbalanced analysis.

This response could have achieved 6/9 marks.

4.3 AQUATIC FOOD PRODUCTION SYSTEMS

You should be able to show the characteristics of aquatic food production systems—capture fisheries and aquaculture

- ✔ A highly diverse range of food webs are supported by photosynthesis by phytoplankton.
- ✔ The most productive parts of the ocean are shallow seas and those with upwelling currents.
- ✔ There is increased demand for aquatic food production.
- ✔ Catching certain species, such as whales, can be controversial, although for some indigenous populations they are an important source of food.
- ✔ Technological developments in the fishing industry have led to the depletion of some fish stocks.
- ✔ It is possible to manage fish stocks more sustainably.
- ✔ The growth of aquaculture has led to increased food resources.
- ✔ There are environmental risks associated with the development of aquaculture.

- **Aquaculture** – raising fish commercially as food for humans and/or animal feed
- **Biorights** – the rights of an endangered species, a unique species or a landscape to be protected
- **Productivity** – the production of energy/organic matter per unit of time
- **Genetically modified organisms** – the addition of certain traits (DNA) to a plant or animal to make it more nutritious, make it grow faster or become more resistant to disease and or pests

Photosynthesis by phytoplankton

Phytoplankton are microscopic organisms that live near the ocean surface and obtain their energy from the Sun. They convert this into food energy (chemical energy) which forms the basis of most oceanic food webs.

The most productive parts of the ocean

The highest rates of **productivity** are found in the upper ocean, especially where upwelling currents bring nutrients to the surface. Higher chlorophyll concentrations and in general higher productivity are observed on the equator, along the coasts (especially eastern margins), and in the high latitude oceans (figure 4.3.2 (a) and (b)). Upwelling currents lead to nutrient-rich zones (figure 4.3.2 (c) and (d)).

Content link

The characteristics of terrestrial food production systems are explored in section 5.2. The impacts of declining numbers of particular species are explored in unit 3. Section 8.4 considers the question of resource depletion.

Test yourself

4.25 Compare and contrast the harvesting techniques of fishing, aquaculture and agriculture. [2]

4.26 Outline reasons for variations in ocean productivity. [2]

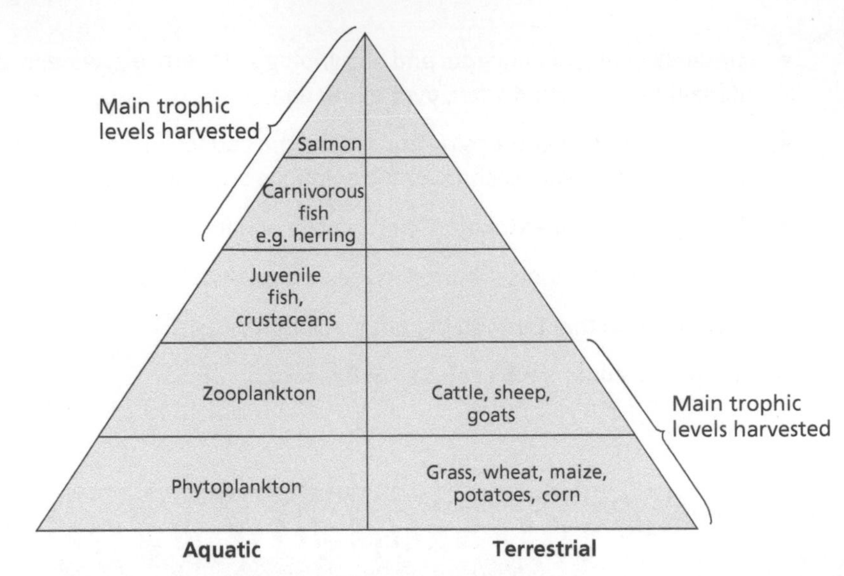

▲ Figure 4.3.1 Fisheries, aquaculture and agriculture – different "harvesting" techniques

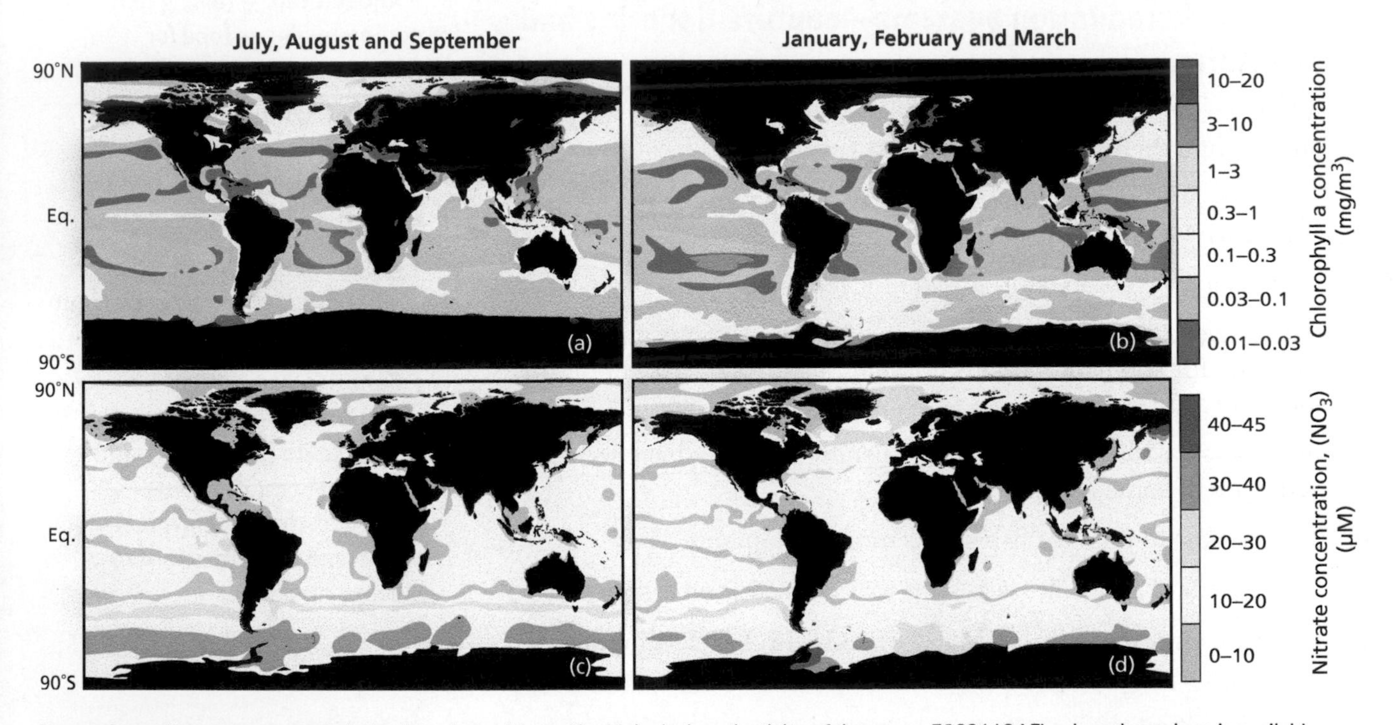

Source: https://www.nature.com/scitable/knowledge/library/the-biological-productivity-of-the-ocean-70631104 The data shown here is available through the NASA's OceanColor and NOAA's National Oceanographic Data Center websites. Sea ice cover impedes measurement of ocean colour from space, reducing the apparent areas of the polar oceans in the winter hemisphere (upper panels).

▲ Figure 4.3.2 Seasonal variations in ocean productivity (a) and (b) and nutrient-rich areas (c) and (d)

Demand for aquatic food resources is increasing

There has been a huge increase in the demand for aquatic food resources as the human population grows and diets change. The change in diet (nutrition transition) that occurs as LEDCs become NICs (newly industrializing countries) results in an increase in consumption of meat, fish and dairy products.

Between 1961 and 2016, the average annual increase in global food fish consumption (3.2%) outpaced population growth (1.6%) and exceeded that of meat from all terrestrial animals combined (2.8%). Food fish consumption grew from around 6.0 kilograms in 1951 to over 20 kilograms in 2015.

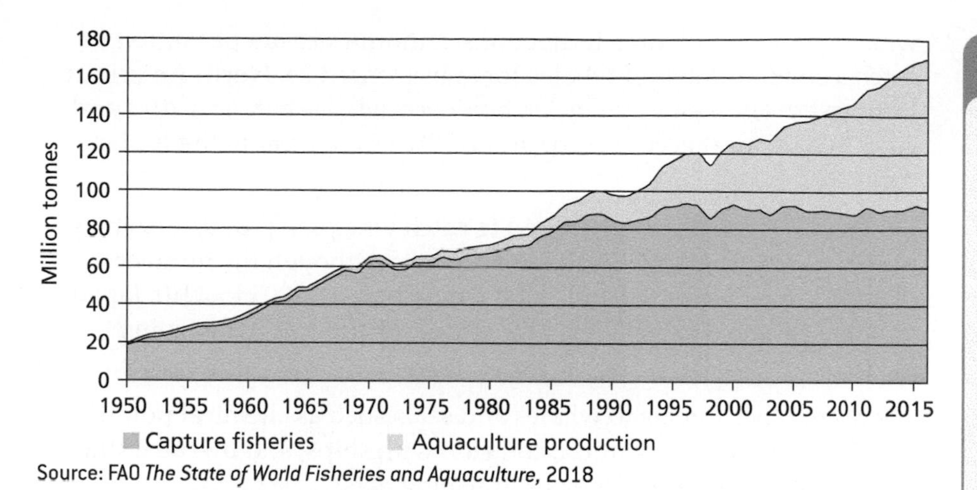

Source: FAO *The State of World Fisheries and Aquaculture*, 2018

▲ **Figure 4.3.3** World capture fisheries and aquaculture production

Of the 171 million tonnes of total fish production in 2016, about 88% (over 151 million tonnes) was utilized for direct human consumption, a share that has increased significantly in recent decades. The greatest part of the 12% used for non-food purposes (about 20 million tonnes) was reduced to fishmeal and fish oil.

Global fish production peaked at about 171 million tonnes in 2016. The value of fisheries and **aquaculture** production in 2016 was estimated at US$362 billion, of which US$232 billion was from aquaculture production. Global capture fisheries production was 90.9 million tonnes in 2016.

Aquaculture continues to grow faster than other major food production sectors. Global aquaculture production in 2016 was 110.2 million tonnes and included 80.0 million tonnes of food fish and 30.2 million tonnes of aquatic plants.

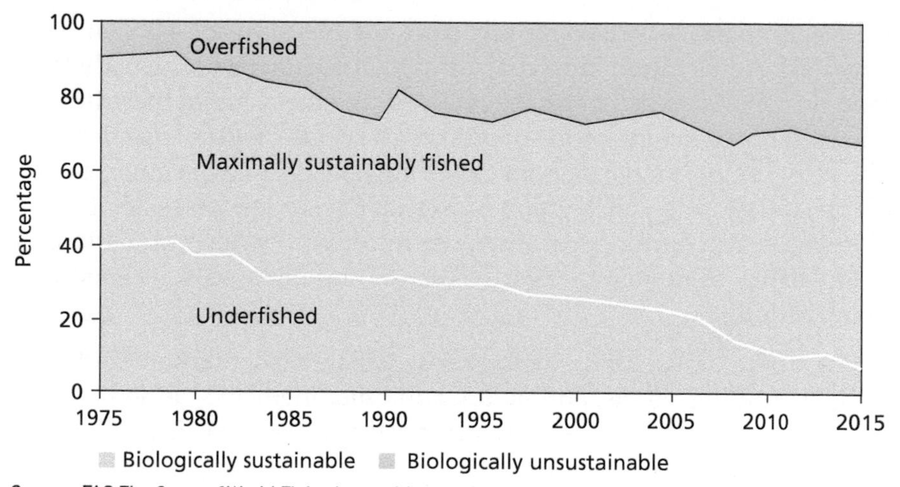

Source: FAO *The State of World Fisheries and Aquaculture*, 2018

▲ **Figure 4.3.4** Global trends in the state of the world's marine fish stocks, 1974–2015

Biorights

The harvesting of certain species, such as whales and seals, is often controversial. Ethical issues arise over **biorights**, the rights of indigenous cultures and international conservation legislation. In the 1930s over 50,000 whales were killed annually. In 1986 the International Whaling Commission introduced a ban on commercial

Assessment tip

Figure 4.3.3 is a compound line graph, i.e. the value of one group (e.g. aquaculture) is placed on top of another (here "capture fisheries") so that it is possible to see the total as well as the relative contribution of each group. Thus, in 2015, total fish production was around 160 million tonnes, with capture fisheries accounting for about 95 million tonnes and aquaculture about 65 million tonnes.

Concept link 

SUSTAINABILITY: The impact of reducing fish numbers is having an impact on biodiversity and the sustainability of resources.

Assessment tip

If answering questions about a graph, make sure you use the scale and try to quantify your answer.

Test yourself

4.27 Describe the trend in over-fished stocks. [2]

4.28 Describe the growth in aquaculture between 1950 and 2015. [3]

whaling. However, some indigenous communities are permitted to catch a small number of whales for subsistence use. North American Inuits were allowed to catch 336 bowhead whales between 2013 and 2018. Whale meat accounts for about 50% of the meat in the Inuits' diet.

Japanese whalers have continued to catch whales. Up until 1986 they were catching some 2,000 whales annually. Although the number dropped to between 500 and 1,200 annually from 2005 to 2010, Japan announced plans to leave the IWC and start commercial whaling in 2019.

There are other threats to whales and seals such as chemical pollution, plastic waste, noise pollution, collision with ships, and by-catch (the unintentional capture of whales in fishing nets).

Concept link 

SUSTAINABILITY and STRATEGY: Managing fisheries requires strategies—such as the Common Fisheries Policy—in order to achieve sustainability.

Managing fisheries

Changes in fishing methods and equipment, e.g. larger nets and smaller mesh sizes, have reduced fish stocks, while practices such as sea-floor dredging have damaged the sea bed. Nevertheless, there is scope for fishing to be managed in ways that are more sustainable.

The Common Fisheries Policy (CFP) of the European Union is an example of fisheries management. The most recent changes to the policy took effect in 2014. The aims of the CFP are to develop an industry that is economically, socially, and environmentally sustainable. The policy states that between 2015 and 2020 catch limits should be set at a sustainable level. According to the CFP, illegal fishing is believed to cost €10 billion per year and yield about 11–26 million tonnes per year, i.e. about 15% of the world's catch.

>> Assessment tip

Fisheries can be managed in a number of ways:

- total allowable catches (quotas for how much can be caught)
- fishing licences can be issued
- boat capacity management (only allow a certain number of boats/ size of boats) into an area
- reducing environmental impact
- specifying a minimum mesh size on nets
- closing fishing grounds at different times of the year.

The CFP has a budget of €6,400 million between 2014 and 2020.

Aquaculture

Aquaculture has increased rapidly since the 1990s (figure 4.3.3 on page 111). Aquaculture involves raising fish commercially, usually for food. In contrast, a fish hatchery releases juvenile fish into the wild for recreational fishing or to supplement a species' natural numbers. The most important fish species raised by fish farms are salmon, carp, tilapia, catfish, and cod. Salmon makes up 85% of the total sale of Norwegian fish farming. Farming was introduced when populations of wild Atlantic salmon in the North Atlantic and Baltic seas declined due to overfishing.

Issues around aquaculture include: loss of habitats, pollution (with feed, antifouling agents, antibiotics, and other medicines added to fish pens), spread of diseases, and escaped species (some involving **genetically modified organisms**).

▲ Figure 4.3.5 Integrated aquaculture

Technological costs are high, and include using drugs such as antibiotics to keep fish healthy, and steroids to improve growth. Breeding programmes are also expensive. Outputs are high per hectare and per farmer, and efficiency is also high.

Environmental effects can be damaging. Salmon are carnivores and so they need to be fed pellets made from other fish. It is possible that farmed salmon actually represent a net loss of protein in the global food supply as it takes 2–5 kilograms of wild fish to grow 1 kilogram of salmon. Other environmental costs include the sea lice and diseases that spread from farmed salmon into wild stocks, and pollution (created by uneaten food, faeces and chemicals used to treat the salmon) contaminating surrounding waters. Organic debris of this type, with steroids and other chemical waste, can contaminate coastal waters.

Accidental escape of fish can affect local wild fish gene pools when escaped fish interbreed with wild populations, reducing their genetic diversity and potentially introducing non-natural genetic variations. In some parts of the world, fish that have escaped from fish farms threaten native wild fish as salmon is an alien species.

QUESTION PRACTICE

Identify two factors which may account for changes in the total capture of Atlantic herring. [2] 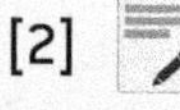 

How do I approach this question?

You must link a correct reason to either an increase or decrease in the amount of fish caught. You will receive 1 mark for each valid reason explained.

SAMPLE STUDENT ANSWER

An increase in catch due to more boats, larger boats, more advanced technology to find the fish.

A decrease in catch due to quotas, total allowable catches, strict regulations, conservation efforts, over-fishing in the past, declining fish stocks.

▲ Valid points highlighted

▲ Valid points highlighted

This response could have achieved 2/2 marks.

4.4 WATER POLLUTION

You should be able to show the causes and consequences of water pollution

- ✔ There are many causes of water pollution.
- ✔ Water pollution can be measured in many ways.
- ✔ Biochemical oxygen demand (BOD) is an indirect method of measuring pollution in water.
- ✔ The presence or absence of certain species can be used to indicate pollution.
- ✔ Eutrophication is a form of pollution caused by excess concentrations of nitrates or phosphates.

- **Anoxia (anoxic conditions)** – a lack of oxygen
- **BOD (biochemical oxygen demand)** – a measure of the amount of dissolved oxygen needed to break down the organic material in a given volume of water through aerobic biological activity
- **Biotic index** – the use of plant or animal species to make conclusions about the level of pollution
- **Trent Biotic Index** – a scale that uses freshwater species to make conclusions about the level of pollution

Causes of water pollution

There are various sources of freshwater and marine pollution. Aquatic pollutants include organic material, floating debris, suspended solids, oil, inorganic plant nutrients (nitrates and phosphates), synthetic compounds, toxic metals, pathogens, hot water, radioactive pollution, noise, light, and biological pollutants (invasive species).

Concept link

Water pollution occurs when natural systems develop a new equilibrium and deviate from the long-term norm. This may result in a loss of biodiversity. Different environmental systems may suggest alternative strategies to manage the situation.

Measuring water quality

A wide range of parameters are used to directly test the quality of aquatic systems. These include temperature, pH, suspended solids (turbidity), nitrates, metals, and phosphates.

Biodegradation of organic material utilizes oxygen, which can lead to anoxic conditions and subsequent anaerobic decomposition, which in turn leads to formation of methane, hydrogen sulfide, and ammonia (toxic gases). Organic pollution can be very serious in the summer due to the solubility of oxygen decreasing as the temperature of the water increases. Aquatic invertebrates and fish can do little to regulate their body temperature; as the water heats up, their rate of respiration increases. Thus they need more oxygen, but less oxygen is available.

Biochemical oxygen demand (BOD)

Biochemical oxygen demand (BOD) measures the amount of dissolved oxygen needed to break down organic material in water through biological activity.

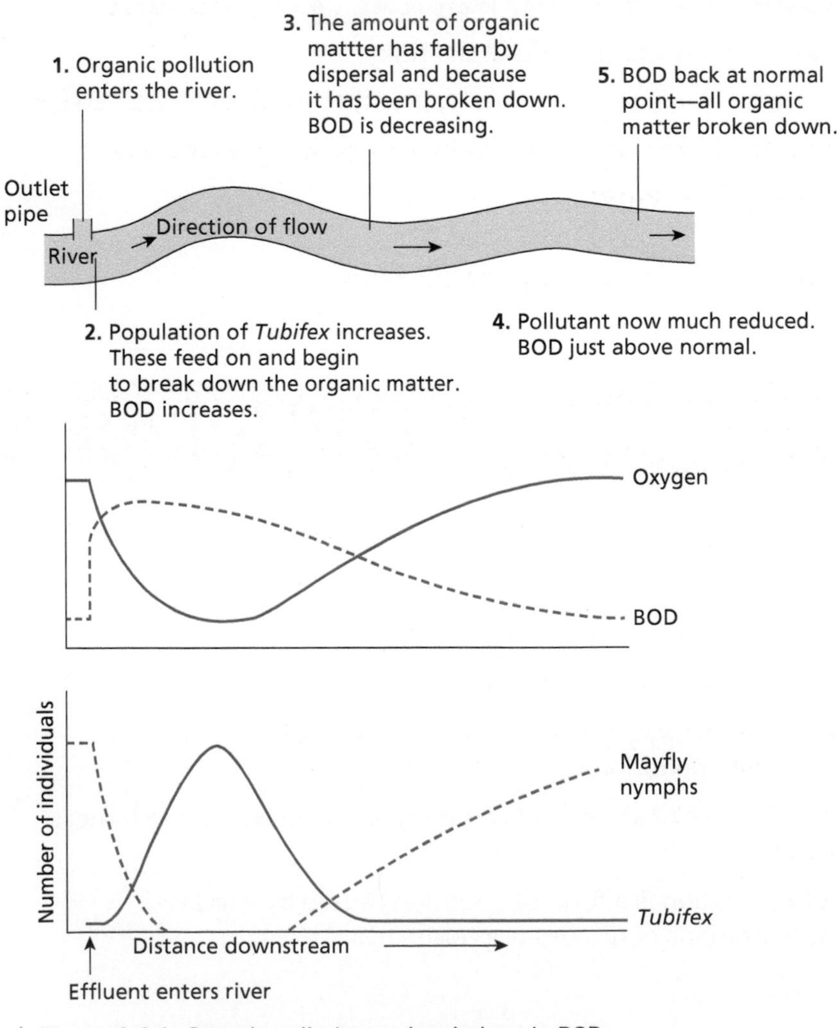

▲ **Figure 4.4.1** Organic pollution and variations in BOD

Assessment tip

Biotic indicators may be more reliable than using dipsticks as they relate to long-term levels of chemicals and other pollutants in water rather than short-term variations.

Indicator species and biotic indices

The presence of some species can be indicative of polluted waters. They can be used as indicator species, e.g. tubifex and mayfly nymphs.

A **biotic index** indirectly measures pollution by examining the diversity of selected species that are present. The **Trent Biotic Index** is based on the disappearance of indicator species as the level of organic pollution increases in a river. This occurs because the species are unable to tolerate changes in their environment such as decreased oxygen levels or lower light levels. Those species best able to tolerate the prevailing conditions become abundant, which can lead to a change in diversity. In extreme environments (e.g. a highly polluted river) diversity is low, although numbers of individuals may be high. Diversity decreases as pollution increases.

Concept link 

EQUILIBRIUM and BIODIVERSITY: In eutrophication, a new equilibrium is achieved—namely increased nitrate loading—leading to an increase in the amount of algae, which leads to a decline in biodiversity.

Eutrophication

Eutrophication leads to increased amounts of nitrogen and/or phosphorus that are carried in streams, lakes and groundwater, causing nutrient enrichment. The increase in nutrients in the system results in an increase in algal biomass. Decomposition of the algae leads to increased nutrients in the system. This is an example of positive feedback. In contrast, the increase in dead organic matter provides more food for decomposers, which increase in number. The increased rate of decomposition leads to a reduction in the amount of dead organic matter—this is an example of negative feedback.

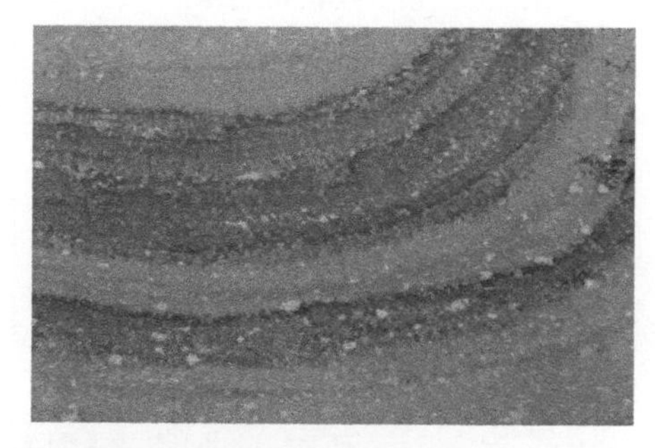

▲ Figure 4.4.2 Eutrophication at the Higher Ley, Slapton, UK

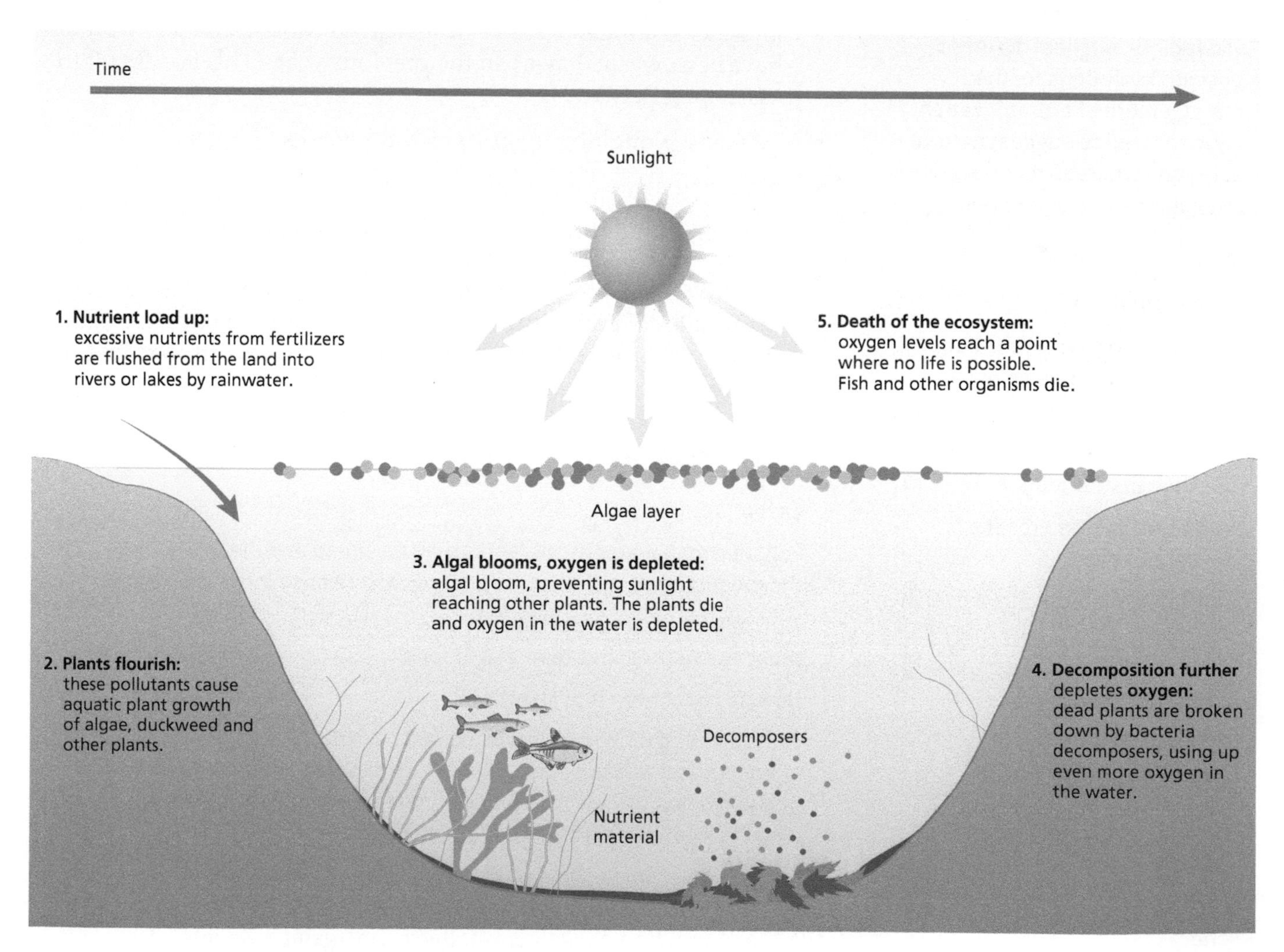

▲ Figure 4.4.3 The process of eutrophication

Dealing with eutrophication

There are three main ways of dealing with eutrophication. These include:

- altering the human activities that produce pollution, for example by using alternative types of fertilizer and detergent
- regulating and reducing pollutants at the point of emission, for example in sewage treatment plants that remove nitrates and phosphates from waste
- restoring water quality by pumping mud from eutrophic lakes.

Possible measures to reduce nitrate loss are as follows:

- avoiding the use of nitrogen fertilizers between autumn and early spring when soils are wet and therefore it is more likely that fertilizer will wash through the soil
- preferring autumn-sown crops—their roots conserve nitrogen in the soil and use up nitrogen left from the previous year
- sowing autumn-sown crops as early as possible, and maintaining crop cover through autumn and winter to conserve nitrogen
- avoiding applying nitrogen to fields next to a stream or lake
- avoiding applying nitrogen just before heavy rain is forecast
- using less nitrogen fertilizer after a dry year because less would have been washed away in the previous year. (This is difficult to assess precisely.)
- avoiding ploughing up grass as this releases nitrogen
- removing algae from eutrophic water
- removing mud from eutrophic streams and ponds
- pumping oxygen into eutrophic water
- using bales of barley straw to prevent the growth of green algae. It uses nitrogen as it decays, with up to 13% less nitrogen lost; it also locks up phosphorus.

Assessment tip

There are three main reasons why the high concentrations of nitrogen in rivers and groundwater are a problem.

- Nitrogen compounds can cause undesirable effects in the aquatic ecosystems, especially excessive growth of algae.
- The loss of fertilizer is an economic loss to the farmer.
- High nitrate concentrations in drinking water may affect human health; for example, methaemoglobinemia (blue baby syndrome) may develop.

Concept link

EVSs: Different environmental value systems, for example ecocentric and technocentric viewpoints, suggest different potential solutions for the management of eutrophication, e.g. ecocentrics suggest natural solutions whereas technocentrics suggest the use of technology.

Test yourself

4.29 Describe the distribution of mayfly nymphs below a source of organic pollution. [3]

4.30 Briefly **explain** the reasons for their distribution. [4]

4.31 Outline the process of eutrophication. [3]

4.32 Outline two ways in which eutrophication can be managed. [2]

QUESTION PRACTICE

Essay

Pollution management strategies may be aimed at either preventing the production of pollutants or limiting their release into ecosystems.

With reference to eutrophication, **evaluate** the relative efficiency of these two approaches to management. [9]

How do I approach this question?

The command term is **evaluate**, so you have to show the relative strengths and weaknesses of strategies to prevent the production of pollutants *and* to limit their release into ecosytems. You will need to come to a conclusion regarding which you think is best, and why.

- You will need to show an understanding of concepts and terminology of eutrophication, distinction between prevention and limiting management strategies, inorganic nutrients, phosphate-free detergents, organic vs inorganic fertilizers, organic farming, domestic and agricultural waste, sewage treatment, nitrate/phosphate stripping, buffer zones, point versus non-point sources, etc.

- You should show breadth in addressing and linking a range of strategies with their effectiveness in reducing impacts of pollutants, from different sources, on different ecosystems, and their relevance and validity for different societies, etc.
- You should provide examples of prevention strategies (changing human activity) e.g. alternative fertilizers, phosphate-free detergents, and limiting strategies e.g. nitrate/phosphate stripping phase in water treatment, use of buffer zones, in named case studies/societies, etc.
- You should offer a balanced analysis of the relative efficiency of the two approaches in reducing impacts on ecosystems; meeting needs of societies, cost and ease of application, etc.
- You will need to provide a conclusion that is consistent with, and supported by, analysis and examples given. For example, generally, prevention strategies are more efficient because they are directed at the root of the problem, but limiting strategies may be seen as more appropriate from an anthropocentric/technocentric perspective as they will be of less hindrance to productivity.

SAMPLE STUDENT ANSWER

Eutrophication is the nutrient enrichment of water bodies due to excessive inputs of nutrients that cause excessive algal growth, resulting in algal blooms. Pollution refers to the addition of a substance into the environment by human activities, at a rate greater than that at which it can be rendered harmless; by the environment, and has an appreciable or negative impact on the organisms in the ecosystem. In the context of eutrophication, pollution would refer to the runoff of fertilizers containing nitrates/phosphates into nearby water bodies, such as lakes, as well as sewage effluent pipes. Fertilizer runoff would be a non-point source of pollution as it originates from numerous, widespread sources. So it would be more difficult to monitor and manage. Discharge from sewage effluent pipes, conversely, is a point-source pollution since it is from a single, clearly identifiable site, so it would be relatively easier to monitor and manage. One pollution management strategy at the level of preventing the pollutants is to ban detergents containing

▲ Good definition

▲ Explanation

▲ Evaluation

▲ Analysis

▲ Evaluation

▲ Strategy

phosphates, which are there to address washing in hard water. Another strategy is to use legislation to promote more eco-friendly detergents, which could be helpful in raising awareness about the environment in the public. Also, using organic fertilizer such as manure instead of artificial fertilizers is beneficial since runoff of phosphates is reduced and organic fertilizer releases nutrient more slowly, helping the crop plants in their uptake of nutrients.

Enforcing legislation to limit the production or use of fertilizers is another strategy, but this may be detrimental to farmers as less fertilizer use would mean reduced crop yields and thus reduced income. Additionally, promoting campaigns and education about using less fertilizer and detergent may be helpful, but they may be limited to a local extent and it should be noted that it is difficult to change human behaviour without financial incentives. Other strategies for preventing production of the pollutant include encouraging farmers to do mixed cropping and crop rotation so that the soil is more fertile and less fertilizer will need to be used, which is economically beneficial to the farmer as well. Pollution management strategies to limit the release of pollutant into aquatic ecosystems include planting a buffer zone between the field and water body to act as a filter and reduce runoff. This is ecologically beneficial as the buffer zones can provide habitats to other species. Also, enforcing legislation to limit the maximum amount of fertilizer that can be sprayed is possible, but would be difficult to enforce and check for compliance. Further, diverting and treating the sewage to decrease slurry can be used, but this diverting is difficult and expensive.

▲ Second strategy

▲ Further strategy

▲ Exemplification – two examples – only one would be needed for the final mark

▲ Very focused answer

▲ Strategy and evaluation

▲ More reasoning

▲ Clear link to question

▲ Explicit link to limitation strategies

▲ Buffer

▲ Evaluation

▲ Evaluation

- Abundant evidence of sound understanding throughout. Distinction of strategies is fully grasped.
- Direct and effective links of statements to both the question and each other.
- Good collection of valid well-explained examples albeit fairly routine.
- The analysis is balanced with many evaluative reflections but lacks overall argument between the two approaches.
- Some intermediary conclusions but no overall conclusion.
- Only these last two factors fail to meet 7–9 mark band criteria.

This response could have achieved 7/9 marks.

5 SOIL SYSTEMS AND TERRESTRIAL FOOD PRODUCTION SYSTEMS AND SOCIETIES

This unit examines the nature of soils and their importance for food production systems. It explores different types of food production systems and the factors that influence them. It looks at ways in which soil can be degraded and also conserved.

You should be able to show:

- ✔ the main features of the soil system;
- ✔ the nature and variety of food production systems;
- ✔ ways in which it is possible to conserve soils.

5.1 INTRODUCTION TO SOIL SYSTEMS

You should be able to show the main features of the soil system

- ✔ Soil profiles contain soil horizons (layers).
- ✔ Soils include weathered rock, organic matter, air, and water.
- ✔ Nutrients are transferred up, down, and sideways through soils.
- ✔ Inputs to the soil system include energy, moisture, and nutrients.
- ✔ There are also transformations such as weathering and decomposition.
- ✔ Soils are often classified according to the proportion of sand, silt, and clay that they contain—this can be shown on a triangular diagram.
- ✔ Fertile soil may be considered a non-renewable resource because of the time needed for soil formation (thousands of years in some places).

- **Inputs** – the factors influencing a system
- **Outputs** – the results (products) from a system
- **Primary productivity** – the biomass gained by producers (plants) in a specific area in a specific amount of time
- **Loam soil** – a mixed soil consisting of sand, silt, and clay
- **Potential evapotranspiration** – the total amount of water that could be lost to evaporation and transpiration if there was an unlimited supply of water

How soils develop over time

Soils form the outermost layer of the Earth's surface, made up of weathered bedrock (regolith), organic matter (both dead and alive), air, and water. The formation of a layer of 30 centimetres of soil takes between 1,000 and 10,000 years. It is formed so slowly that soil can be considered a non-renewable resource. Soils develop from the weathering of bedrock into progressively finer particles. Biological processes in the soil result in further development. The nature of the bedrock, climate, soils, topography, vegetation, and land use influence the extent of weathering and the nature and properties of the soil that develops. Soils can therefore vary significantly over very small areas, for instance within a single field, depending on local changes in parent material, slope and aspect.

Assessment tip

Soils can be observed in most terrestrial places: fields, gardens, and school grounds. Try to study a soil in a place near you, and become familiar with the many characteristics of soils.

Content link

Section 2.2 examines communities and ecosystems. Soil communities form an ecosystem.

▲ **Figure 5.1.1** A rendzina soil showing a litter layer and an organic layer resting on chalk (bedrock/parent material)

Assessment tip

You do not need to know specific soil profile details (i.e. named soils and their profiles) but you do need to understand soil characteristics.

Test yourself

5.1 Describe the main characteristics of the soil shown in figure 5.1.1. [2]

5.2 State the difference between an E horizon and a B horizon. [2]

Content link

Section 4.1 examines the water cycles and makes reference to soil moisture storage in a drainage basin and in the global stores of water.

Test yourself

5.3 Define the terms "field capacity" and "soil moisture deficit". [2]

Soil horizons

A soil profile is a two-dimensional vertical section through a soil and is divided into horizons (or distinguishable layers). These have distinct physical and chemical characteristics, although the boundaries between horizons may be blurred by earthworm activity.

The top layer of vegetation is referred to as the organic (O) horizon. Beneath this is the mixed mineral-organic layer (A horizon). It is generally dark in colour due to the presence of organic matter. An Ap horizon is one that has been mixed by ploughing.

In some soils leaching takes place. This removes material from the horizon. Consequently, the layer is much lighter in colour. In a podzol, where leaching is intense, an ash-coloured Ea horizon is formed. By contrast, in a brown earth, where leaching is less intense, a light brown Eb horizon is found.

The B horizon is the deposited or illuvial horizon. These contain material that has been moved from the E horizon, such as iron (fe) humus (h), and clay (t). Sometimes the B horizon is weathered (w).

At the base of the horizon is the parent material or bedrock. Sometimes labels are given to distinguish rock (r) from unconsolidated loose deposits (u).

Soil has matter in all three states:

- organic and inorganic matter form the solid state
- soil water (from precipitation, groundwater and seepage) forms the liquid state
- soil atmosphere makes up the gaseous state.

Soil atmosphere and water are present in inverse proportions. After a storm there is an increase in the water content at the expense of air. The maximum amount of water that a soil can hold is referred to as its field capacity. When the soil has more water than its field capacity, it is saturated and overland runoff will occur (as will happen if the rainfall intensity exceeds the infiltration rate, i.e. the speed at which water can enter the soil). Soil water depends on a number of factors including texture, organic matter content, and density. During the wet season many soils reach field capacity. With free drainage, excess water drains through the soil. However, impeded drainage results in waterlogging. In the dry season there comes a point where the loss of water by plant uptake and evaporation exceeds the amount of rainfall. The soil begins to dry out. A soil moisture deficit develops and this is measured as the amount of rainfall required to return the soil to field capacity.

Transfers of material

Soil processes involve:

- the gains and losses of material to the profile
- the movement of water between the horizons
- chemical transformations within each individual horizon.

Therefore soils must be considered as open systems in a state of dynamic equilibrium, varying constantly as the factors and processes that influence them alter.

The principal processes include weathering, translocation, organic changes, and biogeochemical transformations (nutrient cycling). The weathering of bedrock gives the soil its C horizon, as well as its initial bases and nutrients (fertility), structure and texture (drainage).

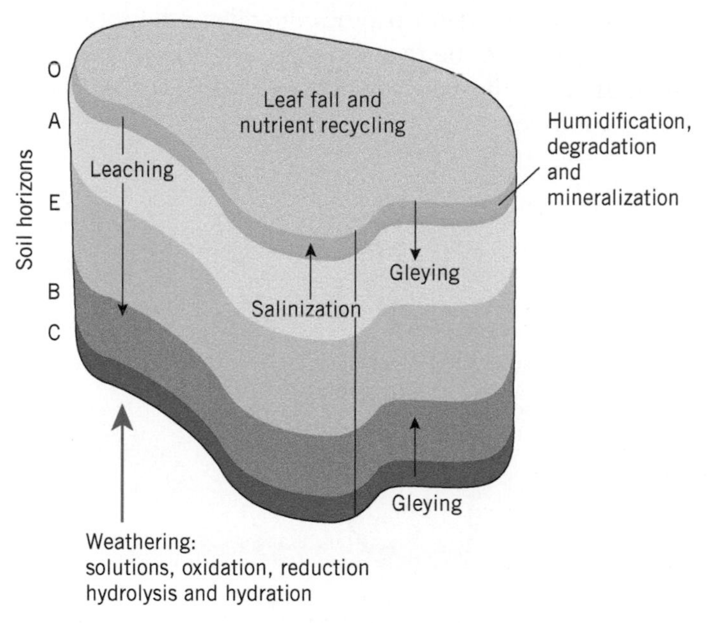

▲ **Figure 5.1.2** The main soil processes

Concept link

EQUILIBRIUM: Soil processes are constantly happening—the soil is in equilibrium with the processes that are taking place.

▲ **Figure 5.1.3** A brown earth showing a gradual decrease in darkness (less organic matter) with depth. The horizons have been blurred by earthworms that mix up the horizons

Organic changes

These are mostly done at or near the surface. Plant litter is decomposed (humified) into a dark amorphous mass. It is also broken down gradually by fungi, algae, small insects, bacteria, and worms. Under very wet conditions humification forms peat. Over a long timescale humus decomposes due to mineralization, which releases nitrogenous compounds. Degradation, humification and mineralization are not separable processes and always accompany each other.

Biological activity is important for the mixing of soils. Figure 5.1.3 shows a soil that has been mixed due to earthworm activity whereas figure 5.1.4 shows an acid soil with no earthworms but very clear horizon development.

▲ **Figure 5.1.4** A soil profile showing different layers (horizons) within the soil – this soil has a dark organic-rich horizon at the top, a leached horizon below, and then a deposited horizon (darker) at its base. The horizons are clear due to the lack of earthworm activity

Water movement in soils

Leaching refers to the downward movement of soluble material and small particles such as clay and humus. Such movements produce eluvial (removal) and illuvial (deposited) horizons.

In arid and semi-arid environments **potential evapotranspiration** exceeds precipitation, so the movement of soil solution is upwards through the soil. Water is drawn to the drying surface by capillary action and leaching is generally ineffective apart from occasional storms. In extreme cases where evapotranspiration is intense, sodium or calcium may form a crust on the surface. This is known as salinization.

Waterlogging

Waterlogged soils are generally found within the water table or in areas of poor drainage. There is limited decomposition, nutrient cycling and root activity.

Test yourself

5.4 Compare and contrast the horizons in figure 5.1.4 (a podzol) with those in figure 5.1.3 (a brown earth). [3]

Test yourself

5.5 Distinguish between leaching and waterlogging. [3]

5.6 Briefly **explain** the term "salinization". [2]

5.7 Describe how nutrients enter into soils. [2]

Transformations

These include decomposition, weathering and nutrient cycling. Decomposition refers to the breakdown of organic matter (see earlier). In contrast, weathering refers to the breakdown of rock (parent material) *in situ*, and is responsible for providing soils with inorganic matter. Nutrient cycling refers to the recycling of nutrients between biomass, litter and soils. The relative size of stores and flows varies with the physical environment.

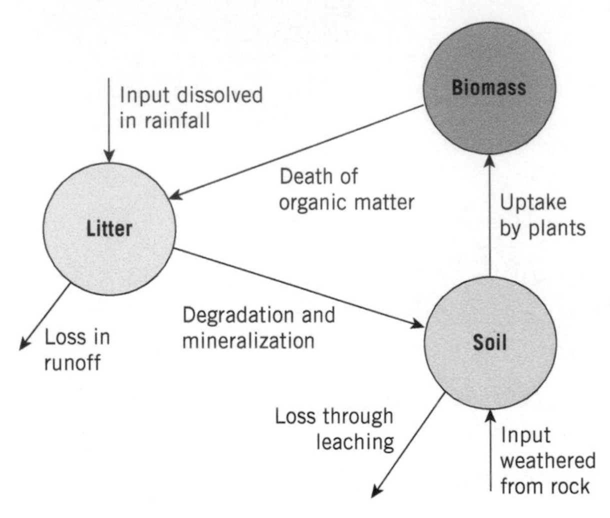

▲ **Figure 5.1.5** A model nutrient cycle

Assessment tip

Make sure you know about the contrasts in clay, sand and loam (mixed) soils, and their potential impacts on crop growth.

Structure and properties of sand, clay and loam soils

Sandy soil particles are generally quite large, 0.02 millimetres to 0.2 millimetres in diameter. This allows water to drain rapidly from them, but they may also lose some nutrients in the process. Sandy soils are considered to be light (easy to work), well aerated, and warm. They are good for the production of early crops, such as potatoes. Overall, sandy soils may have a low potential for **primary productivity** due to the poor water-holding capacity and nutrient status.

In contrast, clay soils have small particles of <0.002 millimetres diameter. The very small pore size means that clay soils can hold lots of water—up to twice as much as a light soil. Clay has a very large chemically active surface area so it tends to retain many nutrients. Clay soils tend to swell when wetted and shrink when dried. Clay soils can be fertile, but unless they are drained, primary productivity may be limited by too much water and poor aeration.

Loam soils are mixed soils—they contain varying proportions of sand, silt and clay. They retain some moisture, allow some water to infiltrate, and have some air and nutrients present. Overall, loam soils support a high primary productivity.

Test yourself

5.8 Define the term "loam soil". [1]

5.9 Suggest why loam soils are good for agriculture. [2]

Soil texture triangular graph

The conventional way of representing texture is by means of a triangular graph. Triangular graphs are used to show data that can be divided into three parts, such as the proportion of sand, silt and clay in a soil. Figure 5.1.6 shows a triangular graph for soil. The data must be shown in percentages and the total must add up to 100%. The main advantages of triangular graphs are that:

- a large amount of data can be shown on one graph (think how many pie charts or bar charts would be used to show all the data on figure 5.1.6)
- groupings are easily recognizable – in the case of soils, groups of soil texture can be identified
- dominant characteristics can be shown easily
- classifications can be drawn up.

Triangular graphs can be difficult to interpret and it is easy to get confused, especially if care is not taken. However, they provide a fast, reliable way of classifying large amounts of data which have three components.

Assessment tip

Triangular graphs show the percentage of sand, silt, and clay in a soil; they do not show any larger particles, soil organic content or moisture content.

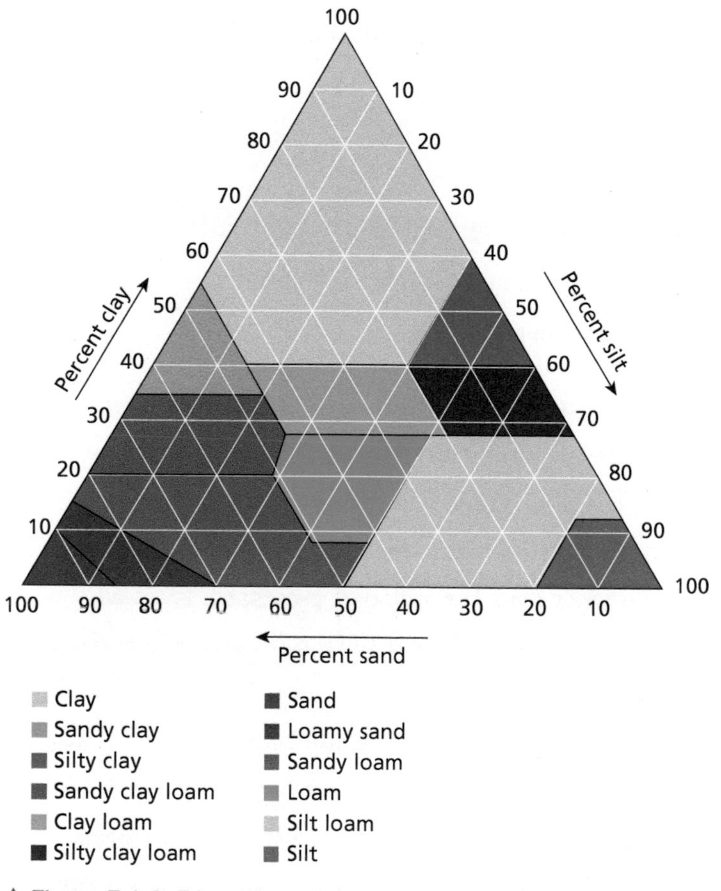

▲ **Figure 5.1.6** Triangular graph to show soil textural groups

Test yourself

5.10 State the type of soil that has:

a) 10% sand, 20% clay and 70% silt; [1]

b) 60% sand, 30% clay and 10% silt. [1]

QUESTION PRACTICE

a) State the soil texture that has the following composition: 20% clay; 55% silt; 25% sand. [1]

b) Describe how the addition of sand to a silty clay loam could alter its characteristics for healthy plant growth. [2]

c) Outline how soil can be viewed as an ecosystem. [4]

How do I approach these questions?

a) Use the triangular graph to locate the point with the characteristics stated in part (a), and state the group that it belongs to.

b) Identify how the silty clay loam will change as more clay is added (use the triangular graph), then describe how this could affect either drainage, temperature or nutrient availability.

c) You only have to give a brief account or summary about how a soil has similar characteristics to an ecosystem, e.g. inputs, outputs, stores and flows.

SAMPLE STUDENT ANSWER

a) Silt loam is composed of the above.

This response could have achieved 1/1 marks.

▲ One valid point made

b) The addition of sand could alter characteristics because sand is very porous and allows for better aeration. Each soil texture has its own qualities that are important to plant growth.

This response could have achieved 1/2 marks.

The answer to question (b) could have mentioned:

- improved drainage
- increased infiltration
- increased flow of nutrients
- easier root penetration
- reduced water holding ability (so less water for plants)
- increase leaching of minerals.

▼ Does not answer the question

▼ Very general, does not answer the question

▼ Outlines how soil can be seen as an ecosystem. Needs to be more specific/precise

▲ Soil as a habitat

▲ Processes e.g. nitrifying the soil

▼ Very general account

c) There are many ecosystems in the world, each one has a purpose, role and different characteristics. We, humans, and our activities affect those ecosystems and the processes that occur in those ecosystems. We can make a positive effect or a negative effect in the ecosystems and the natural processes. Is us who decide to help nature or destabilize it with our overexploitation and pollution.

There are many ecosystems and soil is categorized as one. due to its characteristics and properties. An ecosystem has to have organisms living in it, processes completed in it and has to have a purpose for the equilibrium of nature. And, soil has all those properties. Soil is the habitat of many insects and bacteria which have different roles such as fixing the soil to make it usable for plants and nitrifying the soil. Soil is very important to maintain the equilibrium of the world because the processes completed in it and the plants and animals living in it help complete the food chain and maintain other ecosystems.

This response could have achieved 2/4 marks.

5.2 TERRESTRIAL FOOD PRODUCTION SYSTEMS AND FOOD CHOICES

You should be able to show the nature and variety of food production systems

- ✔ Food production systems vary in terms of sustainability, such as the use of water, fertilizers, pesticides and machinery.
- ✔ Food production and distribution are very uneven throughout the world.
- ✔ Food waste occurs in less economically developed countries (LEDCs) and in more economically developed countries (MEDCs).
- ✔ Many factors affect the choice of food production system, e.g. socio-economic, cultural, ecological, and political factors.
- ✔ Demand for food increases with population growth and with improvements in standards of living.
- ✔ The yield of food from lower trophic levels (primary producers) generally costs less than that from higher trophic levels.
- ✔ Some societies prefer to have food from higher trophic levels.
- ✔ There are variations in terrestrial food production systems in terms of inputs, outputs, system characteristics, socio-economic, and environmental impacts.
- ✔ Sustainability can be achieved through reducing consumption of meat and dairy products, food labelling, government control, and land-use planning.

▲ **Figure 5.2.1** Arable farming, Aero Island, Denmark

- **Arable farming** – the cultivation of crops such as wheat farming in the Great Plains of the USA
- **Pastoral farming** – the rearing of animals, for example sheep farming in New Zealand
- **Intensive farming** – high inputs or yields per unit area, such as battery hen production
- **Extensive farming** – low inputs or yields per unit area, as in free-range chicken production
- **Commercial food production** – the production of food for sale (to make a profit)
- **Subsistence food production** – the production of food to be eaten by the farmer's family
- **Nomadic farming** – farmers moving seasonally with their herds, such as the Pokot, pastoralists in Kenya
- **Sedentary farming** – farmers remaining in the same place throughout the year, for example dairy farmers in Devon and Cornwall
- **Fertilizer** – minerals and nutrients added to a soil to make it more productive (fertile)
- **Transnational companies (corporations)** – large companies that have operations in many countries

▼ **Table 5.2.1** Countries that are the most economically dependent on agriculture (% of GDP from agriculture)

Country	%
Central African Republic	58.2
Sierra Leone	56.0
Chad	52.6
Guinea-Bissau	43.9
Ethiopia	41.9

Source: *Pocket world in figures*, 2017 edition, *The Economist*

▼ **Table 5.2.2** Countries with the largest agricultural output ($ billion)

Country	$ billion
China	950
India	336
USA	226
Indonesia	119
Nigeria	114

Source: *Pocket world in figures*, 2017 edition, *The Economist*

Assessment tip

You should examine two contrasting farming systems, one from a rich country and one from a poor country.

Test yourself

5.11 Explain the terms
(a) capital-intensive
(b) subsistence and
(c) cash crops. [2]

5.12 Suggest why the countries that are most dependent on agriculture are poor. [2]

5.13 Suggest reasons for the world's largest agricultural output. [2]

Sustainability of terrestrial food production systems

Scale

The sustainability of terrestrial food production systems is influenced by several factors, including the degree of dependence on agriculture and the value of agricultural output.

Farming systems in LEDCs (less economically developed countries) and MEDCs (more economically developed countries) are very complex. They are also very different. Hence the following points are generalizations and do not apply to all forms of farming.

Agriculture in MEDCs has more in common with manufacturing industry than it has with farming in LEDCs. For example, much of it is run by companies, and is capital-intensive, highly mechanized, large-scale, and market-orientated, and government involvement is crucial. By contrast, agriculture in LEDCs is typically small-scale, labour-intensive, and subsistence by nature. In addition, MEDCs have considerable control over the price of many products from LEDCs, such as tea, coffee, and cocoa, whereas LEDCs have little influence over price. Countries that are heavily dependent upon one or two crops are particularly vulnerable to price fluctuations and poor harvests.

Although **export production** has been long established in LEDCs, for example in the plantation system, it has frequently been separate from local production. Increasingly, however, agricultural systems in LEDCs are becoming influenced by multinational companies (MNCs), through their:

- ownership of the land
- control of marketing
- supply of **inputs**
- production of crops.

Although many farmers use new techniques, harvest new cash crops and are increasingly commercial, there are limits to which farming systems in LEDCs can adapt to external pressures for change. New farming methods can create problems for the farmers. For example, in cash cropping there is huge uncertainty yet there is little subsistence production to fall back on. Credit is less forthcoming and the result is increased rural poverty and environmental deterioration.

Industrialization

The industrialization of agriculture (also known as agri-industrialization) is associated with large-scale, capital-intensive farming methods aimed at maximizing yields (and profits) often with little regard for the environment or the workers.

Mechanization

Mechanization involves the use of machines such as tractors, combine harvesters and milking machines to increase the amount of work that can be done on a farm, and at a faster rate than using animal or human labour. Mechanization has had a major impact on farming. The use of tractors is correlated with levels of economic development. However, some areas are not suited to large-scale mechanization, such as the terraced paddy fields of South-East Asia. In other areas, such as the

Great Plains in the USA, the number of machines used in agriculture exceeds the number of farm workers.

Fossil fuel use

In developed and emerging countries, where mechanization and industrialization of agriculture have occurred, the use of fossil fuels is extensive. In addition, transport of food to market leads to increased use of fossil fuels, especially when the market for foodstuffs is global. The concept of food miles measures the distance that food travels from the place of production to the place of consumption. It has been estimated that the number of food miles for a festive meal in an MEDC is 24,000 miles (37,000 kilometres).

Content link

Section 2.3 looks at flows of energy and matter. Food miles illustrate how far food has travelled from where it is produced to where it is consumed – this is an energy flow to transport the matter.

Seed, crop and livestock choices

At a very broad level, it is possible to identify diets that are predominantly based on cereals and those which are of a mixed diet. The former is more common in poor countries, the latter in rich and emerging countries. Changes in the demand for chicken illustrate some of the changes in dietary choice that have developed. Since 1990 chicken consumption has increased by 70% globally. It is a relatively cheap meat and has the reputation for being a healthy meat. Western consumers prefer lean white meat whereas Asian and African consumers prefer darker meat (from the legs and thighs). Increasingly, some Western consumers have been choosing to buy meat produced in better environmental conditions e.g. organic meat or free-range eggs rather than battery-produced.

Water use

Agriculture accounts for about 70% of the world's water use. Future global agricultural water consumption is estimated to increase by almost 20% by 2050 due to increased demand for high-quality foods such as meat and dairy products. Much more water (7,500 litres) is required to produce 500 grams of beef compared with 500 grams of cheese (2,500 litres), chicken (1,950 litres), maize (450 litres) or wheat (650 litres). The transition from a diet based on cereals to a diet based on meat and dairy comes with a high demand for water.

Fertilizer use

Fertilizers help to increase crop yield and indirectly livestock yields. China accounts for 30% of world fertilizer and pesticide use on just 9% of global cropland. Intensive use of fertilizers has environmental consequences, e.g. eutrophication of lakes and streams, and the creation of "dead zones" in coastal waters. Nitrogen released into the atmosphere can contribute to global warming, acid rain, and ground level (tropospheric) ozone.

Pesticides

Pesticides are manufactured chemicals used to control weeds, insects, fungi and other small organisms that may damage crops and reduce yields. Pesticide use has risen dramatically since the 1950s. China is the world's largest producer and exporter of pesticides. However, pesticides may accumulate in the food chain (bioaccumulation) and may become more concentrated higher up the food chain (biomagnification). They can also have an impact on human health—an estimated 70,000 agricultural workers die every year due to pesticide poisoning.

Content link

Section 4.4 explores water pollution, and examines eutrophication in more detail.

Antibiotics

There is widespread use of antibiotics in farming. Around 15,400 tonnes a year, about 80% of all antibiotics sold, go to farmers. Chicken farmers use more antibiotics than pig or cattle farmers. A very small proportion is used to cure illness – the main function is to make chickens fatten up more quickly and to act as a prophylactic (preventative medicine) in the cramped conditions in which they live.

Animals receive antibiotics in their feed and water, which may create antibiotic-resistant bacteria in their guts. These bacteria can spread into the environment and may infect animals sold for human consumption. Each year there are some 1.2 million cases of food poisoning due to salmonella, resulting in nearly 20,000 hospital cases and around 400 deaths.

Pollinators

In 2008 Europe's bee population was decimated by the pesticide *clothiandin*, which can be more than 10,000 times more lethal than DDT. It was used to kill the corn rootworm that was destroying crops in the Rhine Valley. In 2013 the European Union banned the use of *clothiandin*, but despite this the bee population has not recovered as well as expected.

Concept link 

BIODIVERSITY: The decline in the bee population due to pesticides is a major threat to biodiversity, and one that may have severe financial consequences for farmers around the world.

Farmers complained that the ban would cost the EU's agricultural sector €880 billion (about £758 billion) a year in reduced crop yields.

According to the Food and Agriculture Organization, pollinators help farmers grow crops worth €577 billion annually. Damage to the bee population is threatening crops around the world. In some parts of China humans have to pollinate crops by hand.

Test yourself

5.14 Briefly **outline** the advantages of fertilizers and pesticides in agriculture. [4]

5.15 Outline the disadvantages of pesticides and antibiotics in farming. [4]

Legislation

An example of government involvement in agriculture is the Common Agricultural Policy (CAP) of the European Union (EU). The main priorities of CAP were to:

- increase agricultural productivity and self-sufficiency
- ensure a fair standard of living for farmers
- stabilize markets
- ensure the availability of supplies
- ensure that food was available to consumers at a fair price.

Between 1958 and 1968 these aims were implemented: a single market existed in agriculture from 1962 and a common set of market rules and prices were introduced by 1968. At the centre of the CAP was the system of guaranteed prices for unlimited production. This encouraged farmers to maximize their production as it provided a guaranteed market.

As a result of over-production, the policy was reformed. The key elements of CAP are price cuts and the withdrawal of land from production. Consequently, there is more incentive to diversify agriculture and to make it less intensive.

Government subsidies

In some MEDCs governments provide subsidies for their farmers. Subsidies encourage farmers to produce more food. Many LEDCs cannot afford to pay their farmers subsidies. MEDCs may also charge import tariffs. This makes it more expensive to import food from overseas. Both of these factors make it difficult for LEDC farmers to sell their products to MEDCs.

Farm size

Some 98% of farms in China are less than 2 hectares compared to an average size of 180 hectares in the USA. Average farm size decreased from the 1980s to 2000 but has increased slightly since then. Strict migration laws within China prevent rural populations from migrating permanently to urban areas, so they tend to keep their farms rather than sell them.

▲ **Figure 5.2.2** Small-scale agricultural production in Antigua

Inequalities in food production

The global distribution of food production is unequal relative to the global distribution of population. International trade generally helps to reduce these inequalities. Exports from the USA, Canada, Argentina, and Malaysia to countries such as China, Japan, and Singapore are especially important in reducing inequalities in food production.

Around one billion people have insufficient dietary intake and a further billion are overweight. Food inequality originates at an international level as well as an intra-national level (i.e. within a country). At the international level, food production is greater in areas that have an adequate water supply and favourable land. At the intra-national level, production of and access to food are influenced by poverty, accessibility, infrastructure, and conflict.

Food production that is traded has increased significantly over recent years due to population growth, rising standards of living, and changes in diet. Food exports of "luxury" foods and "cash crops" rather than staples may increase food inequalities between rich and poor countries.

Test yourself

5.16 Briefly **outline** the main reasons for inequalities in food production. [2]

5.17 Using an example, **explain** how government/multi-government organizations influence food production. [3]

Food waste

In MEDCs, consumerism, excess wealth, and mass marketing lead to wastage. Approximately one-third of food is thrown away in the UK each year. In LEDCs, up to 80% may be wasted before it reaches the market/shops. More efficient farming practices and better transport, storage, and processing facilities ensure that a larger proportion of the food produced reaches markets and consumers. However, produce is often wasted through retail and customer behaviour. Major supermarkets, in meeting consumer expectations, often reject entire crops of perfectly edible fruit and vegetables at the farm because they do not meet exacting marketing standards for their physical characteristics, such as size and appearance. Globally, retailers generate 1.6 million tonnes of food waste annually in this way.

In LEDCs wastage tends to occur primarily at the farmer–producer end of the supply chain. Inefficient harvesting, inadequate local transportation, and poor infrastructure mean that produce is frequently handled inappropriately and stored under unsuitable conditions. As a result, mould and pests (for example rodents) destroy or at least degrade large quantities of food material. Substantial amounts of foodstuffs fall off unsecure loads as they are transported or are bruised as vehicles travel over poorly maintained roads.

Choice of food production systems

In MEDCs most food production is done for commercial reasons. There is very little subsistence farming in MEDCs. Agribusiness refers to running a farm like a business. The main aim is to maximize profit. It is dominated by big businesses such as Bayer, which operates via 420 companies in 90 countries, and Cargill, which operates in 70 countries.

Agribusiness has many impacts. There is large-scale use of fossil fuels. There are losses to habitats and biodiversity. Pollution, including eutrophication and soil degradation, occurs. Small farms have been replaced by very large farms. Farm workers have been replaced by machines.

However, in some MEDCs concern about the environment has led to the growth of organic farming. Sales of organic food have increased since 2010. Concern about animal welfare has led to changes to the way in which animals are farmed. There has been a growth in the number of free-range pigs and chickens being produced.

Concept link

ENVIRONMENTAL VALUE SYSTEMS (EVSs): Different types of farming system represent different EVSs. Agribusiness is a more technocentric form of farming (using technology to increase yields) whereas shifting cultivation is a more ecocentric form of agriculture (working within the bounds of nature).

Shifting cultivation occurs in tropical rainforests where population density is low. It is a form of subsistence farming. The famers grow a variety of crops. This is known as polyculture. The crops grown include cassava and yams. Many of the people who practise shifting cultivation are animists. They believe that everything contains a spirit or soul. This includes trees and places. Even animals have a spirit or soul. Animists respect all living things.

Availability of land

Food production depends, in part, on the amount of land that is available. As the population increases, land availability may decrease. In South Africa, for example, the amount of cultivated land per person reduced from around 0.7 ha per person in 1960 to around 0.2 ha per person in 2016.

Content link

See figure 4.3.1 in section 4.3 for a diagram showing the different trophic layers that aquatic and terrestrial food production systems harvest from.

Trophic layers and yield

In terrestrial food production systems most food is harvested from low trophic levels. These are the producers and herbivores. Terrestrial food production systems that are based on crops (primary producers) are more energy-efficient than those that produce livestock (herbivores). Owing to energy losses between trophic levels, less than 10% of the energy that is made available to plants is available to herbivores.

Comparison of food production systems

Two examples of food production systems include North American cereal farming and subsistence rice farming in South-East Asia. Both are **arable farming** types. In the North American farming system the main crops are wheat and corn. In the South-East Asian farm system the main crop is rice. A second crop may be grown in the dry season.

Most farms in North America are very large. Most farms in South-East Asia are very small. Most farms in North America have large amounts of machinery and much of the work on the farm is done by machine. Most farms in South-East Asia do not have much machinery and so much of the work done on the farm is done by hand.

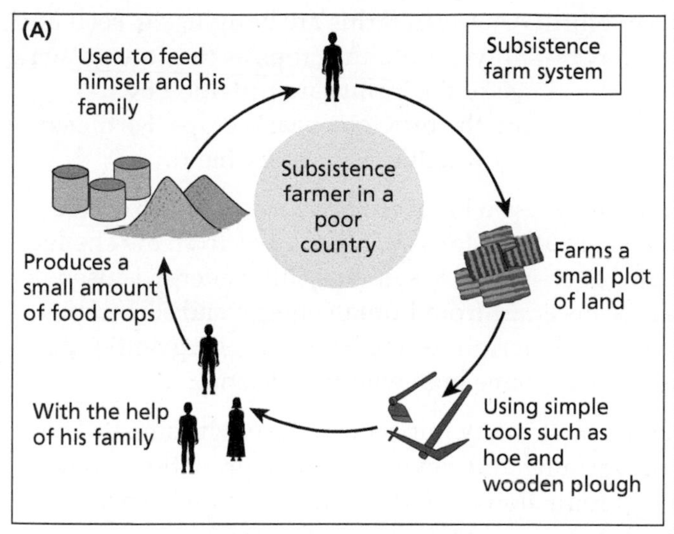

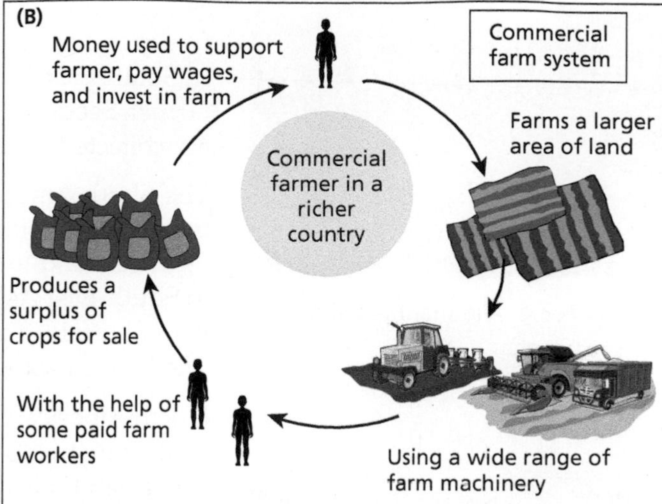

▲ **Figure 5.2.3** A comparison of farming in (A) the poorest LEDCs and in (B) MEDCs

▼ **Table 5.2.3** A comparison of two food production systems

	Cereal production in North America	**Rice production in South-East Asia**
Nature of farming	Commercial	Subsistence
Type of farming	Arable	Arable
Size	Large (c. 200 ha)	Small (c. < 2ha)
Labour	Machinery (capital-intensive)	Human, animal
Climate	Hot summers, cold winters	Hot summers, hot winters
Products	Monoculture	Polyculture
Inputs	Chemical fertilizers, high-yielding varieties of crops	Organic fertilizers
Scale	Extensive	Intensive
Energy use	Energy-intensive	Labour-intensive

North America has hot summers and cold winters. This limits the growing season to about six months. Corn and wheat are the main crops grown. In South-East Asia summers *and* winters are hot. There is a monsoon season in the summer which has heavy rainfall. Two crops are grown. One is grown in the wet season and one is grown in the dry season. Rice is the main wet season crop. Wheat is sometimes grown in the dry season.

Most North American farms are commercial. This means that they sell the majority of their crops for profit. The South-East Asian farm system is subsistence. This means that the crops grown are used to feed the farmer's family. Usually only one type of crop is grown on North American farms. This is called monoculture. Generally, more than one crop is grown on the South-East Asian farms, or a combination of rice and fish in flooded paddy fields. This is called polyculture.

There are many inputs into the North American farms. These include chemical fertilizers and irrigation water. The farms may also use high-yielding varieties of seed which have been genetically modified. In the South-East Asian farm, organic fertilizers are used and water is trapped by using terraces. Farmers may also use high-yielding varieties of rice. Farming in North America is extensive. This means there are low inputs and **outputs** per unit area. Farming in South-East Asia is intensive. This means there are large inputs and outputs per unit area.

▲ **Figure 5.2.4** Allotments on the edge of Shanghai

▲ **Figure 5.2.5** Farming by the Three Gorges reservoir

Test yourself

5.18 Describe the main characteristics of the farming systems shown in figures 5.2.4 and 5.2.5. [2]

Many of the inputs to North American farms are bought, e.g. seed and fertilizer. Pesticides may be sprayed onto the crops as they are growing. In South-East Asian farms most of the inputs are provided by the farmer. Seeds are collected from the previous year's crops. Fertilizers are collected from animals. Any weeding is done by hand.

The North American farms use a lot of energy, for example to drive tractors and combine harvesters. Energy is also used to create chemical fertilizers and to provide irrigation water. Very little energy is used in South-East Asian farms apart from human energy and sometimes water buffalo. The North American farms have low energy efficiency. The South-East Asian farms have high energy efficiency.

North American farms have many impacts on the environment. They contribute to global warming and can cause soil degradation. They also cause habitat loss and reduction of biodiversity. The South-East Asian farms also have impacts on the environment. They cause deforestation and soil exhaustion. Deforestation can lead to habitat loss and reduction of biodiversity.

The sustainability of terrestrial food production systems is influenced by socio-political, economic, and ecological factors. Increased sustainability may be achieved through:

- altering human activity to reduce meat consumption and increase consumption of organically grown and locally produced terrestrial food products
- improved transport and storage in LEDCs
- improving the accuracy of food labels to assist consumers in making informed food choices
- monitoring and control of the standards and practices of multinational and national food corporations by governmental and intergovernmental bodies.

QUESTION PRACTICE

a) Outline how the reasons for food wastage may differ between human societies. [4]

b) Explain how the choice of food production system may influence the ecological footprint of a named human society. [7]

How do I approach these questions?

a) MEDCs: consumers buying too much food and then wasting it; food processing places "best before" date, beyond which food must be disposed of.

LEDCs: poor infrastructure so food cannot be delivered quickly to markets; lack of refrigeration facilities means food rots before reaching market or consumer.

Each correct reason outlined will be awarded 1 mark, up to a maximum of 4.

b) Defining the problem: named society with an associated style of food production; define ecological footprint.

Factors affecting the ecological footprint of the choice of food production systems include productivity of the land available to the society; type of (food) consumption common in the society; type of food production—amount of input required to system, e.g. chemical fertilizers/machinery—oil increases footprint; comment on particular features of food production system, e.g. no-till agriculture increases carbon stores; comment on amount of meat eaten—increases footprint; comment on population size and per capita consumption.

Each correct explanation outlined will be awarded 1 mark, up to a maximum of 7.

SAMPLE STUDENT ANSWER

a) Food waste in MEDCs is most likely to occur in homes and restaurants. People buy too much food - on special offer - then don't get around to eating it. Sandwiches that are bought in supermarkets in MEDCs do not include the end crusts of the loaf - they are thrown away.

In LEDCs food is wasted at the farm or market level. Many farmers have no electricity so food goes off quickly. Also, the state of roads is poor, so some of the produce gets bruised as it is taken to market. People won't buy bruised food.

▲ Valid point

▲ Valid point

▲ Valid point

▲ Valid point

This response could have achieved 4/4 marks.

b) Pig farming in Denmark is a commercial type of livestock farming.

There are very large inputs into the system e.g. energy, food concentrate for the pigs and transport to get the pigs to market.

The farms are small (<30 ha) but intensively used. There is intensive cereal cultivation to produce food for the pigs.

The outputs are also large – 9 million pigs are produced a year, and this amounts for over 40% of Danish agricultural output. 75% of the bacon is exported.

The water footprint of pigs very high (6000 L to produce 1 kg). Denmark has one of the world's largest ecological footprints (the area of land and water required to support a defined human population at a given standard of living) partly due to Denmark's agricultural industry, and the Danes' high meat consumption.

▲ Identifies farming type

▲ Valid point

▲ Another valid point

▲ Good example

▲ More examples

▲ Good use of supporting data

Marks 6/7

- Good points made.
- Could be more structured/ordered.

5.3 SOIL DEGRADATION AND CONSERVATION

- **Soil fertility** – the ability of a soil to enable plant growth
- **Toxification** – the decline in soil quality (fertility) due to a build-up of chemicals in the soil
- **Salinization** – an increase in the amount of salts in the upper soil horizons
- **Desertification** – the spread of deserts into areas previously fertile and productive

You should be able to show ways in which it is possible to conserve soils

✔ Soils change over time.

✔ Human activities may degrade soils.

✔ Food production systems often lead to a decline in the quality of soils.

✔ Reduced soil fertility may lead to environmental degradation.

✔ Soil conservation can help to maintain soil fertility.

Changes to soils

Soils require significant time to develop. However, human activities may reduce **soil fertility** and increase soil erosion in a very short time. Activities such as deforestation, intensive grazing, urbanization and certain agricultural practices (e.g. irrigation and monoculture) can seriously damage soil. Reduced soil fertility may result in soil erosion, **toxification**, **salinization**, and **desertification**. Nevertheless, soil conservation strategies exist and may be used to preserve soil fertility and reduce soil erosion. These measures include soil conditioners (such as organic materials and lime), wind reduction techniques (wind breaks, shelter belts), cultivation techniques (terracing, contour ploughing, strip cultivation), and avoiding the use of marginal lands.

▲ **Figure 5.3.1** Walkers on Croagh Patrick mountain, Ireland

Soil degradation and erosion

Soil degradation and erosion are the decline in the amount and quality of a soil. Soil degradation leads to a reduction in soil fertility. It can be caused by overgrazing, which occurs when too many animals use the land, leading to removal of vegetation and compaction of soil. With less vegetation cover the soil is compacted. This reduces infiltration and increases overland runoff.

Soil degradation and erosion can also be caused by over-cultivation. Over-cultivation results in the removal of specific nutrients from soils. Monoculture can lead to declining soil fertility due to the removal of selected nutrients. This can be overcome through the use of polyculture or crop rotation.

Deforestation can also lead to soil degradation due to a loss of vegetation so that the soil is more easily eroded. Deforestation also removes the protection given to soils by plant roots and canopy.

Excessive irrigation is another cause of soil degradation. Excessive irrigation can lead to salinization which is caused by the evaporation of water from soil. Evaporating water draws salts up towards the surface. Salinization is toxic to plants.

Assessment tip

Some soils take a very long time to form—hundreds, if not thousands, of years—hence they can be considered a non-renewable resource.

Assessment tip

Soil degradation is likely to increase as the human population grows and rising standards of living place greater pressure on soils to produce food.

The removal of hedges and windbreaks can cause soil degradation. When hedges and windbreaks are removed the soil is more vulnerable to wind and water erosion.

Agricultural yields may decline due to reduced soil fertility and the removal of topsoil. To improve yields agricultural systems may use lots of fertilizers. This may cause the soil to be enriched with nitrates. This is a form of eutrophication. In extreme cases eutrophic soils may become toxic to some plants. Salinization may be caused by the excessive use of irrigation on some soils. Salinization is also toxic to plants. It can cause increased stress for plants and can lead to reducing agricultural yields.

In extreme cases, soil degradation and erosion can be linked to desertification. Desertification is the spread of desert-like conditions into previously productive areas. Desertification may be caused by over-population. Over-population forces people to farm in marginal areas. This can cause soil degradation and erosion.

Desertification leads to a decline in plant productivity. Previously fertile areas are no longer able to provide enough food for the people who live there.

Soil conservation

Soil conservation is the protection of soils in terms of their quantity and quality. There are a number of methods of soil conservation.

Soil conditioners include lime and organic materials. Lime is used to make acidic soils less acidic. Crushed limestone can be added as a powder to make acid soils more fertile. Organic matter can also be added to the soil to make it more fertile. This can reduce the effect of wind erosion and raise the temperature of soils.

Wind breaks are linear belts of hedges or trees that reduce wind speed and therefore reduce the potential for wind erosion. There are usually only one or two rows of vegetation in a wind break. Shelter belts are similar to wind belts but generally consist of three or four staggered rows of trees or shrubs that reduce wind speed and provide shelter for the growing vegetation.

Strip cultivation refers to the growing of crops in linear strips. This helps to reduce wind erosion by having varied species of crops in strips rather than having a large-scale exposure of just one crop. Increased friction with the varied vegetation leads to a reduction in wind speed.

Cultivation techniques such as terracing and contour ploughing are important as they reduce the potential for overland flow and wind erosion. Terracing involves the levelling of the land into a series of steps or terraces. These help to reduce soil erosion by having flat surfaces rather than a steep surface. This reduces overland flow. Contour ploughing refers to ploughing around a hill rather than up and down a hill. This also reduces the amount of overland flow and helps to protect the soil.

Marginal lands are those that are not fertile and have very limited agricultural potential. The use of marginal lands for food production may lead to desertification. It is often better to use marginal lands for national parks rather than farming. National parks can derive an income from tourism whereas the farming of marginal lands generally leads to soil degradation.

Assessment tip

Marginal lands are those that are not suited for agriculture. However, as population pressure increases, some of this land may be brought into production for food, especially when people are desperate.

Test yourself

5.19 Suggest how the walkers on Croagh Patrick mountain [figure 5.3.1] have affected the soil quality. [2]

5.20 Suggest how this may change over time. [2]

5.21 Distinguish between overgrazing and over-cultivation. [2]

5.22 Explain how irrigation and salinization are linked. [2]

Concept link 

STRATEGIES: There are many different strategies linked to soil conservation.

Concept link 

SUSTAINABILITY: Conservation techniques aim to restore soils to their natural state and/or prevent them from being degraded.

▲ **Figure 5.3.2** Check dam to reduce soil loss and water loss

Soil management on the Great Plains

Extensive crop production takes place on the Great Plains in the USA and involves the large-scale production of grain using large machinery.

In the 1930s the southern Great Plains experienced severe wind erosion. This created the Dust Bowl. In 1935 the Soil Conservation Act came into force. Contour ploughing was introduced and farmers ploughed around hills rather than up and down them. This reduced the rates of soil erosion. Strip farming has a strip of cropland next to a strip of fallow land. The strips are lined up across the direction of the prevailing wind. This reduces wind speed and soil erosion. Farmers used cover crops to help bind the soil and to reduce the impact of water erosion. Millet grows fast and so is a popular choice of cover crop for many farmers.

Many soil conservation techniques are now practised on the Great Plains. There has not been a repeat of the Dust Bowl of the 1930s so the techniques appear to be working. The USA is one of the world's leading "bread baskets" so it is using its soils and protecting them at the same time.

Test yourself

5.23 Outline the advantages of contour ploughing and cover crops. [2]

5.24 Explain how the check dam in figure 5.3.2 can aid soil conservation. [2]

Soil conservation measures in a subsistence farming system

The Kikuyu are a tribe in Kenya that practise bush fallowing, a type of farming that involves clearing an area of forest or "bush". Farmers may burn some of the trees to provide fertile ash for the soil. The ash contains the nutrients that were contained in the trees. On the cleared land the farmers grow crops such as maize and sweet potatoes. The plot loses its fertility after a few years and is abandoned. The farmers move to a new plot and the process starts over again. The abandoned plot may return to forest or bush after a number of years. In this way, the Kikuyu are able to produce food, and the soils are able to recover over a long period of time.

QUESTION PRACTICE

a) Outline two ways in which human activity may have increased soil erosion in Iceland. [2]

b) With reference to Figures 1 and 2 and Table 1, **explain** the problems associated with land restoration in Iceland. [4]

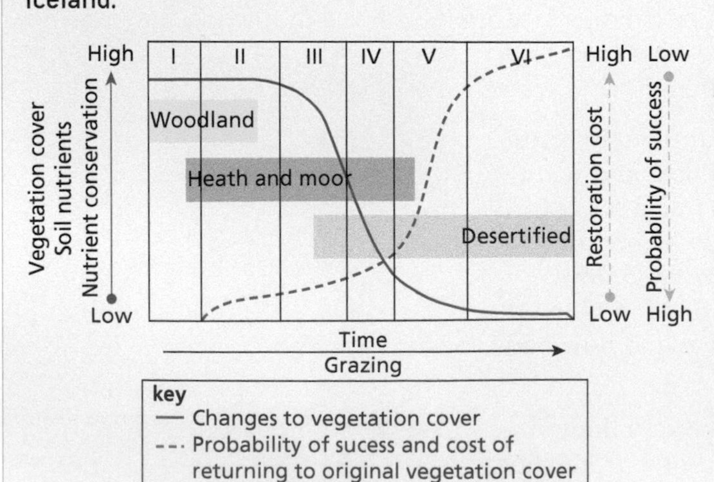

▲ **Figure 1** Model to show changes in vegetation cover during the six stages of soil degradation in Iceland

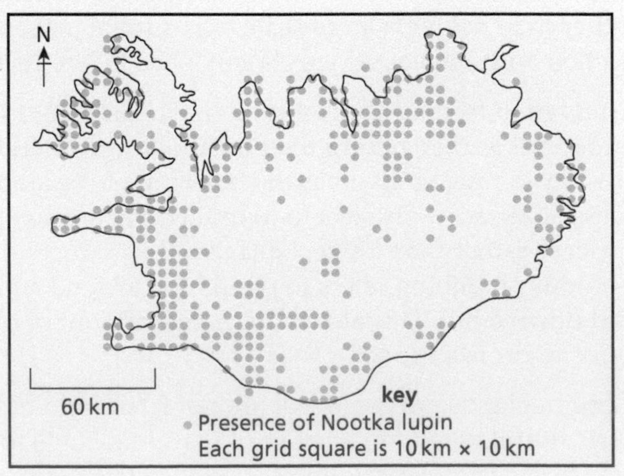

▲ **Figure 2** Known distribution of lupin in 2010

▼ **Table 1** Fact file on Nootka lupin

• Native to North America
• It was tried as an ornamental plant in Icelandic gardens in 1885
• Introduced in Iceland to stop soil erosion in 1945
• Invasive – spread quickly and outcompetes native flora
• Ministry of Environment recommended eradication of the lupin in highlands (above 400 m), national parks, and conservation areas
• Removal methods include: grazing, use of herbicides, pulling up by hand, mowing
• Public participation encouraged to help with the removal of the lupin

How do I approach these questions?

a) Deforestation, overgrazing, climate change, and trampling can expose the soil/lead to soil compaction and prevent the regeneration of vegetation. Maximum of 2 marks available. Marks will only be awarded if the reason given is explained.

b) It can be expensive/labour-intensive; the longer you leave it the more it costs. Strategies may have unforeseen/unpredicted effects. For example, lupins (introduced to control soil erosion) are outcompeting native species; herbicides may cause water pollution; mowing/grazing/pulling up lupins can increase soil erosion; the harsh climate may make land restoration difficult. Cause of the problem may be external/global, e.g. climate change; reduction in or low soil nutrients. Maximum of 4 marks available.

SAMPLE STUDENT ANSWER

a)

• More farming of livestock or clearing of vegetation for agriculture.

▼ Too vague

• Destruction of vegetation for human construction such as roads or human developments (e.g. housing) to be constructed increasing uncovered soil and increasing erosion (soil blows away).

▲ Has explained why soil erosion increases (goes further than first sentence).

This response could have achieved 1/2 marks.

b) As the vegetation, soil nutrients and conservation decrease, restoration cost increases from low to high and the probability of successful restoration decreases as vegetation cover decreases.

▲ Valid point – costs increase as soil nutrients decrease

▲ Probability of success decreases as vegetation cover decreases – correct point

The introduced species aimed at stopping soil erosion has outcompeted native flora and must now be removed – causing further erosion. One of the removal methods includes grazing, mowing and pulling them up, which leads to increased erosion as it loosens the soil and decreases plant cover.

▲ Erosion increases due to grazing and pulling up of trees – another valid statement

Many of the areas are very large and hard to reach, thus there would not be enough support/resources to effectively restore the land in all/ some of the areas.

▲ Identifies that the area are difficult to get to, inaccessible and expensive to restore (implicit)

Most areas are near the coast and distributed far apart – transport and resources would limit restoration in these areas.

This response could have achieved 4/4 marks.

6 ATMOSPHERIC SYSTEMS AND SOCIETIES

This chapter examines the functioning of the atmosphere and its importance to societies. It investigates how variations in atmospheric consumption, structure and functioning impact on ecosystems and people.

You should be able to show:

✔ the composition, structure and functioning of the atmosphere;

✔ the role of stratospheric ozone;

✔ the formation, impact, and management of photochemical smog;

✔ the formation, impact, and management of acid deposition.

6.1 INTRODUCTION TO THE ATMOSPHERE

- **Troposphere** – the lowest layer of the atmosphere extending from the ground's surface to the tropopause (between 10 km and 15 km)
- **Stratosphere** – a layer of the Earth's atmosphere extending from the tropopause to about 50 km
- **Albedo** – the amount of incoming radiation that is reflected by the Earth's surface and atmosphere
- **Greenhouse effect** – the role of certain atmospheric gases (e.g. water vapour, carbon dioxide and methane) to trap outgoing long-wave radiation and raise the Earth's temperature
- **Tipping point (threshold)** – a critical level beyond which change in a system becomes potentially irreversible

Concept link 

EQUILIBRIUM: The Earth's atmosphere is constantly changing its equilibrium as a result of the factors that influence it. Some changes are on a long-term geological scale, some are extremely short-term.

You should be able to show the composition, structure, and functioning of the atmosphere:

✔ The atmosphere changes over time, both long term and short term.

✔ The atmosphere is mainly a mixture of nitrogen and oxygen, with smaller amounts of carbon dioxide, argon, water vapour, and other trace gases.

✔ Human activities have an impact on the atmosphere.

✔ The major processes connected to living systems take place in the troposphere and the stratosphere.

✔ Clouds play an important role in the albedo effect.

✔ The natural greenhouse effect is a good thing.

▲ **Figure 6.1.1** Radiation fog over the European Alps

The atmosphere as a dynamic system

The atmosphere is a dynamic system (with inputs, outputs, flows, and storages) that has undergone changes throughout geological time. The Earth's temperature changes for a number of reasons, one of the most obvious of which is a change in the output of energy from the Sun.

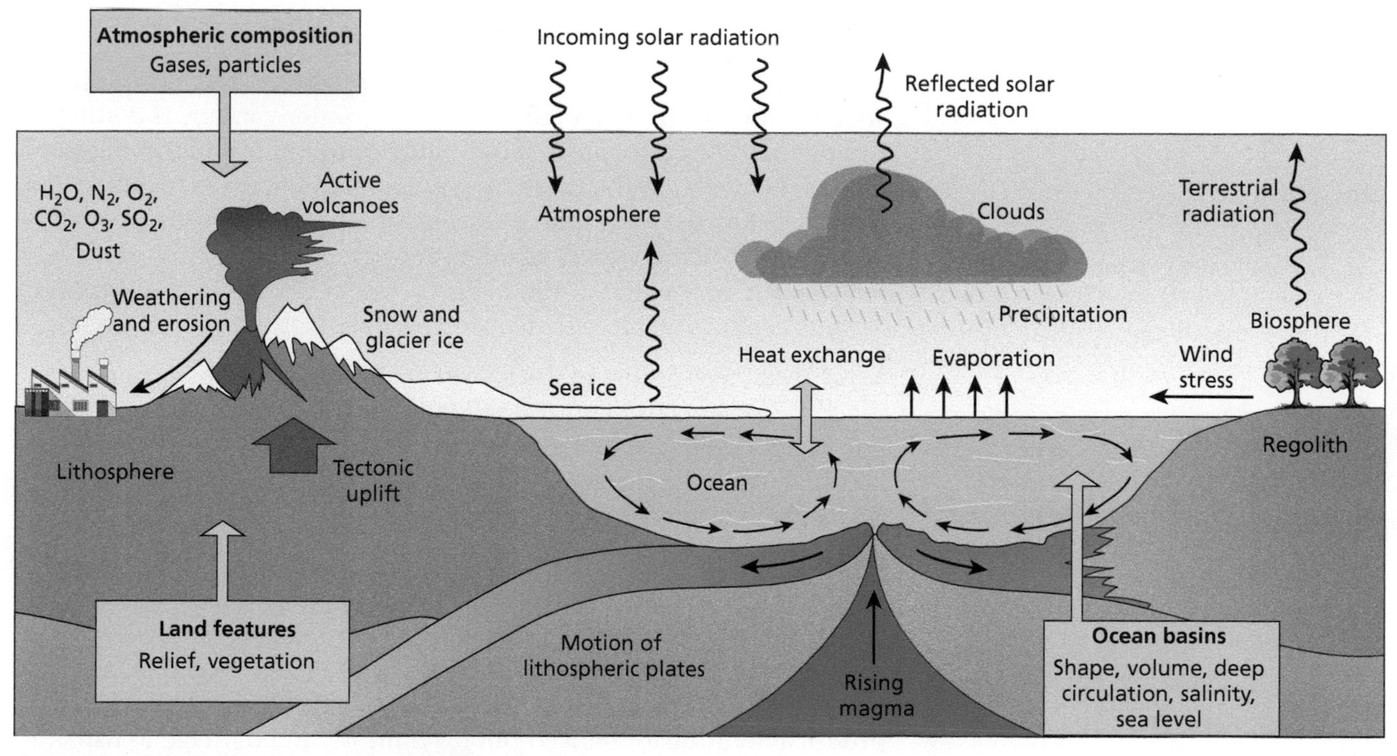

▲ **Figure 6.1.2** The Earth's climate system

The Earth's climate system consists of a number of interacting subsystems including atmosphere, hydrosphere, biosphere, and lithosphere. Solar energy drives the climate system, while the clouds, oceans, continents, and vegetation all influence the Earth's climate.

Test yourself

6.1 State the main input into the Earth's climate system. [1]

6.2 Identify two outputs from the Earth's climate system. [1]

6.3 Outline two flows in the Earth's climate system. [2]

6.4 Identify two stores of freshwater in figure 6.1.1. [2]

6.5 Outline how human activity may influence the Earth's climate system. [1]

Changes to the atmosphere over time: carbon dioxide (CO_2) levels

There have been considerable changes in the levels of carbon dioxide in the geological past. Generally, higher levels of carbon dioxide are associated with higher temperatures. Lower levels of carbon dioxide are associated with lower temperatures and with glacial periods.

Atmospheric concentrations of carbon dioxide in the Early Carboniferous period (350 million years ago) were very high at around 1,500 ppm. By the Middle Carboniferous period (300 million years ago) carbon dioxide had declined to about 350 parts per million (ppm) – comparable to today's figure. Average global temperatures in the Early Carboniferous period were approximately 20 °C. Thus high temperatures correlate with high carbon dioxide levels. Cooling during the Middle Carboniferous period reduced average global temperatures to about 12 °C. This is similar to today's levels.

In the last 600 million years of the Earth's history, carbon dioxide levels have generally been higher than 400 ppm. It is only during the Carboniferous Period and our present age (the Quaternary Period) that carbon dioxide levels have been less than 400 ppm.

Content link

Section 7.2 discusses causes and impacts of recent changes in global climate change.

Assessment tip

Many scientists believe that humanity has entered a new geological era – the Anthropocene – in which most of the world's "natural" processes are heavily influenced by human activities.

Test yourself

6.6 State the geological eras in which global temperatures were higher than today. [2]

Content link

Section 4.1 discusses how much of the world's water is stored in the atmosphere.

Levels of carbon dioxide are currently rising and the increase is thought to be mainly due to human activities. Carbon dioxide levels have risen from 280 ppm in 1850 to over 400 ppm today. This is a significant rise in 170 years. Temperatures have been rising over this same period.

The nature of the atmosphere

The atmosphere is mainly a mixture of nitrogen and oxygen, with smaller amounts of carbon dioxide, argon, water vapour, and other trace gases. The atmosphere is a mixture of solids, liquids, and gases that are held near to the Earth by gravitational force. Up to a height of around 80 km, the atmosphere consists of nitrogen (78%), oxygen (21%), argon (0.9%), and a variety of other trace gases such as carbon dioxide, helium, and ozone. In addition, there is water vapour and solids (in the form of aerosols) such as dust, ash, and soot.

Most "weather" occurs in the lowest 16–17 km, the **troposphere**. In the troposphere, temperatures fall with height (on average 6.5 °C per km). Certain gases are concentrated at height. Most water vapour is contained in the lowest 15 km of the atmosphere. Above this, the atmosphere is too cold to hold water vapour.

Human activities impact atmospheric composition through altering inputs and outputs of the system. Changes in the concentrations of atmospheric gases—such as ozone, carbon dioxide, and water vapour—have significant effects on ecosystems. For example, a decrease in stratospheric ozone can lead to a decline in marine productivity (due to reduced photosynthesis) and damage to fish larvae and eggs. Increased levels of carbon dioxide are linked to global warming and may push some ecosystems past their **tipping point**. Too much or too little water may also have the same effect.

Pollution and global dimming

Assessment tip

Global dimming may be reducing the worst impacts of global temperature rise, but it is causing a large increase in the number of deaths due to poor air quality.

Content link

Section 4.1 discusses flows and stores in the hydrological cycle.

Scientists have shown that from the 1950s to the early 1990s, the level of solar energy reaching the Earth's surface reduced by 9% in Antarctica, 10% in the USA, 16% in parts of the UK, and almost 30% in Russia. This was all due to high levels of pollution at that time. Natural particles in clean air provide condensation nuclei for water. Polluted air contains far more particles than clean air (for example ash, soot, sulfur dioxide) and therefore provides many more sites for water to bind to. The droplets formed tend to be smaller than natural droplets, which means that polluted clouds contain many much smaller water droplets than naturally occurring clouds. Many small water droplets reflect more sunlight than fewer larger droplets, so polluted clouds reflect far more light back into space, thus preventing the Sun's heat from getting through to the Earth's surface.

Water distributed around large natural particles forms a few large droplets with moderate reflectivity, which eventually fall as rain. The same amount of water distributed around small polluting particles forms many small droplets with increased reflectivity. This water does not fall as rain.

Global dimming is where a few large droplets in the atmosphere hold the same amount of water as a greater number of smaller ones, but the total surface area is less for the larger particles, so they reflect less sunlight than the smaller pollution particles.

▲ **Figure 6.1.3** Stratocumulus clouds

At high altitude there are significant concentrations of gases, such as ozone between 25 km and 35 km, nitrogen between 100 km and 200 km, and oxygen between 200 km and 1,100 km. Most processes connected with living systems occur in the lower layers of the atmosphere, i.e. the troposphere.

Clouds and albedo

The proportion of insolation that is reflected back to space is known as the **albedo**. Cloud types affect the albedo. The albedo for a complete cover of cirrostratus clouds is about 44–50% whereas for a cumulonimbus cloud it is about 90%.

The greenhouse effect

The **greenhouse effect** is the process by which certain gases (greenhouse gases) allow short-wave radiation from the Sun to pass through the atmosphere but trap an increasing proportion of outgoing long-wave radiation from the Earth. This radiation leads to a warming of the atmosphere. The greenhouse effect is a good thing, for without it there would be no life on Earth. For example, the Moon is an airless planet that is almost the same distance from the Sun as the Earth. However, daytime temperatures on the Moon may get as high as 100 °C, whereas by night they may be −150 °C. Average temperatures on the Moon are about −18 °C compared with about 15 °C on Earth. The Earth's atmosphere therefore raises temperatures by about 33 °C.

There are a number of greenhouse gases. **Water vapour** accounts for about 95% of greenhouse gases by volume and for about 50% of the greenhouse effect. However, the gases mainly implicated in global warming are carbon dioxide, methane and chlorofluorocarbons.

Carbon dioxide (CO_2) levels have risen from about 315 parts per million (ppm) in 1950 to over 400 ppm in 2015, and are expected to reach 600 ppm by 2050. The increase is due to human activities: burning fossil fuel (coal, oil and natural gas) and land-use changes such as deforestation. Deforestation of the tropical rainforest is a double blow, since it not only increases atmospheric CO_2 levels but it also removes the trees that produce oxygen. Carbon dioxide accounts for about 20% of the greenhouse effect but an increased proportion of the enhanced greenhouse effect.

Methane (CH_4) is the second-largest contributor to global warming, and its presence in the atmosphere is increasing at a rate of 1% per annum. It is estimated that cattle convert up to 10% of the food they eat into methane and emit 100 million tonnes of methane into the atmosphere each year. Natural wetlands and paddy fields are other important sources: paddy fields emit up to 150 million tonnes of methane annually, while, as global warming increases, bogs trapped in permafrost will melt and release vast quantities of methane. This is an example of positive feedback.

Chlorofluorocarbons (CFCs) are synthetic chemicals that destroy ozone as well as absorbing long-wave radiation. Some CFCs are up to 10,000 times more efficient at trapping heat than CO_2. HCFCs (hydrochlorofluorocarbons) have increased rapidly from the 1990s (in contrast to CFCs) and contribute to the enhanced greenhouse effect.

▲ **Figure 6.1.4** Cumulonimbus clouds

Test yourself

6.7 Briefly **explain** how polluted air leads to global dimming. [3]

6.8 Define the term "albedo". [1]

6.9 Suggest why cumulonimbus clouds and stratocumulus clouds have a high albedo. [2]

▼ **Table 6.1.1** Average albedo of various surfaces

Surface	Albedo
Earth systems (average)	0.31
Earth surface	0.14–0.16
Global cloud	0.23
Cumulonimbus	0.9
Stratocumulus	0.6
Cirrus	0.4–0.5
Fresh snow	0.8–0.9
Melting snow	0.4–0.6
Sand	0.30–0.35
Grass, cereal crops	0.18–0.25
Deciduous forest	0.15–0.18
Coniferous forest	0.09–0.15
Tropical rainforest	0.07–0.15
Water bodies*	0.06– 0.10

*Increases when the sun is at a low angle

Source: Barry, R., and Chorley, R., 1998, *Atmosphere, Weather and Climate*, Routledge

Concept link

EQUILIBRIUM: The greenhouse effect is a natural phenomenon. The enhanced greenhouse effect is the change in greenhouse gases and represents a changing equilibrium as a result of human activities.

Test yourself

6.10 Outline the functioning of the natural greenhouse effect. [4]

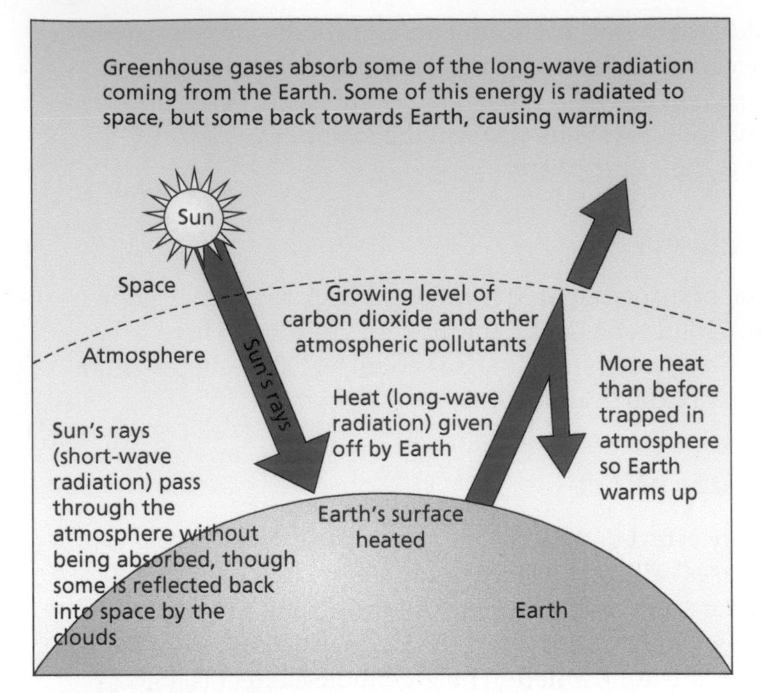

▲ **Figure 6.1.5** The greenhouse effect

QUESTION PRACTICE

Identify two ways in which the climate of Iceland may limit productivity. [2]

▼ **Table 1** Climate data for Reykjavik, Iceland

	Jan	Feb	Mar	Apr	May	Jun	Jul	Aug	Sep	Oct	Nov	Dec
Mean temp. (°C)	−0.6	0.1	0.3	2.8	6.3	8.9	10.7	10.3	7.5	4.3	1.3	−0.1
Rainfall (mm)	35	36	34	15	19	10	14	21	23	20	29	37

How do I approach this question?

You are required to give two separate reasons—no explanation is needed. For example, short growing season, low temperatures, a number of days with temperatures below 0 °C.

SAMPLE STUDENT ANSWERS

▲ Correct point

▲ Correct point

The cold temperatures throughout the year act as a limiting factor, especially between October and April. This limits the rate of reaction needed to sustain plants and animal life such as photosynthesis. Also, the small number of hours of sunshine during most of the year, especially when only for 2 hours per day, will limit photosynthesis, needed for primary productivity.

This response could have achieved 2/2 marks.

6.2 STRATOSPHERIC OZONE

You should be able to show the role of stratospheric ozone:

✔ The destruction and re-formation of ozone is an example of dynamic equilibrium.

✔ Ozone-depleting substances are used in various products such as aerosols, flame retardants, etc.

✔ Ultraviolet radiation damages living human tissues and biological productivity.

✔ Pollution management strategies are in place to deal with stratospheric ozone.

✔ There is an illegal market for ozone-depleting substances.

✔ The Montreal Protocol is the most significant and successful international agreement relating to an environmental issue.

- **Stratospheric ozone** – "good" ozone found at an altitude of between 10 km and 50 km, which blocks out some ultraviolet radiation (short-wave radiation) which is harmful to humans, some animals and some plants
- **CFCs (chlorofluorocarbons)** – non-toxic, non-flammable chemicals containing atoms of chlorine, fluorine, and carbon. They are used in the manufacture of aerosol sprays, foams and packing materials and as refrigerants
- **Ultraviolet radiation** – also known as short-wave radiation, i.e. radiation at the wavelength 0.1–4.0 μm
- **The "ozone hole"** – the thinning of the concentration of ozone in the stratosphere. Ozone is a trace gas and is measured in Dobson Units
- **Dobson Units (DU)** – the unit of measurement for ozone
- **Halocarbons** – also known as halogenated organic gases and first identified as depleting the ozone layer in the stratosphere. Now known to be potent greenhouse gases. The most well known are chlorofluorocarbons (CFCs)
- **Halogen** – any of a group of five non-metallic elements with a similar chemical bonding: fluorine, chlorine, bromine, iodine, and astatine. They react with metals to produce a salt. Halogenated means a halogen atom has been added
- **Ozone-depleting substances (ODSs)** – halogenated organic gases (such as chlorofluorocarbons) and halogen atoms (such as chlorine) that can destroy ozone

Depletion of stratospheric ozone

Ozone occurs in the **stratosphere**. It is mostly found around at 10–50 km. It is important for the filtering of harmful **ultraviolet radiation**. Ozone is essential for sustaining life. The ozone layer shields the Earth from harmful radiation that would otherwise destroy most life on the planet.

Assessment tip

Do not confuse **stratospheric ozone** with tropospheric ozone (ground-level ozone). Stratospherioc ozone is good and protects us from ultraviolet radiation, whereas tropospheric ozone is a pollutant that causes air pollution and aggravates respiratory problems.

Some ultraviolet radiation is absorbed by ozone. This happens during the formation and destruction of ozone. Ozone also absorbs some outgoing long-wave radiation, so it is a greenhouse gas too. Ozone is created by oxygen rising up from the top of the troposphere reacting with sunlight.

Ultraviolet radiation (short-wave radiation) breaks down oxygen molecules into two separate oxygen atoms. The oxygen atoms (O) combine with oxygen molecules (O_2) to form ozone (O_3).

Human activities can also lead to the depletion of ozone. Human production of **chlorofluorocarbons** (CFCs) is linked to a decrease in ozone. Figure 6.2.1 shows how natural processes and human processes combine to destroy ozone.

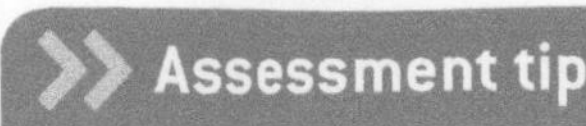

You do not need to memorize chemical equations relating to ozone and ultraviolet radiation.

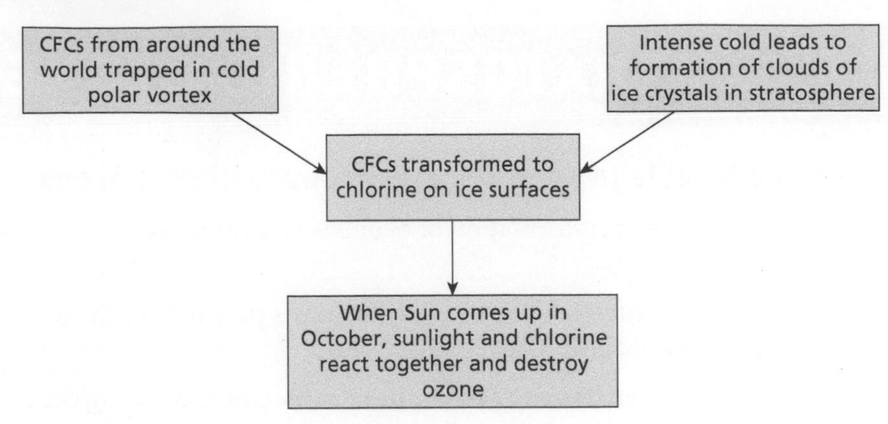

Source: Nagle, G., 2005, Ozone—too little up there, too much down here, GeoFactSheets

▲ **Figure 6.2.1** Human and natural processes and the destruction of ozone

Content link

Look at section 1.5 Humans and pollution to see how human activity can cause pollution, as well as managing pollution.

Ozone-depleting substances

The chemicals that cause stratospheric ozone depletion include **halocarbons**. Halogens include chlorofluorocarbons (CFCs). These are found in many products, including aerosols and refrigerators. They can also be found in air conditioners and foamed plastics. They are sometimes found in pesticides and are also used for fire extinguishers and solvents.

Halogenated organic gases are very stable under normal conditions but can liberate **halogen** atoms when exposed to ultraviolet radiation in the stratosphere. Halogen atoms react with oxygen and slow the rate of ozone re-formation.

Pollutants increase the destruction of ozone. They change the equilibrium of the ozone production system. They cause "holes" in ozone layer. The **ozone hole** is a thinning of the concentration of ozone in the stratosphere.

The ozone "hole" allows more ultraviolet radiation to pass through the Earth's atmosphere. This can be very damaging. There is a clear seasonal pattern to the concentration of ozone. Each spring there is a decrease in the amount of ozone over Antarctica. The ozone layer recovers when summer comes. This is because in winter, air over Antarctica becomes cut off from the rest of the atmosphere. The intense cold allows the formation of clouds of ice particles. Chemical reactions occur on these ice particles. The chemical reactions involve chlorine. The chlorine may come from CFCs. The chlorine atoms destroy ozone each spring. By summer the ice clouds have disappeared. Thus there is less destruction of ozone in the summer months.

(a)

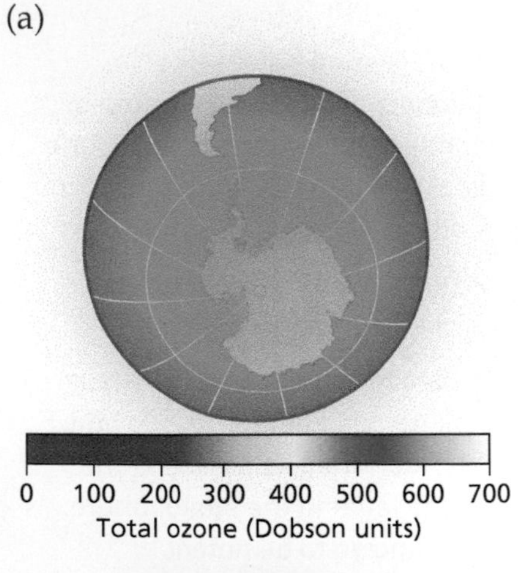

(b)

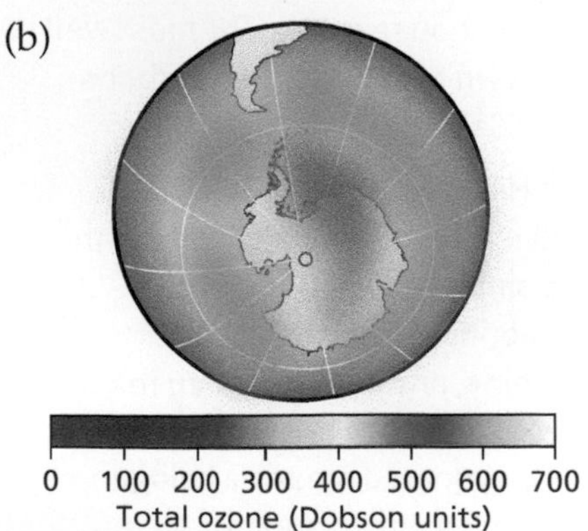

▲ **Figure 6.2.2** Ozone hole over Antarctica, (a) January 1979 and (b) March 2018

The effects of ultraviolet radiation on living tissues and biological productivity

Increased ultraviolet radiation is damaging to ecosystems that contribute significantly to global biodiversity as it damages plant tissues and plankton.

Ultraviolet radiation damages marine phytoplankton, which is one of the major primary producers of the biosphere. It causes reduced rates of photosynthesis. In aquatic ecosystems the organisms that live in the upper part of the water are most affected. Organisms that live at the surface of the water during their early life stages are also affected. These include phytoplankton and fish eggs and larvae. Most adult fish are protected from UV radiation since they inhabit deep waters. UV radiation can also cause genetic mutations in DNA. There are also negative impacts on reproduction.

Ultraviolet radiation is damaging to human populations around the world. It causes eye damage and cataracts. The effects of long-term exposure are irreversible and can cause blindness. It can also cause sunburn and eventually skin cancer. Research suggests annual loss of **ozone** is about 1% and there has been an increase in skin cancers of 4%.

Concept link 

BIODIVERSITY: Ultraviolet radiation damages living organisms and reduces ecosystem productivity. This could lead to a decline in biodiversity.

Test yourself

6.11 Outline the role of stratospheric ozone. [1]

6.12 Explain the meaning of the term "ozone hole". [1]

6.13 State two products that contained ozone-depleting substances in the past. [2]

Pollution management strategies

Recycling refrigerants

The use of CFCs in refrigeration was one of their most important uses. Now, a combination of HFCs (hydrofluorocarbons) and hydrocarbon refrigerants has largely replaced the CFCs in fridges. Fridges with **ozone-depleting substances (ODSs)** (halogenated organic gases (such as chlorofluorocarbons) and halogen atoms (such as chlorine), that can destroy ozone) can be replaced with "greenfreeze" technology that uses propane and/or butane. Old CFC coolants in fridges and air conditioning units can be recycled.

Alternatives to gas-blown plastics and propellants

Huge quantities of CFCs were used as propellants in aerosol sprays. Alternatives to aerosols can be used. A good example is using soap instead of shaving foam. Pump-action sprays and trigger sprays can be used instead of aerosols.

The phase-out of methyl bromide

Methyl bromide gas has been used to control pests. It is an ODS. Its production and import in the USA and Europe was phased out in 2005. There are some exceptions. It can be used to eliminate quarantine pests and it can be used in farming where there are no alternatives.

There are alternative chemicals to methyl bromide. Some of these react in ultraviolet radiation to have an impact on germs. Other non-chemical alternatives include biofumigation and crop rotation. These are examples of organic farming. Cultivation of plants in water (hydroponics) can also reduce the risk of pests.

The United Nations Environment Programme (UNEP) has played a key role in providing information, and creating and evaluating international agreements, for the protection of stratospheric ozone. In 1987 it brought 28 countries together to sign the first Montreal Protocol. It is working to end production of HFCs by 2040.

An illegal market for ozone-depleting substances persists and requires consistent monitoring. It exists because ODS substitutes are often more expensive than CFCs, as is changing equipment to enable the use of alternative chemicals.

Assessment tip

Ozone depletion/UV radiation has no significant **direct** impact on global warming. However, many ODSs not only deplete ozone but are greenhouse gases contributing to global warming. Ozone depletion/UV radiation does impact biological productivity, thus reducing carbon sinks and so **indirectly** contributes to global warming.

National and international approaches to reducing the emissions of ozone-depleting substances

The 1987 **Montreal Protocol on Substances that Deplete the Ozone Layer** is the most significant and successful international agreement relating to an environmental issue. By 2012 the world had phased

Test yourself

6.14 Describe the impact of increased UV radiation on living tissue and biological productivity. [3]

6.15 Outline the main methods to reduce ozone-depleting substances. [3]

6.16 Explain why the Montreal Protocol has been described as "the most successful environmental management scheme". [2]

out 98% of the ozone-depleting substances contained in nearly 100 hazardous chemicals worldwide

Nearly 200 governments have signed up and implemented the agreed changes according to the Montreal Protocol. It is believed that ozone could recover by 2050 as a result of the Montreal Protocol. The Protocol has been revised seven times since it was first introduced in 1987. Subsequent revisions have reduced the phasing out timescale because of success—phase-out in Europe was achieved by 2000. Total global phase-out is expected by 2030. The Protocol provided an incentive for countries to find alternatives. It also raised public awareness of the use of CFCs. Technology has been transferred to LEDCs to allow them to replace ozone-depleting substances.

Some of the substances that replaced CFCs, such as HFCs, are powerful **greenhouse gases**. The long life of the chemicals in the atmosphere means that damage will continue for some time—some argue until 2100. It is harder for LEDCs to make the changes. The second-hand appliance market means old fridges are still in circulation. The success of the protocol depends on national governments agreeing to its requirements.

QUESTION PRACTICE

Explain how changes in the concentration of stratospheric (and tropospheric) ozone in the atmosphere can affect global biodiversity. [7]

How do I approach this question?

There is 1 mark available for the following point: stratospheric ozone has decreased, and (production of) tropospheric ozone has increased.

There are then up to 4 marks for points relating to stratospheric ozone and tropospheric ozone, up to a combined maximum of 6 marks. Points relating to stratospheric ozone include:

- change in stratospheric ozone allows more UV radiation to reach Earth
- causes mutations/damage to DNA/cancers
- this may result in death of organisms/reduction in biodiversity
- it may also reduce plant growth/NPP/especially phytoplankton/damages chlorophyll
- this affects populations all along the food chain/reducing diversity of the food web.

SAMPLE STUDENT ANSWER

▲ Intro which shows changes in ozone allowing more UV radiation to reach Earth. (1 mark for stratospheric ozone)

Stratospheric ozone is beneficial as it absorbs and prevents harmful short wavelength UV rays from reaching the Earth's surface. Depletion of this ozone layer caused by ozone depleting substances (ODSs) such as OFCs from aerosols, refrigerants and air conditioning as well as HCFCs and halons from fire retardants and pesticides which release halogen atoms that react with the ozone and break it into oxygen and molecular oxygen, as well as preventing oxygen from reacting to ozone in a catalytic process.

SAMPLE STUDENT ANSWER

UV radiation can promote vitamin D synthesis in animals, which is beneficial. But incoming UV radiation due to depletion of ozone layer leads to increased skin cancer and cataracts in humans. Also, UV rays cause mutations in DNA, which may be harmful to the organisms and thus decrease biodiversity. UV also degrades the chlorophyll in the leaves of plants, decreasing primary productivity thus secondary productivity, resulting in decreased global biodiversity. Further, UV radiation due to depletion of ozone layer kills the mutualistic algae in coral reefs, leading to coral leaching and death of corals, thus decreasing productivity and global biodiversity. UV radiation further damages plants, reducing photosynthesis thus oxygen production and carbon fixation, so increased amount of CO_2 not fixed.

▲ Valid point. (1 mark for stratospheric ozone)

▲ Valid point. (1 mark for stratospheric ozone)

▲ Relates to biodiversity – chlorophyll point developed. (1 mark for stratospheric ozone)

▲ Valid, but maximum 4 marks reached for stratospheric ozone

This response could have achieved 4/7 marks.

6.3 PHOTOCHEMICAL SMOG

You should be able to show the formation, impact and management of photochemical smog:

- ✔ The main source of primary pollutants is road transport.
- ✔ Primary pollutants are converted into secondary pollutants in the presence of sunlight.
- ✔ Nitrogen oxides interact with others in the presence of sunlight to produce tropospheric ozone.
- ✔ Tropospheric ozone damages plants and can cause breathing difficulties and eye irritation in humans.
- ✔ Local conditions that intensify tropospheric ozone include climate, topography, population density, and the use of fossil fuels.
- ✔ Temperature inversions may trap smog in valleys.
- ✔ Deforestation and burning may also contribute to smog.
- ✔ There are economic losses that are caused by tropospheric ozone.
- ✔ Several pollution management strategies are in place to deal with tropospheric ozone.

- **Tropospheric ozone** – “bad” ozone which is found at ground level i.e. within the troposphere, and is formed when oxygen molecules react with molecules from nitrogen dioxide in the presence of sunlight
- **Primary pollutants** – these are the emissions that are active when emitted, e.g. nitrogen monoxide
- **Secondary pollutants** – pollutants that arise from the physical or chemical change of primary pollutants, e.g. nitrogen oxides reacting with sunlight to form photochemical smog
- **Thermal (temperature) inversion** – an atmospheric situation in which cold air is found at low altitudes and warm air at higher altitudes, inverting the normal pattern of a decrease in temperature with increasing altitude

(a)

(b)

▲ **Figure 6.3.1** Hong Kong (a) clear view and (b) with smog

Primary and secondary pollutants

The main cause of photochemical smog (ground level/**tropospheric ozone**) is the volume of road transport concentrated in cities. Important pollutants are released when fossil fuels are burned, including hydrocarbons, e.g. carbon monoxide, carbon dioxide (from unburned fuel) and nitrogen monoxide (also called nitric oxide or NO). These are **primary pollutants**. Tropospheric ozone—or ground-level ozone—is a secondary pollutant because it is formed by reactions involving oxides of nitrogen (NO_x). Nitrogen monoxide reacts with oxygen in the presence of sunlight to form nitrogen dioxide (NO_2). This is a brown gas that contributes to urban haze. It can also absorb sunlight and break up to release oxygen atoms that combine with oxygen in the air to form ozone.

Effects of tropospheric ozone

Ozone is a toxic gas and an oxidizing agent. It irritates the eyes and can cause breathing difficulties in humans. It may also increase susceptibility to infection. Ground-level ozone reduces plant photosynthesis and can reduce crop yields significantly. It damages crops and forests. Ozone pollution has been suggested as a possible cause of the dieback of German forests (previously it was believed these had died as a result of acidification). Ozone is highly reactive and can attack fabrics and rubber materials.

Nitrogen oxides interact with others in the presence of sunlight to produce tropospheric ozone. Ozone formation can take a number of hours, so the polluted air may have drifted into suburban and surrounding areas. Smog is more likely under high-pressure (calm) conditions. Rain cleans the air and winds disperse the smog—these are associated with low-pressure conditions.

Content link
See sections 1.5 and 8.2 for more information about how human activities are causing widespread environmental degradation, and affecting atmospheric, aquatic, and terrestrial systems.

Frequency and severity of smog

The frequency and severity of photochemical smogs depends on local factors. These factors include topography and climate as well as population density and use of fossil fuel. **Temperature inversions** may trap smog in valleys. A temperature inversion refers to an atmospheric situation in which cold air is found at low altitudes and warm air at higher altitudes, inverting the normal pattern of a decrease in temperature with increasing altitude.

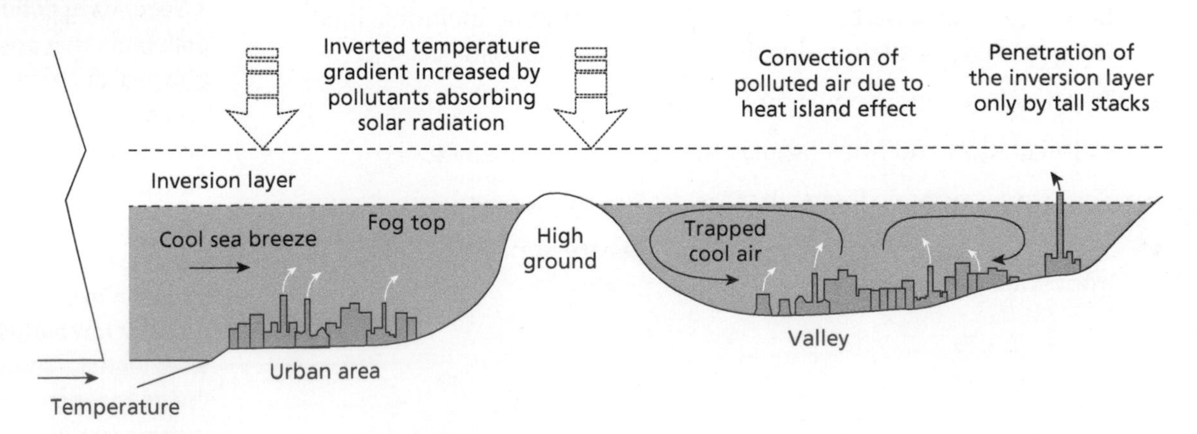

▲ **Figure 6.3.2** Thermal (temperature) inversion

Temperature inversions happen regularly in Los Angeles and Mexico City. The air is unable to disperse since cold air from the surrounding mountains and hills prevents the warm air from rising. Cold air is denser than warm air and so it traps the warm air below. Concentrations of air pollutants can build up to harmful levels.

Urban microclimates also affect the production of ground-level ozone. Urban areas generally have less vegetation than surrounding rural areas. Urban areas also have a greater concentration of buildings and industries that generate a lot of heat.

Deforestation and burning

Deforestation and burning may also contribute to smog. Fire is used to:

- clear land for farming and other uses (deforestation)
- remove dead vegetation and improve soil quality.

The South Asian brown haze of the late 1990s was caused by large-scale burning of rainforests in Malaysia and Indonesia to develop land for plantations and other developments.

Economic loss

The economic losses of photochemical smog include clean-up costs, loss of tourism, cost of health care, reduced productivity of workers, and the decline in ecological services provided by ecosystems. In China it has been estimated that the economic cost of air pollution amounts to about 4% of GDP.

Pollution management strategies

There are many ways in which urban air pollution can be managed. The first level of pollution management involves measures that lead to a change in human activity. Some measures may lead to a reduction in the use of fossil fuels, such as reducing demand for electricity and switching to renewable energy. Increased use of public transport can reduce emissions of fossil fuels. The promotion of clean technology/ hybrid cars also reduces the use of fossil fuels. Preventing cars from entering parts of the city can result in improved air quality. In Mexico City cars with odd-numbered registration plates are allowed into the city centre on certain days and even-numbered cars on other days. By increasing the provision of bus lanes and cycle lanes more people are encouraged to cycle or travel by bus. Tolls for private cars to enter the city centre have reduced the amount of vehicles entering the city centre, e.g. the London Congestion Charge.

Reducing the consumption of fossil fuel through urban design (e.g. south-facing windows and triple-glazed windows) can lead to an improvement in air quality. Reducing fossil fuel combustion by switching to renewable energy methods also improves air quality. The relocation of industries and power stations away from centres of population leads to an improvement in air quality.

The second level of strategies focuses on the controlled release of the pollutant. These strategies include the development of catalytic convertors in cars to reduce emissions of NO_x. The chimneys used in industries and power stations should be tall to help disperse pollutants. It is possible to filter and catch pollutants at the point of

Assessment tip

Ground-level ozone generally occurs during high-pressure (calm) conditions in which pollutants may build up for hours. In low pressure conditions (windy), the pollutants are dispersed by the wind. In addition, low-pressure conditions are cloudy, so there is less photochemical reaction occurring (less sunlight).

Test yourself

6.17 Distinguish between primary pollutants and secondary pollutants. [2]

6.18 Define the term "temperature inversion". [2]

Concept link

STRATEGIES: Different pollution management strategies may be influenced by different EVSs. For example, the use of catalytic converters is a technocentric approach whereas the use of green space (open areas) and reforestation is a more ecocentric form of pollution management.

emission. Fuel quality can be regulated by governments. Urban design can be made more sustainable. Open space and water courses help to reduce the temperature and allow evaporative cooling.

The third level of pollution management strategies includes clean-up strategies such as reforestation and regreening, and conservation of areas to sequester carbon dioxide can also improve the quality of the environment.

Weaknesses of pollution management strategies

Most urban pollution comes from cars. Old cars tend to be more polluting than new cars. Vehicles using diesel fuel produce emissions of particulate matter. The use of catalytic converters reduces fuel efficiency and increases CO_2 emissions. Hybrid cars and electric cars are less polluting, but they only account for a small percentage of the vehicles currently in use. Public transport can be expensive and may be inconvenient. Sustainable urban design is expensive.

Test yourself

6.19 Explain how different pollution management strategies deal with photochemical smog. [3]

QUESTION PRACTICE

1) **Explain** how changes in the concentration of (stratospheric) and tropospheric ozone in the atmosphere can affect global biodiversity. [7]

2) With reference to named examples, **distinguish between** a primary and a secondary pollutant. [4]

3) a) **State one** human factor that contributes to photochemical smog. [1]

b) **State one** natural factor that contributes to photochemical smog. [1]

c) **Explain** why the formation of photochemical smog may have harmful effects on the environment of cities. [4]

▲ **Figure 1** A layer of smog covering the Chilean city of Santiago

How do I approach these questions?

1) You will be awarded 1 mark for the following point: production of tropospheric ozone has increased.

There are then up to 4 marks for points relating to tropospheric ozone, up to a combined maximum of 6. Points relating to tropospheric ozone include:

- change in tropospheric ozone in urban areas may lead to photochemical smog
- this is toxic (to humans and/or other species)
- it damages plant leaves, reducing NPP of ecosystems/food chains
- tropospheric ozone is also a greenhouse gas, and contributes to global warming/climate change
- this may lead to population declines/death/reduction in biodiversity.

2) You will be awarded 1 mark for each correct point, e.g.

- **Primary pollutants** are the emissions that are active when emitted, e.g. nitrogen monoxide.
- **Secondary pollutants** are the pollutants that develop from the physical or chemical change of primary pollutants, e.g. nitrogen oxides reacting in sunlight to form photochemical smog.

3) a) You need to provide a specific human factor (economic, social, cultural or political) that can contribute to photochemical smog—no explanation is needed.

b) You need to provide a specific physical factor (natural environment) that can contribute to photochemical smog—no explanation is needed.

c) You need to provide a detailed account, including reasons or causes.

SAMPLE STUDENT ANSWER

1) Tropospheric ozone levels have increased due to the combustion of fossil fuels and emissions of nitrogen oxides, carbon dioxide, hydrocarbons and VOCs. Tropospheric ozone caused breathing problems and may lead to lung cancer. People's growing need for medicine may harm biodiversity as plants or animals are killed for medical purposes. Tropospheric ozone is also a greenhouse gas that traps long-wave radiation reflected from the Earth and reflects it back to the surface, increasing global temperatures. Species and genetic diversity of animals such as polar bears or penguins may reduce due to their inability to adjust or relocate to colder climates due to rising global temperatures. Organisms and plants favouring warmer climates, however, may increase in species and genetic diversity as they are able to live in larger areas of the planet.

▲ Change in tropospheric ozone toxicity (1 mark for tropospheric ozone)

▲ Role as a greenhouse gas and effect. (1 mark for tropospheric ozone)

▲ Impact on certain organisms. Genetic diversity (2 marks for tropospheric ozone)

▲ Maximum 4 marks for tropospheric ozone

This response could have achieved 4/7 marks.

2) Primary pollutants are those that directly contribute to the enhanced greenhouse effect, without needing to react with anything. An example of a primary pollutant is CO_2. Secondary pollutants are those that are harmful only once they react with a certain molecule. An example would be how SOx reacts with water in the atmosphere to form H_2SO_3 or H_2SO_4 which contributes to acid deposition.

▲ Well expressed – clear

▲ Good example

▲ Good point – reacts with something

▲ Good example

This response could have achieved 4/4 marks.

3) a) One human factor that contributes to photochemical smog is the use of fossil fuels in cars and factories, leading to the increase of the greenhouse effect.

▲ Valid reason

▼ Irrelevant

This response could have achieved 1/1 marks.

b) The ventilation of the area that has smog. If the cities don't have good ventilation, because they are surrounded by mountains, this would contribute to the photochemical smog.

▲ Valid point – could have topography, light winds, thermal inversion

This response could have achieved 1/1 marks.

▲ Three impacts although very brief – suggesting that a decline in air quality leads to these impacts

c) Some cities are very vulnerable to smog, which decreases the air quality, creates lung cancer and respiratory problems, harms tree leaves, etc.

This response could have achieved 3/4 marks.

Another point that could be given for the final mark is that it harms the long-term educational development/mental development of children.

6.4 ACID DEPOSITION

- **Acid precipitation** – the increased acidity of rainfall, largely as a result of human activity
- **Acid deposition** – the increased amount of acid in rainfall and the placing/leaving of acid on rocks and in soil, largely as a result of human activity
- **Dry deposition** – fallout of particulates of oxides of sulfur and nitrogen
- **Wet deposition** – rain and snow that have become acidified through emissions of sulfur dioxide and oxides of nitrogen

You should be able to show the formation, impacts and management of acid deposition

✔ Primary pollutants, such as sulfur dioxide and oxides of nitrogen, can be converted into secondary pollutants such as dry and wet acid deposition.

✔ Acid deposition has major impacts on soil, water, vegetation and buildings.

✔ Acid deposition affects large-scale areas downwind from major coal-burning and/or industrial areas.

✔ Different pollution management strategies are in place to deal with acid deposition.

▲ **Figure 6.4.1** The effect of acid rain on buildings

Content link

See section 7.1 to investigate the main coal-producing nations.

Formation of acid deposition

Acid deposition is the increased acidity of rainfall and dry deposition. This is largely as a result of human activity and is caused by carbon dioxide in the atmosphere combining with moisture in the atmosphere. Rainfall is naturally acidic with a pH of about 5.5. Acid rain in Europe has been declining since many countries moved away from burning coal and moved some of their industries to LEDCs.

The major causes of acid rain are the sulfur dioxide and nitrogen oxides produced when fossil fuels are burned. Sulfur dioxide and nitrogen oxides are released into the atmosphere. There they are absorbed by the moisture and become weak sulfuric and nitric acids. The pH can be as low as 3.

Dry deposition typically occurs close to the source of emission and causes damage to buildings and structures. **Wet deposition** occurs when the acids are dissolved in precipitation and fall at great distances from the sources. Wet deposition has been called a "transfrontier" pollution, as it crosses international boundaries with disregard. Figure 6.4.2 shows the formation and impacts of acid rain.

Impact of acid deposition on soils

When levels of acidity increase, many nutrients become unavailable to plants. These include nitrogen, phosphorus, molybdenum, and boron (Figure 6.4.3). Essential nutrients such as calcium and magnesium

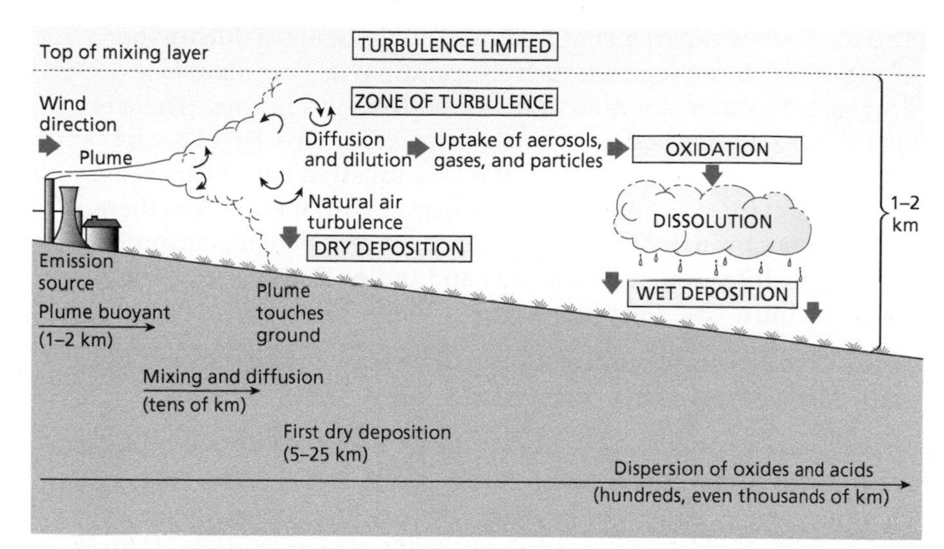

▲ **Figure 6.4.2** The formation of acid rain

can be **leached** from a soil as it becomes more acidic, and this can be detrimental to plant growth. Other nutrients become more common and can be toxic to living organisms. Copper becomes more available in acidic soils. Iron and aluminium may be **mobilized** when soil pH becomes lower than 4.5.

Impact on coniferous trees

Acid rain has many impacts on **coniferous trees**. Coniferous trees are more at risk than deciduous trees. This is because coniferous trees do not shed their leaves at the end of the year, so they are exposed to acid deposition all year round. The trees may also take up toxic aluminium ions from the soil. The trees fail to grow because of a lack of nutrients and the presence of too many toxins. Root hairs may be damaged and so there is less uptake of water from the soil. Needles are lost and there may be dieback of the crown.

Impact on living organisms

Increasing acidity leads to falling numbers of fungi, bacteria and earthworms. Aluminium and mercury reduce the number of soil microorganisms. Earthworms cannot tolerate soils with a pH below 4.5. Aluminium damages fish gills by causing a mucus to build up, making breathing difficult.

Impact on water

Iron and aluminium are washed from the soils into streams and lakes. The water may become too acidic to support fish. In Canada and the eastern USA there are over 48,000 lakes that are too acidic to support fish.

Areas affected by acidification

Acid deposition has been called a "transfrontier" pollution because the area producing the acid deposition is not the same as the regions receiving it. Areas experiencing acid deposition are downwind from the main sources of sulfur dioxide and nitrogen dioxide. Coal-fired power stations and heavy concentrations of vehicles emit vast quantities of sulfur dioxide and nitrogen dioxide.

Test yourself

6.20 Distinguish between wet deposition and dry deposition. [2]

6.21 Identify the two chemicals responsible for acidification. [1]

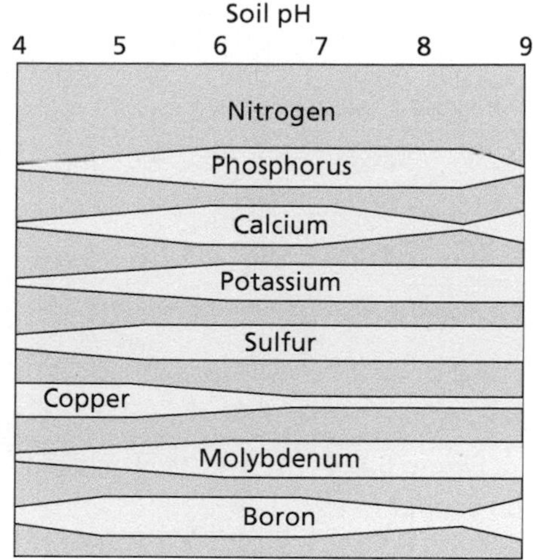

▲ **Figure 6.4.3** The effect of soil pH on nutrient availability

Source: *University of Missouri Extension*

Concept link 

BIODIVERSITY: As acid deposition reduces the growth of some plants, the food supply for secondary consumers is reduced, and biodiversity is threatened.

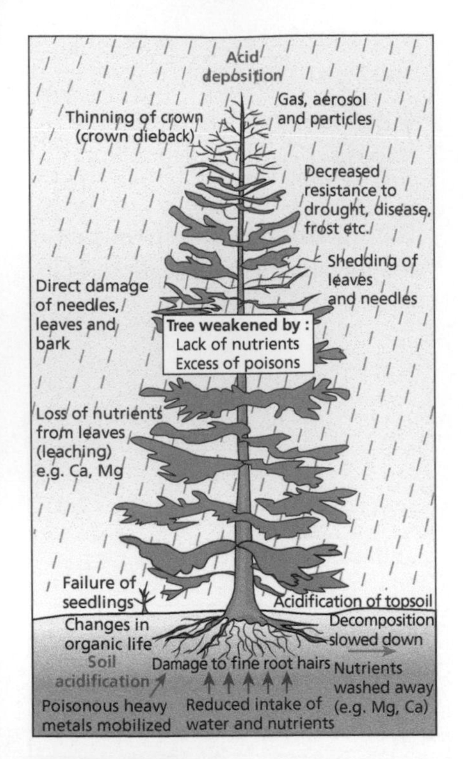

▲ **Figure 6.4.4** The effects of acid rain on conifer trees

Test yourself

6.22 Identify the chemical that becomes available when soil pH is less than 5. [1]

6.23 Outline the impact of acidification on coniferous trees. [3]

Concept link 

EVSs: Some pollution management strategies are technocentric, such as the use of limestone scrubbers, whereas others are anthropocentric, such as legislation, e.g. the 1979 Convention on Long-range Transboundary Air Pollution.

Test yourself

6.24 Outline how human activity may be altered to reduce acidification. [2]

6.25 Explain how pollutants can be regulated and/or monitored to reduce the threat of acidification. [2]

The main areas experiencing acid rain are those areas downwind of major industrial regions, such as Scandinavia. Scandinavia is downwind from major industrial belts in Western Europe. There is also much acid deposition in north-east USA and eastern Canada. These areas are downwind from the US industrial belt. There is less acidification in Scandinavia now compared with the 1980s as there is less **heavy industry** in Western Europe. Areas that are currently causing acidification include China and India. This is because both countries burn vast amounts of coal.

Areas experiencing acidification usually have high rainfall and thin soils. Many of them have forests and lakes.

Some environments are able to neutralize the effects of acid rain. This is called the buffering capacity. Chalk and limestone areas are very alkaline and can neutralize acids effectively. The underlying rocks over much of Scandinavia and northern Canada are granite. They are naturally acidic and have a very low buffering capacity. These areas have had the worst damage from acid rain.

Pollution management strategies for acid deposition

There are many ways to try to reduce the damaging effects of **acid deposition**. The first main type of strategy is to alter human activity and to reduce the production of pollutants

The most effective long-term treatment is to reduce the emissions of SO_x and NO_x. This can be achieved in a variety of ways, such as by:

- reducing the demand for electricity
- increasing the use of public transport and reducing the use of private cars
- using alternative energy sources that do not produce nitrate or sulfate gases.

There are a number of international agreements concerning acidification. The first agreement was the 1979 Convention on Long-Range Transboundary Air Pollution. This was important for the clean-up of acid rain in Europe. The 1999 Gothenburg Protocol commits countries to reducing emissions of sulfur dioxide and oxides of nitrogen in an attempt to reduce acidification and other forms of pollution. In North America the 1991 Air Quality Agreement between the USA and Canada focused on reducing the impacts of acid rain and smog.

The second method is to control the release of pollutants. Methods to achieve this include:

- using **limestone scrubbers** in the chimneys of power stations to neutralize the acid
- removing the pollutants before they reach the atmosphere.

Clean-up strategies (the third level of pollution management) include the liming of acidified lakes and soils. By adding powdered limestone to lakes and soils, acidity levels are reduced. This was used in Swedish lakes in the early 1970s and was very cost-effective.

Uncertainties

Increased SO_2 and NO_2 are not the only causes of acidification. Critics of acidification stress the uncertainties. Rainfall is naturally acidic and could cause some of the damage. No single industry/country is the sole emitter of SO_2/NO_x, so it is impossible to pinpoint the polluter, i.e. it is an example of non-point source pollution. Car owners with **catalytic converters** have reduced emissions of NO_x. Different types of coal have variable sulfur content – some coal is "cleaner" than others.

QUESTION PRACTICE

Essay

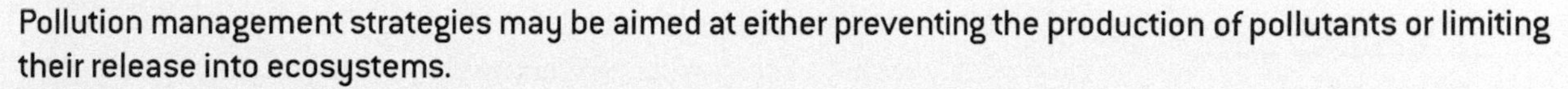

Pollution management strategies may be aimed at either preventing the production of pollutants or limiting their release into ecosystems.

With reference to either acid deposition or eutrophication, **evaluate** the relative efficiency of these two approaches to management. [9]

How do I approach this question?

The command term is "evaluate" which requires you to make an appraisal/judgment by weighing up strengths and weaknesses of both approaches. The guidelines below show certain features that you may include in your answer. The five headings coincide with the criteria given in each of the mark bands (shown in bold below). This guide simply provides some possible inclusions and should not be seen as the only way to answer this question. It outlines the kind of elements that you could include.

Acid deposition:

- **Understanding concepts and terminology** of acid deposition, distinction between prevention and limiting management strategies, NO_x and SO_x, atmospheric emissions, fossil fuels, biodiesel, alternative energy, cement/pulp and paper industries, etc.
- **Breadth** in addressing and linking a range of strategies with their effectiveness in reducing impacts of pollutants, from different sources, on different ecosystems, and their relevance and validity for different societies, etc.
- **Examples** of prevention strategies (changing human activity), such as using alternative energy sources, e.g. solar, hydro, wind, etc., energy-saving technology, transport bans/public transport, paper recycling, and limiting strategies, e.g. scrubbers, catalytic converters, etc.
- **Balanced analysis** of the relative efficiency of the two approaches in reducing impacts on ecosystems, meeting needs of societies, cost and ease of application, etc.
- A **conclusion** that is consistent with, and supported by, analysis and examples given, e.g. "Generally, prevention strategies are more efficient because reducing the use of fossil fuels will simultaneously resolve many other environmental impacts of using this resource which ultimately will become unavailable."

SAMPLE STUDENT ANSWER

Pollution is the addition of a substance/agent at such a rate that it cannot be rendered harmless by the environment and has appreciable negative effects on the ecosystem. Acid deposition, whether wet or dry, is a form of secondary pollution caused when nitrogen and sulfur oxides released from exhaust pipes and coal-burning industries, fall as dry particles or oxidize to form nitric and sulfuric acid and fall as precipitation or acid rain, a secondary pollutant.

▲ Definition of acid rain and scene setting

▲ Sound link to question and distinction of strategies

As a form of pollution, acid deposition can be managed by either reducing/preventing the production of the culprit substances, i.e. NO_x and SO_x, or by limiting the release of these substances despite their production. Each approach can be evaluated.

▲ Prevention – alternative to coal. Explanation and some evaluation

Prevention of deposition can be done in a variety of ways. Firstly, instead of coal, industries can use less-polluting energies, such as renewable, for example, Iceland is majority powered by geothermal energy, which minimizes otherwise large emissions of SO_x, CO_2 and so on.

▲ Some depth to analysis

This has the advantage of being sustainable, but the disadvantage of being expensive and unavailable in several countries. The UK has unreliable/limited geothermal potential, for example. This reduces the efficiency of this solution which can be defined more generally as the convenience of solutions and their relative effectiveness compared to effort required.

▲ Sulfur removal

Further, NO_x and SO_x cannot be produced in the first place if sulfur is removed from gas before combustion, meaning sulfur dioxide is never produced. This solution is more efficieint in some ways, as it would allow continued dependence on oil, merely a further step in the process of its manufacture and use.

▲ Evaluation – expensive

However, in some ways it is inefficient as it would be difficult to implement in less economically developed countries where a desire for economic development and a lack of technology would likely lead to illegal combustion of regular oil, meaning that additional time and effort must be expended to regulate illegal activity.

▼ Oil not coal

On the other hand, there are ways available to simply prevent SO_x and NO_x from being released into the environment even if they are created.

▲ Flue-gas desulfurization (FGD)

Firstly, flue-gas desulfurization allows sulfur to be removed after combustion. This would decrease overall release but be costly and inefficient for many countries without the necessary infrastructure already in place as it is a somewhat recent technology.

▲ Evaluation

If not flue-gas desulfurization, catalytic converters implemented on exhuast pipes of vehicles prevent the release of NO_x and SO_x from the process of engine combustion with chemical sorbents and scrubbers.

▲ Evaluation

This is extremely efficient in allowing vehicle-dependent activity to continue but it would be costly and time consuming to not only instal converters on all vehicles but then check up on them regularly.

Broadly, preventing the production of SO_x and NO_x is the more efficient and sensible solution. Limiting the release of the pollutant is never perfectly efficive, and the possible consequences of acid deposition are considerable. Over half of German spruce trees in areas have been killed. In Sweden 40% or more of wells have a pH that renders them undrinkable. Conifer trees and the ecosystem they support in Norway are at great risk. Their needles are dying as their membranes and fatty tissues break down, hindering photosynthesis. Simple solutions that involve a change in habits and mindset but are sustainable, effective and efficient once implemented include biking, carpooling, hybrid cars and the complete phaseout of coal in favour of renewable energy. Denmark is one of the least impacted Scandinavion countries by acid deposition because of an effective combination of bicycle-heavy transport and wind energy.

▼ Conclusion needs to be based on balanced analysis, and so this does not achieve full marks – this is a mix of conclusion and new materials (analysis)

▼ Impacts of acid rain

▲ Some reference to alternative energy and transport

- Generally extremely well covered.
- Some drift in the conclusion.

This response could have achieved 8/9 marks.

7 CLIMATE CHANGE AND ENERGY PRODUCTION

This chapter examines climate change and energy production. It looks at the range of energy sources that are available and the choices that are made. It also looks at the causes and impacts of climate change, and the schemes for mitigation and adaptation.

You should be able to show:

- ✔ The energy choices that societies have and the security it gives them;
- ✔ The causes and impacts of climate change;
- ✔ The potential for climate change mitigation and adaptation.

7.1 ENERGY CHOICES AND SECURITY

- **Energy security** – having access to clean, reliable and affordable energy sources for cooking, heating, lighting, communications, and production
- **Fossil fuels** – coal, gas, and natural gas, the use of which results in a depletion of natural capital (the reserves)
- **Renewable energy** – energy sources such as solar, tidal, wind, and hydroelectric schemes whose use does not deplete natural capital

▲ **Figure 7.1.1** A "nodding donkey" used in oil extraction

Content link

Look at section 7.2 to assess the relative importance of fossil fuels in global climate change.

You should be able to show the energy choices that societies have and the security these give them

- ✔ Fossil fuels are the main source of energy in most countries.
- ✔ There are other sources of energy that have a lower carbon content.
- ✔ Energy security depends on reliable and affordable energy sources.
- ✔ The choice of energy source depends also on sustainability.
- ✔ Improvements in energy efficiency could reduce energy use.

Energy sources

There is a range of energy resources available to society. Energy resources are usually divided into two main types: **non-renewable energy resources** and **renewable energy resources**.

Non-renewable energy resources cannot be replenished within a similar timescale to that at which they are extracted. Oil and coal are good examples of non-renewable energy resources. They are also called **fossil fuels**. Their use is unsustainable.

In contrast, renewable energy resources are replenished within a similar timescale to that at which they are extracted. Examples include biofuels and hydroelectric power. These can be used in a sustainable way, although they may be used unsustainably, e.g. if the rate of deforestation for fuelwood exceeds the rate of annual renewal.

Fossil fuels contribute to most of humankind's energy supply. The impacts of their production and their emissions vary considerably. Their use is expected to increase to meet global energy demand.

New sources of modern energy

Biofuels are a type of modern energy source. They are made from plants grown today, whereas fossil fuels are formed from plants and animals that died millions of years ago. For decades, Brazil has turned sugar cane into ethanol, and some cars there can run on pure ethanol or use it as an additive to fossil fuels. In 2016, United Airlines announced a new initiative to integrate biofuel into its energy supply in the hope of reducing greenhouse gas emissions by 60%. However, biofuels do not provide as much energy per kg as oil, for example.

Test yourself

7.1 Distinguish between renewable and non-renewable energy sources. [2]

7.2 Identify the type of energy production shown in Figure 7.1.1. [1]

Low carbon energy sources

There are several energy sources that have lower carbon dioxide emissions than fossil fuels. These include hydroelectric power (HEP), solar, wind, tidal, and geothermal. However, there may be carbon emissions in the construction of energy facilities (e.g. HEP schemes). Biofuel is a renewable energy resource but it does emit carbon. It has relatively low set-up costs but production costs may be high. This is because it takes lots of effort to cut and carry the biofuel back to the house or farm. Nevertheless, Brazil has produced biofuels on a large scale for over 40 years.

Patterns and trends in energy consumption

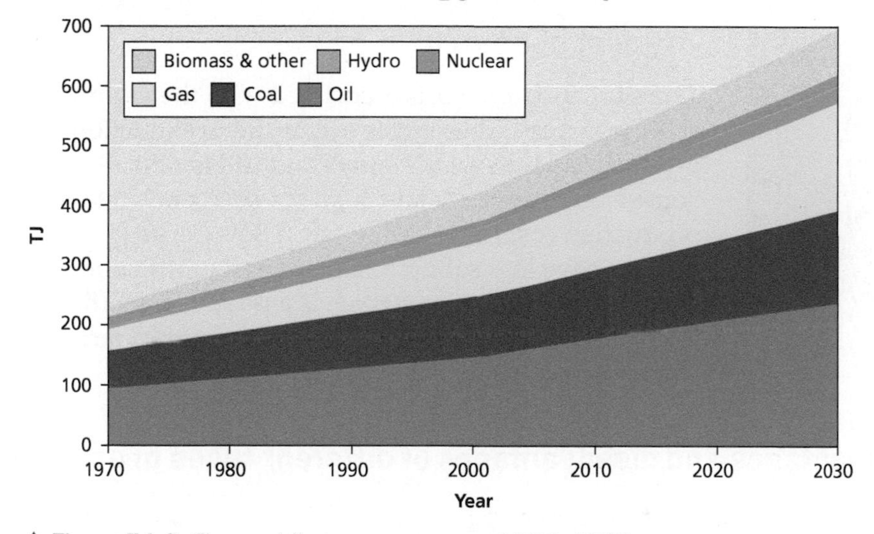

▲ **Figure 7.1.2** The world's energy sources, 1970–2030

Assessment tip

Figure 7.1.2 is an example of a compound line graph. The projected value for biomass in 2030 is about 80TJ (the thickness of the pink line) not 700 TJ (the top of the pink line).

1 TJ = 1000 gigajoules

Overall global oil production increased from around 65 million barrels per day in 1991 to over 90 million barrels per day by 2016. The largest relative increase was in the Middle East and South and Central America, but there were relative falls in Europe/Eurasia and North America, despite an absolute increase in the amount of oil produced.

A report in *The Economist* in 2019 suggested that the global demand for oil and natural gas could increase by 13% by 2030, and that all the major oil companies are expected to expand their output. Exxon Mobil plans to pump 25% more oil and gas in 2025 compared with 2017. To limit global temperature rise to 1.5°C oil and gas production would need to fall by 20% by 2030 and 55% by 2050. It is predicted that in 2030 85% of cars will still use the internal combustion engine, and will continue to depend on fossil fuels.

Nuclear energy provides a relatively small amount of the world's energy. In 1991 it accounted for less than 500 mtoe (million tonnes of

Test yourself

7.3 Estimate the projected growth in consumption of oil and gas between 2019 and 2030. [1]

7.4 Identify the energy source that has the highest absolute increase (projected) for 2020–2030. [1]

7.5 Identify the two energy sources that have the lowest absolute increase (projected) for 2020–2030. [2]

oil equivalent) out of a global total of approximately 8000 mtoe (which is around 6%). By 2016, less than 5% of world energy consumption came from nuclear energy. Nuclear energy peaked in 2006 but then fell around 2011 (possibly reflecting reaction to the Fukushima–Daichii nuclear explosion in Japan).

Changes in demand

The major consumers of energy are the MEDCs, although demand for and use of energy resources by NICs has been increasing rapidly. Energy resources are used in large quantities for manufacturing and transport. The world continues to use fossil fuels—mainly oil, natural gas, and coal—despite the growth in renewable energy sources. This is partly because there are still resources available, the infrastructure is already in place and, in some cases, the energy companies are important sources of revenue for governments, as well as being powerful influencers. The geographical pattern of demand is projected to shift from the OECD region to NICs.

LEDCs and NICs will continue to grow faster than MEDCs, but their consumption remains low by comparison. However, as economic growth rates slow down as economies mature, there may be a small decline in the use of energy.

Concept link

SUSTAINABILITY: Energy emissions account for the major share of greenhouse gas emissions, and attempts to manage global climate change will require more sustainable forms of energy.

▲ **Figure 7.1.3** Peat cutting and afforestation – balancing the climate effect?

The cost of energy

There are different costs associated with the various energy sources. These costs occur due to extraction, production, transport, storage, and utilization. Some energy sources have high set-up costs but low production costs, e.g. solar energy. Some energy sources have low set-up costs but high production costs, e.g. peat burning, which is a relatively inefficient form of energy and releases a large amount of carbon when burned.

Advantages and disadvantages of different types of energy

▼ **Table 7.1.1** The advantages and disadvantages of three different types of energy

Advantages of oil	Disadvantages of oil
• It is a very efficient source of energy. • It can be used as a fuel and to generate electricity. • It is relatively cheap.	• It is a finite resource and will run out. • Burning oil releases carbon dioxide, which is a greenhouse gas. • Oil spills damage ecosystems.
Advantages of hydroelectric power	**Disadvantages of hydroelectric power**
• It is a renewable source of energy so will not run out in the future. • Reservoirs can be multi-purpose and have other benefits such as recreation and fishing. • It allows nations to control their own energy supplies. • Once built, HEP schemes are relatively cheap to run.	• Reservoirs flood habitats and displace people. • Dams act as barriers for species that migrate up and down rivers.
Advantages of nuclear power	**Disadvantages of nuclear power**
• It is a very efficient form of energy requiring only a small amount of uranium to generate a large amount of energy. • There are plentiful supplies of uranium. • It does not produce greenhouse gases when producing electricity. • Over a long timescale the cost of producing electricity reduces.	• Vast amounts of carbon are released during the construction of nuclear power stations. • It is very expensive to build nuclear power stations. • It is very expensive to make nuclear power stations safe (decommissioning) when they have stopped producing energy. • There have been several nuclear disasters, e.g. Chernobyl and Fukushima–Daiichi.

Energy security

According to the analyst Chris Ruppel, the period from 1985 to 2003 was an era of **energy security**, and since 2004 there has been an era of energy insecurity. Ruppel claims that following the energy crisis of 1973 and the Iraq War (1990–91), there were periods of low oil prices and energy security. However, insecurity has increased for many reasons, including:

- increased demand, especially from newly industrializing countries (NICs)
- decreased reserves as supplies are used up
- geopolitical developments: countries with oil resources such as Russia have been able to "flex their economic muscles" in response to the decreasing resources in the Middle East and the North Sea
- global warming and natural disasters such as Hurricane Katrina (2005), which have increased awareness about the misuse of energy resources
- terrorist activity such as in Syria
- the conflict between Russia and Ukraine.

For most consumers, a diversified energy mix offers the most energy security. Depending on a single source, especially from a single supplier, is more likely to lead to energy insecurity.

Test yourself

7.6 Outline one advantage and one disadvantage of biofuels. [2]

7.7 Suggest two reasons why the burning of peat as a fuel has ceased in some countries. [2]

7.8 Briefly **explain** why the burning of fossil fuels is projected to increase by 2030. [3]

Factors that affect the choice of energy sources adopted by different societies

Factors affecting the choice of energy include availability, cost, and environmental, cultural and political factors. Availability affects the type of energy used by societies. The Middle East has vast reserves of oil and consequently it relies on it as a source of energy. In contrast, Japan does not have any oil reserves so it has had to develop other sources of energy, notably nuclear power and hydroelectric power.

- Economic factors affect the type of energy used by societies. Coal is a relatively cheap form of energy and so it is used a lot in developing countries. Nuclear power is a very expensive form of energy so only rich countries can afford it.
- Cultural factors also affect the type of energy used by societies. Many people believe that the risks associated with nuclear power outweigh the benefits. **Fuelwood** is used in many poor countries for cooking and for heating.
- Environmental factors affect the type of energy used by societies. Burning fossil fuels gives off **greenhouse gases** which are linked to **global warming**. Burning coal may also result in **acid rain**. Disposing of nuclear waste is very dangerous. Attempts to extract natural gas from **shale** rocks can lead to earthquakes and **groundwater** pollution.
- Political factors can affect energy sources. For example, Russia has considerable gas and oil reserves, and can choose which countries it wishes to supply. Following the nuclear disaster at Fukushima–Daiichi, a number of governments (e.g. Germany, Switzerland, Spain, and Italy) decided to phase out the development of nuclear power.

Content link

Read section 6.4 to examine the role of coal burning in acid deposition.

Test yourself

7.9 Identify two methods of energy conservation. [1]

7.10 Outline how zero carbon emissions can be achieved. [2]

Energy efficiency and conservation

Energy conservation includes the use of smart meters (for monitoring energy use) and improved environmental/building standards. Improved design could include reduced energy demand due to improved insulation of walls and roofs, triple glazing of windows, larger windows to allow more sunshine/heat in, and energy-efficient appliances. Reduced carbon emissions could be achieved by using more public transport, car sharing or electric-hybrid vehicles that use energy from renewable sources. Increasing energy efficiency and/or conservation contributes towards energy security.

QUESTION PRACTICE

1) a) **Identify** two environmental impacts of hydro-electricity generation in Brazil. [2]

b) The use of renewable resources is not always sustainable due to the activities involved in their production.

Justify this statement for a named source of renewable energy. [7]

Essay

2) Increasing concern for energy security is likely to lead to more sustainable energy choices.

Discuss the validity of this statement, with reference to named countries. [9]

How do I approach this question?

1) a) To gain full marks you need to provide cause and impact. The answer must focus on environmental impacts and can be either positive or negative.

b) You will be awarded 3 marks if the example of the energy source is non-renewable or unnamed. You will be awarded 1 mark for each correct point.

Essay

2) Good answers will include:

- understanding of concepts and terminology—energy security, sustainability, environmental value systems (EVSs), fossil fuels, renewable resources, non-renewable resources
- breadth in addressing and linking different countries' issues of energy security and energy sources—MEDCs, NICs, LEDCs, energy security, sustainability, fossil fuels, renewable resources, non-renewable resources—argument and counter-argument
- named examples (different countries, energy sources, energy security)
- balanced analysis
- a conclusion that is consistent with, and supported by, analysis and examples.

SAMPLE STUDENT ANSWER

▲ Valid environmental point

▼ Adds nothing

▼ Unclear. Increased evaporation would be in the lake area

▼ Adds nothing

1) a) Less fossil fuels will be used, reducing carbon emissions into the atmosphere that can potentially cause global warming. This is a positive effect. Also, this type of energy generation requires the building of dams which can hurt aquatic and terrestrial ecosystems due to increased evaporation in rivers. This is a negative impact.

This response could have achieved 1/2 marks:

b) Electricity can be generated through wind power. Windmills are a source of renewable energy that have been set in place in many countries including Spain, France and the United States. Even though windmills are a source of renewable energy their production is not always sustainable. The production of windmills is costly, and involves the release of various greenhouse gases which aid the enhanced greenhouse effect. The production of windmills also generates inorganic wastes that can be unsustainable if not dealt with correctly. As a result the production of windmills can be unsustainable.
In addition, when windmills are placed they fragment ecosystems which leads to a loss of biodiversity. Lastly, if a windmill malfunctions and catches fire the gases released will damage the surrounding populations and contribute to the enhanced greenhouse effect.

▲ General point

▼ Why is wind renewable?

▲ General point

▲ Valid point

▲ Valid point, but say which wastes

▲ Valid impact on ecosystem/ biodiversity

▼ Unclear point

This response could have achieved 3/7 marks.

- The first two sentences do not answer the question.
- The candidate then makes three valid points "production is costly", "produces wastes" and "impacts on biodiversity".
- The final point about fires is unclear, i.e. what gases are released, how do they harm local populations (people or fauna/flora)?
- The candidate could have scored more had they explained about the use of greenhouse gases produced in the transport of materials, construction of access roads and for maintenance to gain access.

Essay

2) Energy security is very important, especially if a country is using nuclear energy. As the Chernobyl accident proved, nuclear energy can be catastrophic if not properly controlled. Nuclear plants are also prone to terrorist attacks and must be heavily guarded. Countries such as France, Spain and China have shifted from non-renewable resources such as nuclear energy to more renewable energy. The increasing concern for energy security has led to more sustainable energy choices to a large extent.
France and Spain are very similar when it comes to energy. They both have various nuclear plants around the country, and both have implemented windmills as the main source of renewable energy. Spain is looking to close at least 1 of its 5 nuclear power plants in the following 10 years. This is a result of increased fear of the dangers of nuclear energy. Similarly, France recently closed one of its nuclear plants.

▲ General point, focuses on nuclear power

▲ Problems related to nuclear power. Named countries

▼ Overall still quite general; note that very general answers often get about half marks

▲ Wind power as an alternative to nuclear power

▼ All very general

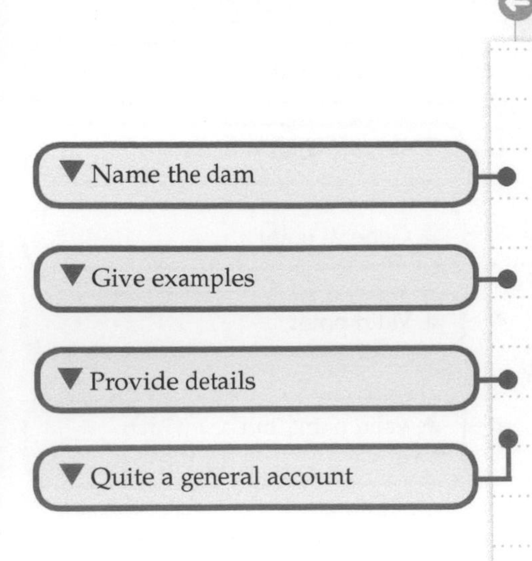

In China, the increasing concern for energy security has led them to find ways of making more sustainable energy. This led the Chinese government to construct the huge dam that could supply hydroelectric energy to many people in the surrounding areas. Moreover major cities in China have started implementing electric cars, to reduce carbon emissions in major cities. Lastly, China has invested in solar energy to shift away from fossil fuels.

In conclusion, the increasing concern for energy security has led to more sustainable energy choices to a large extent.

This response could have achieved 4/9 marks.

- There needs to be greater use of concepts and terminology, e.g. energy security, sustainability, etc.
- The answer looks at four countries, including MEDCs and an NIC. However, the level of detail is limited.
- The analysis is very general and it needs to be more detailed.
- The conclusion should be based on the detailed analysis.

7.2 CLIMATE CHANGE—CAUSES AND IMPACTS

- **Climate** – the state of the atmosphere over a period of not less than 30 years
- **Weather** – the state of the atmosphere over a short timescale (normally less than one week)
- **Positive feedback** – change in a system that leads to increasing movement away from equilibrium and contributes to instability
- **Negative feedback** – change in a system that counteracts any movement away from equilibrium and contributes to stability
- **Ecosystem services** – the benefits that are provided to societies by the environment, e.g. climate regulation, natural resources and health benefits

You should be able to show the causes and impacts of climate change

- ✔ Climate change has been a feature of the Earth's history.
- ✔ Human activity has accelerated climate change.
- ✔ Climate change has many impacts.
- ✔ Climate is the state of the atmosphere over a period of not less than 30 years.
- ✔ Weather and climate are affected by atmospheric and oceanic circulation systems.
- ✔ Human activities are increasing greenhouse gases.
- ✔ The greenhouse effect has led to increased temperatures, increased frequency of extreme events, rising sea levels and an increased potential for changes in climate and weather.
- ✔ The potential for climate change varies from place to place.
- ✔ Feedback mechanisms are involved in global climate change.
- ✔ Different environmental value systems view global climate change differently.
- ✔ There is uncertainty regarding global climate change.

▲ **Figure 7.2.1** Iceland could experience melting glaciers and an increase in agricultural productivity due to climate change

Climate and weather

The term "**climate**" refers to the state of the atmosphere over a period of not less than 30 years. It includes the averages and extremes of temperature, precipitation (rain, snow, dew), humidity, cloud cover, wind, and pressure. The term "**weather**" refers to the state of the atmosphere at a given time—normally we look at it over a period of less than one week. It includes all of the same aspects as climate, but on a much shorter timescale.

Test yourself

7.11 Outline one way in which the people living in the settlement shown in figure 7.2.1 could benefit from global warming and one way in which they may be disadvantaged by global warming. [2]

Weather and climate and the effect of oceanic and atmospheric circulatory systems

There are important variations in the receipt of solar radiation according to latitude and season. The result is an imbalance: excesses or a positive budget in the tropics, deficit or a negative budget in temperate regions and towards the poles. However, neither region is getting progressively hotter or colder. To achieve a balance the horizontal transfer of energy from the equator to the poles takes place by winds and ocean currents. This gives rise to an important second energy budget in the atmosphere—the horizontal transfer between low latitudes and high latitudes to compensate for differences in global insolation.

The excess of net radiation in lower latitudes leads to a transfer of energy towards the poles from tropical latitudes by ocean currents and wind systems. This is in the form of sensible heat (warm air masses/ocean water) and latent heat (atmospheric water vapour, e.g. in hurricanes).

Content link

Look at section 6.1 for a background to the atmosphere as a dynamic system with inputs, outputs, stores and flows.

Assessment tip

The atmosphere is an open energy system receiving energy from both the Sun and the Earth. Although the Earth is very small compared to the Sun, it has an important local effect, as in the case of urban climates. Incoming solar radiation is referred to as insolation.

Oceanic systems

The effect of ocean currents on temperatures depends upon whether the current is cold or warm. Warm currents from equatorial regions raise the temperatures of polar areas (with the aid of prevailing westerly winds). However, the effect is only noticeable in winter. For example, the Gulf Stream transports heat northwards and then eastwards across the North Atlantic. The Gulf Stream is the main reason why north-west Europe has mild winters and relatively cool summers. By contrast, there are other areas that are made colder by ocean currents. Cold currents, such as the Labrador Current off the north-east coast of North America, can reduce summer temperatures, but only if the wind blows from the sea to the land.

Atmospheric systems

The world's major wind systems are largely determined by variations in temperature and pressure.

- Trade winds blow from sub-tropical high-pressure (STHP) belts towards the Equator. Owing to the strength of the STHP, trade winds are regular and predictable.
- Mid-latitude westerlies blow from the STHP belts towards the poles. Those in the southern hemisphere are stronger and more persistent due to the relative lack of large land masses.
- Polar easterlies blow from the polar high-pressure zone towards the mid-latitudes.

Human activities and greenhouse gases

Human activities have led to increased levels of greenhouse gases in the atmosphere, which has caused:

- increased global temperatures
- changes in the frequency and intensity of extreme weather events
- long-term changes in climate and weather
- rising sea level.

The main greenhouse gases include water vapour, CO_2, methane and chlorofluorocarbons (CFCs). Human activities are increasing levels of CO_2, methane and CFCs in the atmosphere. Figure 7.2.2 shows the increase in CO_2 from 1960 to 2012.

Permafrost contains vast amounts of carbon stored as frozen methane or organic materials, unable to decay while frozen. If permafrost continues to melt, some 190 gigatonnes of carbon could be released into the atmosphere by 2200, in turn leading to an increased melting of permafrost, i.e. positive feedback.

Content link

Section 6.1 provides a discussion of the greenhouse effect.

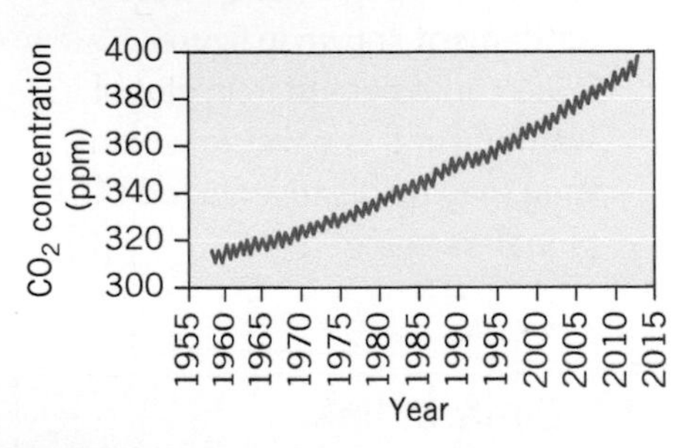

▲ **Figure 7.2.2** Keeling curve to show the change in atmospheric CO_2, 1960–2012

Concept link

EQUILIBRIUM: Human activities have pushed the natural greenhouse effect to such an extent that that it has reached a new equilibrium associated with global climate change.

Test yourself

7.12 Estimate the (a) absolute and (b) relative change in atmospheric CO_2 between 1960 and 2012. [2]

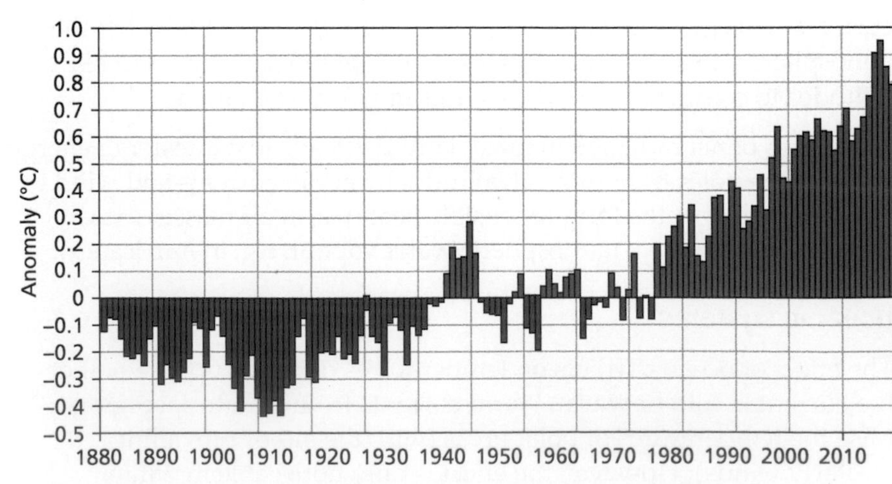

▲ **Figure 7.2.3** Departure of global temperature from the average, 1880–2018

Extremes of climate and weather

Extremes of climate and weather are predicted to occur in many parts of the world. In 2019 the USA experienced a "polar vortex" with temperatures as low as −30°C, while at the same time Australia was experiencing temperatures of over 40°C. The number of deaths from extreme weather is expected to rise.

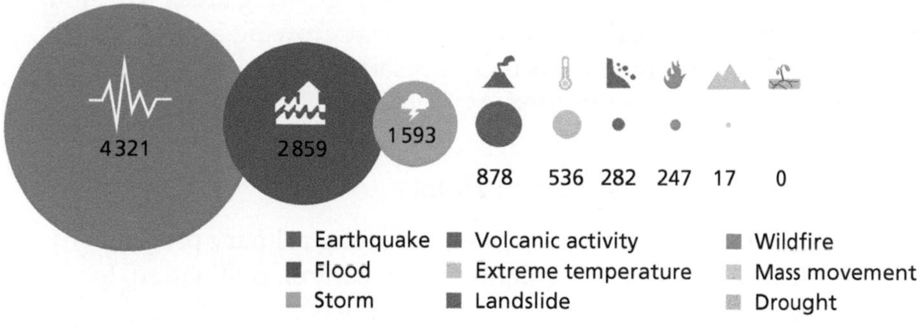

▲ **Figure 7.2.4** Number of deaths per disaster type, 2018

The potential for long-term climate change is extremely high. A report by NASA in 2018 suggests that climate change will continue throughout this century and into the next. Temperatures will continue to rise, the frost-free season (and growing season) in the USA will lengthen, there will be more droughts, heatwaves, and changes in precipitation patterns, and hurricanes will become stronger and more intense.

NASA also suggested that global sea levels will rise by between 30 cm and 120 cm by 2100, which is higher than suggested in Figure 7.2.6. Coastal flooding will occur as global warming leads to thermal expansion of water and melting of glaciers and ice caps. These will contribute to a rise in sea level. This could have many impacts such as increased coastal erosion. The intrusion of salt water could cause the contamination of soils and the decline of agricultural production. Ecosystems could also be affected. Coastal flooding could cause a reduction of mangrove forests and prevent coral reefs from obtaining enough light. Many of the world's megacities are located in low-lying coastal areas, e.g. Shanghai and Tokyo, and these will be vulnerable to coastal flooding and storm surges in a warmer world.

There are many potential impacts of increased mean global temperature. These include some impacts on the natural environment and some impacts on the human environment. Potential effects on the natural environment include the distribution of biomes and changes to weather patterns. Potential effects on the human environment include changes to global agriculture and the spread of tropical diseases.

Changes in water availability

As temperatures rise, water is going to become scarce in some areas. Cape Town almost lost all of its water supply in early 2018 (referred to as "Day Zero"), and there have been major droughts in the USA, Australia, and southern Africa. The Middle East and North Africa are likely to experience an increase in water stress in the coming decades.

Changes in distribution of biomes

One impact of increased mean global temperatures is **biome** shifting. This means that biomes could shift by latitude or by altitude. Alpine (mountain) species are particularly at risk, because zonation will move up

Test yourself

7.13 Describe the variations in global temperature between 1880 and 2018. [3]

7.14 Compare the average prediction for temperature with that of the worst-case scenario. [2]

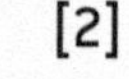

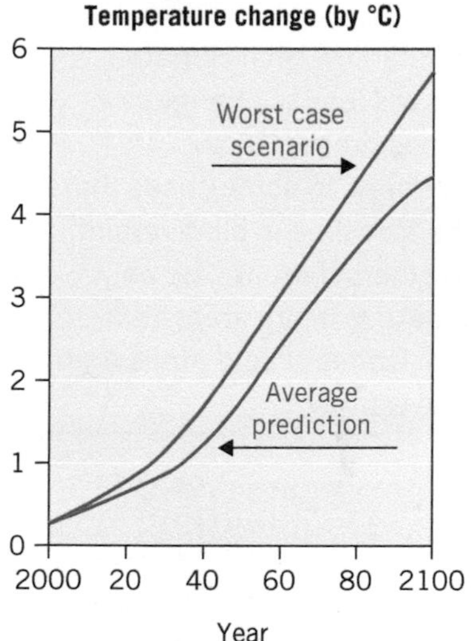

▲ **Figure 7.2.5** Predicted temperature changes, 2000–2100

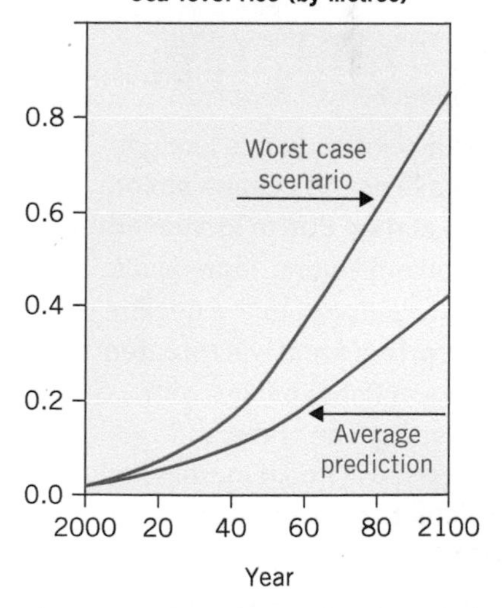

▲ **Figure 7.2.6** Predicted sea level rise (metres), 2000–2100

Test yourself

7.15 Estimate the total number of deaths that were caused by weather events. [2]

7.16 Suggest why this figure may not be accurate. [2]

Test yourself

7.17 Suggest why the number of deaths due to drought is difficult to measure. [2]

7.18 Describe the projected changes in sea level, 2000–2100. [2]

Concept link

BIODIVERSITY: Global climate change will lead to changes in species present in an environment. In some cases there may be a decline in biodiversity, and in others there will be an increase, e.g. in dry areas that become wetter or cold areas that become warmer.

Assessment tip

Ecosystems provide vital services for humanity, e.g. climate regulation and flood control. These will change as global climate changes.

Assessment tip

Ocean acidification is likely to increase as the oceans absorb more carbon due to increased carbon emissions. More acidic oceans will lead to the decline of coral reefs and will threaten stocks of fish. Species with shells, such as crabs and lobsters, will find it increasingly difficult to grow shells in acidic waters.

This is not caused by climate change but is caused by an increase in CO_2—it is a parallel impact of increased levels of CO_2.

the mountain. There may be some loss of **species diversity** as species that are unable to adapt or have limited scope for shifting may become extinct. Animals can migrate but plants shift their range more slowly.

Change in location of crop-growing areas

There may be an increase in primary **productivity** as a result of global warming. Shifting biomes mean crop growing areas will shift. Scientists project a northward shift of wheat-growing areas in North America. In some areas a shortage of resources could lead to increased conflict, e.g. over water or food.

Changes to ecosystem services

There are many **ecosystem services**, including primary productivity, nutrient cycling, climate regulation, flood control, pollination, food, and water.

As climate changes, some ecosystems will alter and be unable to provide the services they do currently. One suggestion is that the tropical rainforest may become a savanna, and that would lead to a massive decline in the production of oxygen and carbon storage and provision of rainfall.

Changes in weather patterns

Researchers have considered the effect of a doubling of CO_2 from the 270 ppm (parts per million) in pre-industrial times to 540 ppm. They believe this will lead to temperatures rising by about 2°C. Increased warming is likely to be greater at the poles rather than at the Equator. There are also likely to be changes to prevailing winds and to precipitation patterns. Continental areas, i.e. areas that are far from a coastal/maritime influence (e.g. the great plains of the USA or central Asia), will become drier.

Impact on human health

Global warming can have various impacts on human health. An increase in the amount of stagnant water may lead to more mosquitoes and they could cause diseases. Changes in the distribution of organisms may cause new diseases to occur in an area. The risk of heat stroke will increase in a warmer world. Salt water intrusion and a decline in food production could lead to hunger and malnutrition.

Negative and positive feedback mechanisms

There are many **feedback mechanisms** (i.e. knock-on effects or impacts) that are associated with an increase in mean global temperatures. Some of these involve **positive feedback** and some involve **negative feedback**.

Negative feedback may involve increased evaporation in tropical latitudes, leading to increased snowfall on the polar ice caps. The surface of snow and ice is very reflective so the albedo is increased. Increased reflectivity reduces the amount of solar radiation received and so lowers temperatures.

In contrast, positive feedback may involve greater thawing of permafrost, leading to an escalation in methane levels, which increases the mean global temperature. As methane is a greenhouse gas, it has the potential to increase temperatures, thereby reinforcing the rise in temperature. This is shown in Figure 7.2.8.

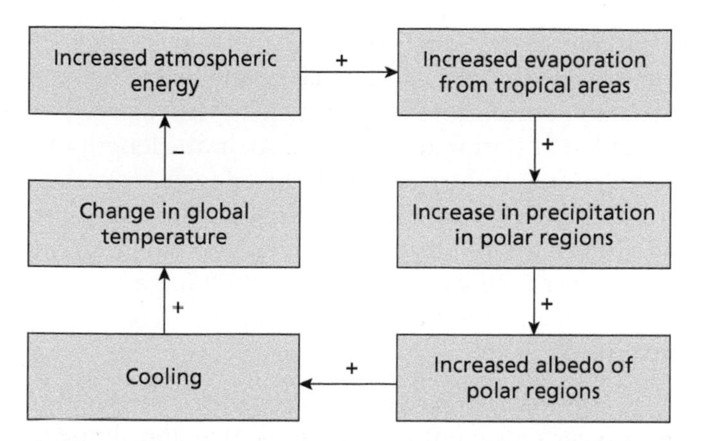

▲ **Figure 7.2.7** Negative feedback and global climate change

Concept link 

EQUILIBRIUM: Negative feedback maintains the equilibrium in natural systems.

Any feedback mechanisms associated with global warming may involve very long time lags.

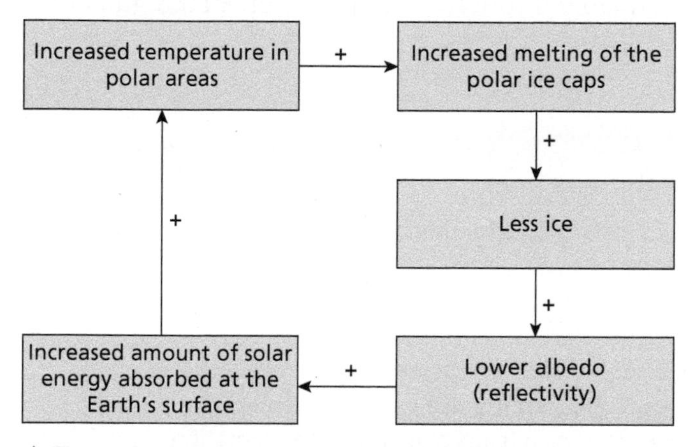

▲ **Figure 7.2.8** Positive feedback in polar areas

Concept link 

EQUILIBRIUM: Positive feedback mechanisms cause natural systems to deviate away from pre-existing equilibria.

Test yourself

7.19 Outline the difference between negative and positive feedback. [2]

Environmental value systems (EVSs) and global warming

Climate change is the subject of debate, and some people state that it is not happening. Some critics state that even if humans are causing climate change, the Earth will correct itself. This is in part the view stated in the Gaia hypothesis, i.e. that the Earth is a self-regulating entity.

Those who believe that human-induced global warming is real

Some people claim that scientific data proves that the climate is warming. Scientific data shows that carbon dioxide levels and greenhouse gas levels are increasing. Data from a variety of sources and times indicates warming. These people claim that human activities and/or fossil fuel combustion are known to increase carbon dioxide/ greenhouse gas levels. They also insist that carbon dioxide and other greenhouse gases are known to impact global temperatures. Therefore, it is likely that human activities are resulting in global climate change. Moreover, the rapid rate of increase in CO_2 implies a human link.

Those who dismiss global warming

Other people claim that natural fluctuations occur, so changes in climate could still be a short-term trend. They argue that the only technologically verifiable data has been collected over a short period

Assessment tip

Global climate models

These are very complex as they cover the whole of the Earth (and beyond) and try to integrate atmospheric, oceanic, terrestrial, lithospheric, hydrological, and human systems. In addition, the further forward in time scientists try to predict, the less accurate their estimates are.

Much depends on whether humans will make the effort to control climate change or whether they will just carry on regardless of the consequences.

There are also natural events, such as volcanic eruptions, which may have some impact on climate change.

of time. They also state that other aspects of climate change are not fully understood. They argue that the climate has changed in the past. This is in part due to natural fluctuations such as Milankovitch cycles (variations in the Earth's orbit around the Sun, in the length of seasons, and in the orientation of the poles towards or away from the Sun). Moreover, current carbon dioxide levels and global temperature fluctuations are moderate compared to geologic history. Therefore, it is not conclusive that humans are causing global climate change.

Other views on climate change

Ecocentrics stress the importance of nature for humanity. They see the reduction of greenhouse gas emissions as vital for controlling climate change, even if this led to reduced economic growth.

In contrast, technocentrics assume that people could find solutions to the problem of global warming, such as geoengineering.

Finally, anthropocentrics hold the view that people must manage global warming sustainably through taxes, environmental regulation, and legislation.

Test yourself

7.20 State one way in which (a) a technocentrist (b) an anthropocentrist and (c) an ecocentrist would view climate change. [3]

QUESTION PRACTICE

a) Outline how CO_2 emissions can cause a change in the global climate. [2]

b) Identify two possible reasons for the projected change in CO_2 emissions for China. [2]

c) Evaluate the potential impacts of climate change on Iceland. [5]

How do I approach these questions?

a) You will get 1 mark for each correct reason that outlines how CO_2 emissions can lead to a change in global climate, e.g. emissions (lead to higher concentration) of CO_2 which is a greenhouse gas; this causes a greater absorption of infrared/heat radiation and rise in global temperature, which leads to increased evaporation, changing winds, changing rainfall patterns, drought, extreme weather events, etc.

b) You will receive 1 mark for each correct reason identified, up to a maximum of 2, e.g. growing number of fossil-fuelled vehicles; rapidly developing economy; increased standard of living; increase in fossil-fuelled power plants; increased industrialization; increase in intensive or mechanized farming systems; burning of forests to clear land for agriculture.

c) There is a maximum of 3 marks for either positive or negative impacts. You should explain impacts, e.g. why each is a benefit or a problem. There is a maximum of 4 marks if there is no evaluation/opinion/appraisal.

SAMPLE STUDENT ANSWERS

▲ Valid point – greenhouse gases

▼ This is the opposite of the first point – less CO_2 and less global warming

a) CO_2 emissions may cause a change in the global climate as increase in CO_2 will increase the global warming because more greenhouse gases will be released and if there is less CO_2 then there will be less global warming as less greenhouse gases will be released.

This response could have achieved 1/2 marks.

b) One possible reason is the amount of fossil fuel burning vehicles that will be present in 2030. Another possible reason could be the amount of people present in China using electricity through the burning of coal

▲ Valid point

▲ Valid point

This response could have achieved 2/2 marks.

Identifies two correct answers—fossil-fuelled vehicles and burning coal in power stations.

c) Climate change, and particularly global warming, is likely to have a number of impacts on Iceland, both beneficial and detrimental.

▲ Generic introduction

Beneficial

- Temperature would increase and more crops would be able to be grown – increasing agricultural output.
- Temperature increase would increase vegetation growing (higher productivity and biomass produced) increasing vegetation cover and decreasing erosion.
- Number of ecosystems may be eroded, increasing biodiversity.
- Higher temperature would reduce heating thus reduce emissions from energy production.
- Productivity is likely to increase – increasing vegetation cover and thus increasing ecosystem diversity/biodiversity and reducing erosion.

▲ Cause and effect

▲ Vegetation and erosion

▲ Max of 3 for positive impacts achieved

▼ Valid but no credit

Detrimental/negative impacts

- Increased temperature may cause sea level rise – leading to flooding in some areas of the island.
- Some ice sheets may melt – reducing range of ecosystems and decreasing tourist attractions – economic and social impact.
- The distribution/types of ecosystem may change and endemic species may become extinct/biodiversity may be lowered.

▲ Valid negative impact

▼ Max of 4 reached, no credit given

Though climate change may have a number of positive impacts, the way in which the environment in and around the island may respond is unknown, and they may in fact be greatly damaging environmentally, socially and economically particularly regarding rare/vulnerable species and the economic reliance on tourism.

▲ Good point

▲ Valid evaluation

This response could have achieved 5/5 marks.

7.3 CLIMATE CHANGE—MITIGATION AND ADAPTATION

- **Mitigation** – in respect of climate change, this relates to attempts to reduce the causes of climate change
- **Adaptation** – in respect of climate change, this relates to attempts to manage the impacts of climate change
- **Carbon capture and storage (CCS)** – capture of CO_2 on emission (from power stations) and storage deep underground, for example
- **Geoengineering** – large-scale engineering scheme that alters natural processes
- **Biomass** – living matter formed of biological molecules
- **UNREDD** – United Nations Initiative on Reducing Emissions from Deforestation and Forest Degradation in low-income countries.

You should be able to show the potential for climate change mitigation and adaptation

- ✔ Mitigation refers to attempts to prevent global climate change, whereas adaptation refers to measures to adapt to climate change.
- ✔ Mitigation strategies include reduced energy consumption, reduced emissions of nitrogen and methane and increased use of alternative energy and geoengineering.
- ✔ Mitigation includes CO_2 removal, using biomass as an alternative energy source and using carbon capture and storage.
- ✔ Adaptation refers to living with climate change, e.g. flood defences, vaccination programmes, desalinization, and alternative crops.
- ✔ Adaptation varies from place to place.
- ✔ International cooperation may help develop schemes for mitigation and adaptation.

▲ **Figure 7.3.1** The Thames Barrier which protects London from tidal flooding. Rising sea levels will mean the Barrier is closed more frequently

Assessment tip

Some climate change "solutions" may be an example of both mitigation and adaptation. Afforestation is an example of mitigation (use as a renewable energy source with carbon capture to reduce the emissions of carbon) *and* a form of adaptation (to store increased atmospheric carbon).

Test yourself

7.21 Does figure 7.3.1 show climate mitigation or adaptation? **Justify** your answer. [2]

Mitigation

Mitigation refers to attempts to reduce the causes of climate change. Some of the methods are shown in table 7.3.1.

▼ **Table 7.3.1** Methods of climate change mitigation

National and international methods	Individual methods
• Control the amount of atmospheric pollution • Geoengineering • Develop carbon capture schemes • Develop renewable energy sources • Ocean fertilization • Incentives/campaigns/education to promote individual methods	• Use public transport • Use locally produced foods • Use energy-efficient products • Turn off appliances when not in use • Reduce heating by insulating buildings • Use double- or triple-glazed windows

Mitigation strategies to reduce GHGs

Reduction of energy consumption

Energy consumption could be reduced in many ways, e.g. by using public transport, using locally produced foods, turning off appliances when not in use, becoming vegetarian/vegan, installing better insulation in homes.

Reduction of emissions of oxides of nitrogen and methane from agriculture

In a report in *Science* in 2018, the authors argued that the biggest single way to reduce personal environmental impact is to avoid meat and dairy products. A diet rich in meat and dairy products is harmful to the environment as cattle are a major source of methane.

Use of alternative energy sources

The use of alternative energy sources would reduce emissions of carbon dioxide. These alternative energy sources include hydroelectric power and solar power. They do not produce carbon dioxide when they are operating although carbon dioxide is released during the construction of the facilities. However, only certain places have the potential to produce alternative energy supplies.

Geoengineering

Some scientists have suggested using the sulfate aerosol particles in the air to dim the incoming sunlight and thereby cool the planet to offset the warming effects of carbon dioxide. Another idea is to place giant mirrors in space to deflect some of the incoming solar radiation. These are radical, and perhaps unworkable, ideas.

Concept link 

ENVIRONMENTAL VALUE SYSTEMS: Some forms of climate change management are technocentric (such as forms of geoengineering) whereas others are ecocentric (such as afforestation and the use of open spaces).

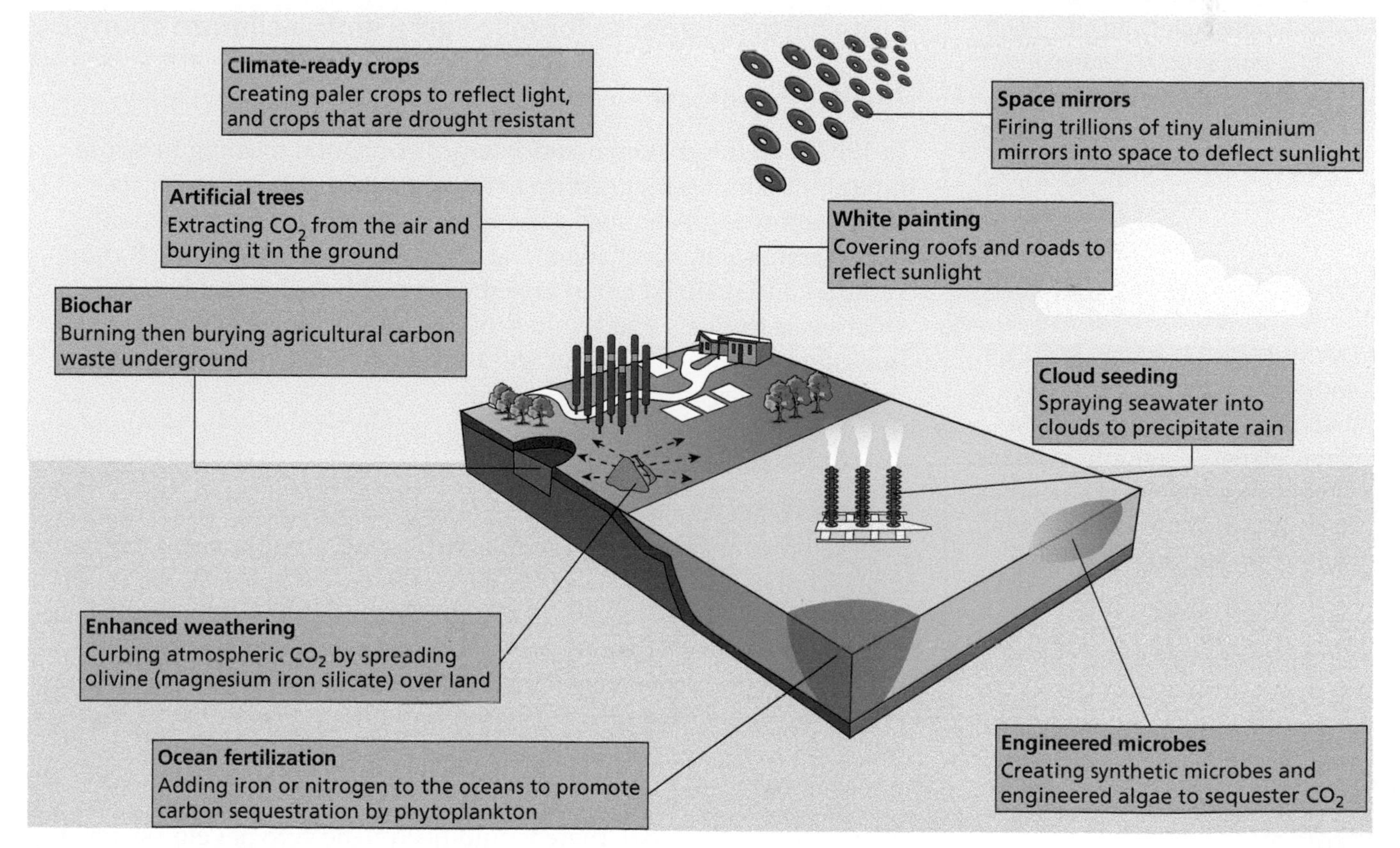

▲ **Figure 7.3.2** Geoengineering strategies

Mitigation strategies for carbon dioxide removal

Protecting and enhancing carbon sinks

The United Nations Initiative on Reducing Emissions from Deforestation and Forest Degradation in developing countries **(UNREDD)** was launched in 2008 and aims to improve conservation and sustainable use of forests and increase the amount of carbon stored in forests.

Using biomass as a fuel source

Biomass is potentially a renewable source of energy although it emits greenhouse gases. It is not a very efficient form of energy, but if coupled with **carbon capture** it could be renewable and clean.

Carbon capture and storage (CCS)

Currently, when fossil fuels are burned, the CO_2 enters the atmosphere, where it may reside for decades or centuries. A potential solution is to capture the CO_2 instead of allowing it to accumulate in the atmosphere.

>> Assessment tip

CCS

Two main ways to do this have been proposed:

- Capture the CO_2 at the site where it is produced (the power plant) and then store it underground in a geologic deposit (for example an abandoned oil reservoir).
- Allow the CO_2 to enter the atmosphere but then remove it using specially designed removal processes (for example collecting the CO_2 with special chemical sorbents that attract the CO_2). This approach is called "direct air capture" of CO_2.

Ocean fertilization

Carbon dioxide absorption can be increased by fertilizing the ocean with compounds of iron, nitrogen, and phosphorus. This introduces nutrients to the upper layer of the oceans, increases marine food production, and removes carbon dioxide from the atmosphere. In some cases, it may trigger an algal bloom. The algae trap carbon dioxide and sink to the ocean floor. In some locations, upwelling currents bring nutrients to the surface, for example off the coast of Peru. These support large-scale fisheries and help to absorb and store carbon.

Long-term prospects for managing global climate change

The Kyoto Protocol

In 1997 at an international and intergovernmental meeting in Kyoto, Japan, 183 countries signed up to an agreement that called for the stabilization of greenhouse gas emissions at safe levels that would avoid serious climate change. The agreement is known as the Kyoto Protocol and it aimed to cut greenhouse gas emissions by 5% of their 1990 levels by 2012. The Kyoto Protocol came into force in 2005 and was due to expire in 2012 but was extended to 2015, when the Paris Agreement was adopted.

Concept link 

ENVIRONMENTAL VALUE SYSTEMS: Anthropocentric EVSs suggest that humans must manage the global climate system sustainably through regulation and legislation. The Kyoto Protocol and the Paris Agreement are good examples of this.

The Paris Agreement, 2015

The 2015 UN Climate Change Conference was held in Paris. France was taken as an example of an MEDC that had decarbonized its energy production—it generates over 90% of its energy from nuclear power, hydroelectric power, and wind energy. The conference resulted in the Paris Agreement on the reduction of climate change. One hundred and seventy-four countries signed the agreement. The key objective is to limit global warming to 2°C compared with pre-industrial levels. It also seeks for zero net anthropogenic greenhouse gas emissions between 2050 and 2100.

To achieve a 1.5°C goal, there would need to be zero net emissions by 2030–50.

Unlike the Kyoto Protocol, there are no country-specific goals and there is no detailed timetable for achieving the goals. Countries are expected to reduce their carbon usage "as soon as possible". However, there is no mechanism to force a country to set a specific target, nor is there any measure to penalize countries if their targets are not met.

The 2015 conference wanted to achieve a binding and universal agreement on climate change from all of the world's countries. The USA and China both agreed to limit greenhouse emissions.

Assessment tip

A report in 2018, "Trajectories of the Earth System in the Anthropocene", suggested that the Earth is warming between 2.5°C and 4.5°C, a situation which could cause the Earth to release more carbon rather than absorb it.

Adaptation strategies in response to climate change

Adaptation strategies can be used to reduce adverse effects and maximize any positive effects of climate change. Examples of adaptations include flood defences, vaccination programmes, desalinization plants, and planting of crops in previously unsuitable climates.

Adaptation refers to the initiatives and measures to reduce the vulnerability of human and natural systems to climate change.

There are many problems related to climate change and many possible ways of adapting to them. Some of these are shown in table 7.3.2.

Test yourself

7.22 Outline how increased reflectivity will reduce global warming. [2]

7.23 Explain how afforestation will decrease global warming. [2]

7.24 Outline how ocean fertilization occurs and why it reduces global warming. [3]

7.25 Identify whether the Kyoto Protocol is an example of climate adaptation or mitigation. Justify your answer. [1]

7.26 State by how much the Earth's temperature will rise according to Hothouse Earth. [1]

▼ **Table 7.3.2** Risks of climate change and possible adaptation strategies

Climate change risks	Potential adaptation strategies
• Flooding	• Early warning systems
• Disease	• Emergency shelters
• Sea level rise	• New forms of agriculture
• Contaminated water	• Genetic engineering/high-yielding varieties (HYVs)
• Dehydration	• Irrigation
• Drought	• Sea walls
• Famine/food shortages	• Mosquito nets
• Over-heating	• Desalination
	• Migration

It is possible to reduce human emissions of greenhouse gases substantially. The technologies are within reach, and measures such as energy efficiency, low-carbon electricity, and fuel switching (for example electrification of buildings and vehicles) are all possible. Nevertheless, even with these, CO_2 will continue to increase for several decades. By the time the oceans stop warming they are likely to add a further 0.6°C to global temperatures. Thus, as well as trying to mitigate climate change, humanity will need to adapt to climate change.

Test yourself

7.27 Identify the adaptation strategies that are likely to deal with (a) sea level rise and (b) drought. [2]

Adaptive capacity varies from place to place and can depend on financial and technological resources. MEDCs can provide economic and technological support to MICs and LEDCs.

In agriculture, crop varieties must be made more resilient to higher temperatures and more frequent floods and droughts. Cities need to be protected against rising ocean levels and the greater likelihood of storm surges and flooding. The geographic range of some diseases, such as malaria, will spread as temperatures rise, so more widespread vaccination programmes will be needed. Demand for water will increase in some areas, and so more desalination plants will be required. These are expensive and some LEDCs may struggle to meet the demand for fresh water.

Test yourself

7.28 Suggest why adaptation capacity varies from place to place. [2]

7.29 Outline some problems of trying to develop efficient irrigation and water management schemes. [2]

International efforts to address climate change

Intergovernmental Panel on Climate Change (IPCC)

The IPCC is the United Nations' unit for assessing the science related to climate change. It was created to provide policymakers with regular scientific assessments on climate change, its implications and potential risks, as well as putting forward adaptation and mitigation options. It was created by the United Nations Environment Programme and the World Meteorological Organization in 1988. It has 195 member countries. The IPCC has three working groups (WGs). WG I examines the Physical Science Basis, WG II looks at Impacts, Adaptation and Vulnerability, and WG III examines Mitigation of Climate Change.

United Nations Framework Convention on Climate Change (UNFCCC)

The UNFCCC's objective is to "stabilize greenhouse gas concentrations in the atmosphere at a level that would prevent dangerous anthropocentric interference with the climate system". The UNFCCC has 197 member parties. The parties have met annually since 1995. In 1997 they established legally binding targets for MEDCs at the Kyoto meeting. While in Paris, in 2015, they established a target of reducing temperature increases to 1.5°C (originally this was 2°C).

National Adaptation Programmes of Action (NAPAs)

A NAPA is a plan submitted by an LEDC to the UNFCCC to outline the country's perception of its urgent and immediate needs to adapt to climate change. Completing a NAPA makes LEDCs eligible for project funding from the Global Environment Facility (GEF).

Test yourself

7.30 Suggest why Nepal's forest resources are under threat. [2]

The NAP process

The National Adaptation Plan (NAP) process helps developing countries conduct comprehensive medium- and long-term planning to reduce vulnerability to climate change and facilitates integration of climate change adaptation in different sectors. NAPs are not linked to a funding source, but technical advice and support are available from the GEF. Table 7.3.3 outlines some of the needs identified in Nepal's NAP process.

▼ **Table 7.3.3** Part of Nepal's adaptation pathways

Theme	Adaptation pathway
Agriculture and food security	• Development of efficient irrigation and water management systems
Forests and biodiversity	• Determine the vulnerability of forest ecosystems, forest communities, and society • Monitor to determine the state of the forest and identify when critical thresholds are reached
Water resources and energy	• Analyse climate change trends and future scenarios for water resources and energy
Climate-induced disasters	• Enhancing the early warning systems for climatic hazards throughout the country
Public health and WASH	• Reaching the unreached and the most vulnerable populations and settlements with health services
Tourism, natural and cultural heritage	• Analyse current and future impacts of climate change on tourism, including the economic impact
Urban settlements and infrastructure	• Enforce land-use planning and bylaws to reduce construction in highly exposed areas such as floodplains and landslide-prone areas
Gender and social inclusion (marginalized groups)	• Addressing resource access issues related to forests, water, and energy for women and marginalized groups
Livelihoods	• Support diversification of livelihoods, including non-farm-based strategies
Governance	• Establish climate change sections in all development ministries

Source: *MoFE, 2018. Nepal's National Adaptation Plan (NAP) Process: Reflecting on lessons learned and the way forward. Ministry of Forests and Environment (MoFE), Government of Nepal, the NAP Global Network, Action on Climate Today (ACT) and Practical Action Nepal. Retrieved from http://napglobalnetwork.org/*

QUESTION PRACTICE

1) a) **Identify** one reduction strategy that the United States might use to achieve its projected change in CO_2 emissions. [1]

b) **Explain** how the ability to implement mitigation and adaptation strategies may vary from one country to another. [4]

Essay

2) **To what extent** do anthropocentric value systems dominate the international efforts to address climate change? [9]

How do I approach these questions?

1) a) You should show how your suggested strategy will lead to a reduction in CO_2 emissions.

b) You will receive 1 mark for each correct explanation, up to a maximum of 4.

Essay

2) Good answers are likely to:

- show understanding of concepts and terminology, e.g. anthropocentric/technocentric/ecocentric values, sustainability, climate change, global warming, carbon emission, international NGOs, international agreements/protocols, mitigation, adaptation, MEDCs/LEDCs
- show breadth in addressing and linking international strategies addressing climate change with relevant environmental value systems (EVS), e.g. anthropocentric with carbon storage, alternative energies, flood defences, e.g. ecocentric with afforestation, energy reduction, reduced consumerism
- include examples of international strategies, e.g. Kyoto protocol, Paris Agreement, REDD (Reduced Emissions from Deforestation & Degradation), and range of strategies employed internationally e.g. desalinization in areas of water scarcity, flood defences in coastal regions
- show balanced analysis of the extent to which international efforts are dominated by anthropocentric values, acknowledging relevant counter-arguments/alternative viewpoints
- include a conclusion that is consistent with, and supported by, analysis and examples given.

SAMPLE STUDENT ANSWER

1) a) The United States may develop wind farms in order to generate clean electricity.

▲ Valid suggestion

This response could have achieved 1/1 marks.

b) Every country has a different government that adheres to different needs so placing an international expectation on climate change won't work. Some countries are fully invested in stopping climate change, while other countries aren't convinced it even exists.

▼ Not enough

▲ Valid point

This response could have achieved 1/4 marks.

Could have referred to: political will, pressure from public; other, more immediate problems e.g. civil war; finance may be a problem for some countries; some countries may be at increased risk due to location (islands, low-lying coastal areas)

Essay

2) In our world today, there is a variety of environmental value systems which shape the way in which international efforts view climate change. Climate change can be examined as the nation's changing weather patterns.

An anthropocentric viewpoint offers the concept that humans can solve environmental issues such as climate change. In regard to climate change, the anthropocentrism view can be examined through legislation, taxes and governmental viewpoints and actions into how to manage and reduce the impacts of climate change. An example of an anthropocentric approach which impacts international efforts to reduce climate change is the Kyoto Protocol. The Kyoto Protocol is an agreement signed by 153 countries which came into place on 11th December 1997. The aim of this was to reduce emissions which are negatively contributing to climate change and can be examined as an anthropocentric viewpoint as it was signed by government officials and has the aim of humans reducing the issue of climate change through lowering human emissions. This is seen as international as many countries signed it. The Protocol introduced the aim that each country had limits to the amount of CO_2 they were able to emit in attempts to address climate change. This has successfully worked as the amount of CO_2 has decreased globally. However, illegal trading has occurred in LEDCs and larger time-scales will be needed to reach their goals leading to a decrease in emissions. Anthropocentric actions such as the UNEP, the Rio Summit and the international panel of climate change are all anthropocentric viewpoints which dominate international efforts to address climate change. Thus, it can be examined that to a great extent anthropocentric viewpoints dominate ways to address climate change.

In opposition the technocentric viewpoint can dominate international efforts in regard to addressing climate change. A technocentric viewpoint is one which believes that technology can solve issues regarding climate change. In Masdar City in Dubai, the town is fully functional on solar energy, a technocentric approach to reducing the impacts of climate change.

▼ Scene setting

▲ Range of strategies identified

▲ Covers the anthropocentric view with some supporting details

▼ Should be 192 countries

▲ Some detail on Kyoto Protocol

▲ Some evaluation – success but illegal emissions

▲ Different agreements

▲ Balanced account – begins to look at alternative viewpoints

▲ Masdar City and solar energy example

In Masdar they have a recreation centre which half yearly hold seminars for people from around the world to illustrate ways in which technology can influence ways to address climate change. Moreover, through the BBC producing a documentary on Masdar City, the programme and use of solar has become internationally known in a way to address climate change through technology. Moreover, technocentrics have created wind power, geothermal energy and hydroelectric power which is used globally to address climate change. Also technocentrists have created adaptation such as air conditioning to address climate change. Thus it can be examined that technocentrists influence the international efforts on climate change.

In conclusion, although technocentrics can influence international effects/efforts on climate change through international agreements, talks and legislation the anthropocentric viewpoint dominates.

▲ Poorly expressed, but a good example of a strategy to help more people become aware of new technological innovations

▲ Range of ways in which technocentrics influence climate change

▼ Very brief conclusion

This response could have achieved 7/9 marks.

8 HUMAN SYSTEMS AND RESOURCE USE

This unit examines the main features of human population change. Human use of resources and the production of waste materials are considered. Finally, the question of whether the Earth is over-populated is discussed.

You should be able to show:

- ✔ how human population dynamics vary;
- ✔ how resource use in society varies;
- ✔ the range of solid domestic waste and options for its disposal;
- ✔ how the ecological footprints of societies impact on the carrying capacity of the environment.

8.1 HUMAN POPULATION DYNAMICS

- **Crude birth rate** – the number of live births per 1,000 population per year
- **Crude death rate** – the number of deaths per 1,000 population per year
- **Total fertility rate** – the average number of births per woman of childbearing age
- **Doubling time** – the time (in years) for a population to double in size
- **Demographic transition model** – a model that describes how countries change from having high birth and death rates to eventually having low birth and death rates (although in the most developed countries death rates may rise in the final stages, due to an aging population)
- **Natural increase** – the increase resulting from the crude birth rate being higher than the crude death rate
- **Population policies** – the strategies introduced by governments to try to increase/decrease population size in a country

You should be able to show how human population dynamics vary

- ✔ There are many indicators of human population change, including birth rates, death rates, total fertility rate, and natural increase.
- ✔ Global population growth was rapid in the 20th century but it is slowing down.
- ✔ Population growth puts a strain on the Earth's resources.
- ✔ Population composition is often shown by the use of age–sex pyramids.
- ✔ The demographic transition model shows how birth and death rates change with the level of development.
- ✔ Many factors affect population growth, including economic, social, and political factors.
- ✔ Increasingly, governments and non-government organizations are influencing population policies.

Demographic tools

Crude birth rate (CBR)

The **crude birth rate** (CBR) is defined as the number of live births per thousand people in a population. The CBR is easy to calculate and the data are readily available. For example:

$$\text{Crude birth rate (CBR)} = \frac{\text{Total no. of births}}{\text{Total population}} \times 1000$$

In China in 2018 there were 15.23 million births out of a population of 1,415 million. This gives a birth rate of 10.94‰.

However, the crude birth rate does not take into account the age and sex structure of the population.

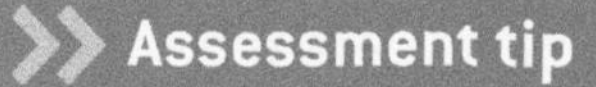

Assessment tip

Birth rates and death rates are expressed in rates per thousand (‰).

Population mortality

The **crude death rate** (CDR) is the number of deaths per thousand people in a population.

$$\text{Crude death rate (CDR)} = \frac{\text{Total number of deaths}}{\text{Total population}} \times 1000$$

In China in 2018 there were 9.93 million deaths, giving it a crude death rate of 7.13 ‰. However, the CDR is a poor indicator of mortality trends—populations with a large number of aged, as in most MEDCs, will have a higher CDR than countries with more youthful populations.

Assessment tip

Many MEDCs have higher death rates than LEDCs because they have aging populations (i.e. a higher proportion of elderly people).

Total fertility rate (TFR)

The **total fertility rate** is the average number of births per women of childbearing age. In 2018 China had a TFR of 1.6 children for every woman (i.e. every 100 women would give birth to 160 children).

In general, the highest fertility rates are found among the poorest countries, and few LEDCs have made the transition from high birth rates to low birth rates. By contrast, the birth rates have reduced in most MEDCs.

Natural increase rate (NIR)

Natural increase is the increase in population as a result of the birth rate exceeding the death rate. In China in 2018 the CBR was 10.94‰ and the CDR 7.13‰, giving a natural increase of 3.81‰. However, natural increase is expressed as a percentage, so this makes China's natural increase 0.38%.

Assessment tip

Natural increase is given as a percentage (%) although birth rates and death rates are expressed in rates per thousand (‰).

Doubling time (DT)

The **doubling time** refers to the length of time it takes for a population to double in size, assuming its natural growth rate remains constant. Approximate values for it can be obtained using the following formula:

$$\text{Doubling time (years)} = \frac{70}{\text{growth rate in percentage}}$$

For China this is $\frac{70}{0.38}$, i.e. 184 years to double its population.

▲ **Figure 8.1.1** South Africa has a youthful population and a fertility rate above replacement level (2.1)

Human population growth

The world's population is growing rapidly. Most of this growth is quite recent. The world's population doubled between 1804 and 1922, 1922 and 1959, 1959 and 1974. It is thus taking less time for the population

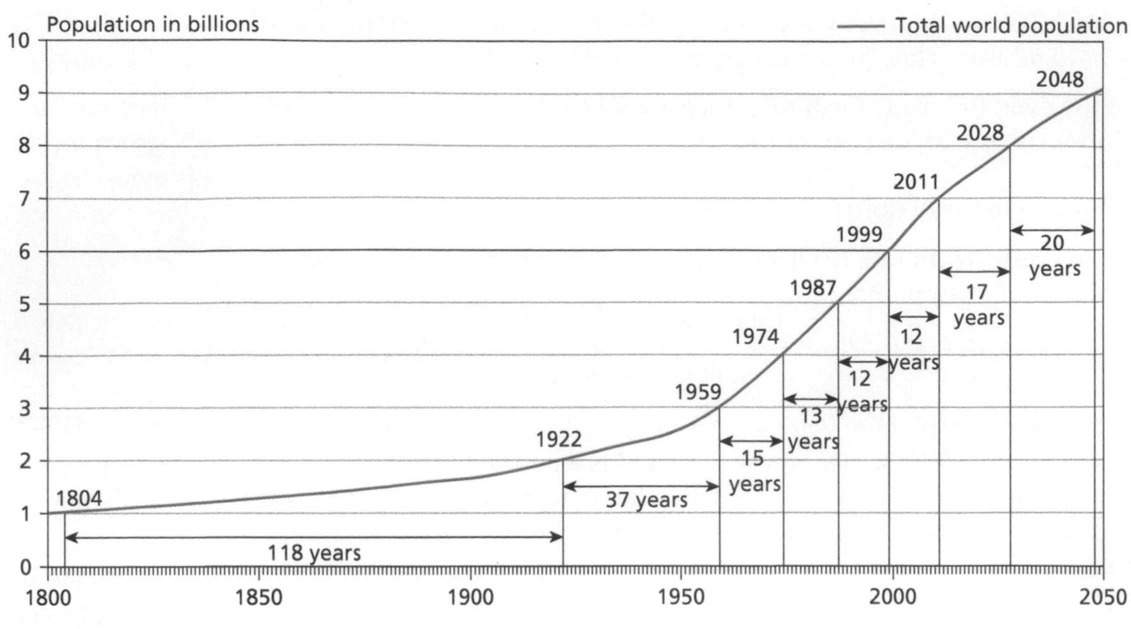

▲ **Figure 8.1.2** Exponential growth

Test yourself

Mexico's population in 2018 was about 133 million. There were around 2,535,000 births and 625,000 deaths that year. (Source: https://countrymeters.info/en/Mexico#population_2018)

8.1 Calculate the crude birth rate and crude death rate for Mexico. [4]

8.2 Calculate the natural increase and doubling time for Mexico. [4]

to double, although growth has slowed down since 1999 (figure 8.1.2). Such growth is referred to as exponential growth. Up to 95% of population growth is taking place in less economically developed countries (LEDCs). However, the world's population is expected to stabilize at about 8.5 billion, having peaked at about 9 billion.

However, the range of predictions for population growth by 2050 is very large and reveals uncertainty as to how much the world's population will grow. The high prediction is 11 billion whereas the low prediction is 7.5 billion. The medium prediction is 9 billion. The further into the future that scientists try to predict, the less certainty there is.

Predictions are important as they enable planners to plan for the future in order to meet needs. These include a need for more food and more health care. More schools and homes will also be required. Knowing whether there will be 11 billion people or 7.5 billion people will affect what facilities and services need to be provided.

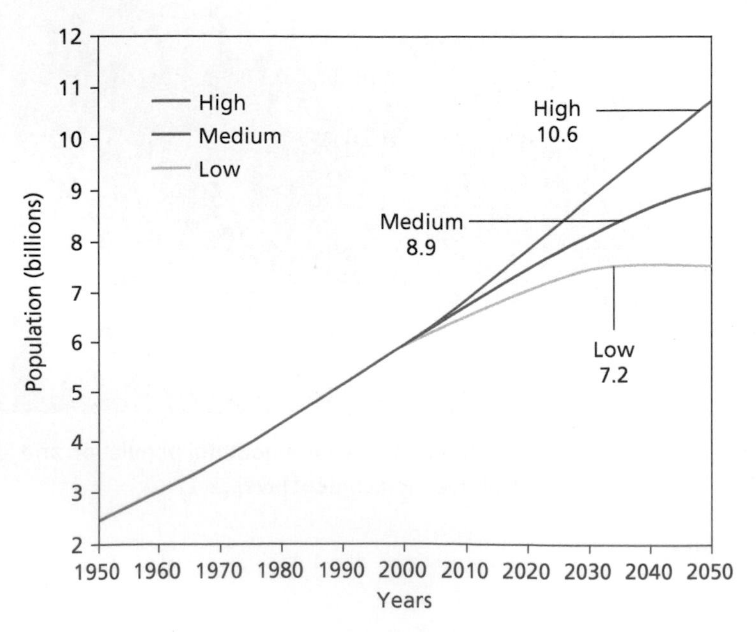

▲ **Figure 8.1.3** Population projections for the world in 2050

The impact of exponential growth is that a huge amount of extra resources are needed to feed, house, clothe, and look after the increasing number of people. However, it can be argued that the resource consumption of much of the world's poor population (i.e. those in LEDCs) is much less than the resource consumption of populations in MEDCs, where population growth rates are much lower.

Global population is estimated to reach 9 billion by 2075, and there could be an extra three billion mouths to feed by the end of the century, a period in which substantial changes are anticipated in the wealth, calorific intake, and dietary preferences of people in developing countries across the world. As their level of wealth increases, many people are eating more meat and dairy products. These are higher up the food chain and require more land to produce the food. The increased need for food will require more land, more water, and more fossil fuels for chemical fertilizers and for transport.

Content link

Section 5.2 examines inequalities in food production and distribution.

Age–gender pyramids

A population pyramid is a bar or line graph on its side showing variations in the **age structure** and **sex structure** of a population. Figure 8.1.4 shows population pyramids for China in 1950, 1979 and 2020.

Population pyramids tell us a great deal of information about the age and sex structure of a population:

- A wide base indicates a high **birth rate**
- A narrowing base suggests falling birth rate
- Straight or near vertical sides reveal a low **death rate**
- Concave slopes characterize high death rates
- Bulges in the slope suggest high rates of **in-migration**
- Indentations in the pyramid may indicate **out-migration** or **age-specific deaths.**

Test yourself

8.3 Calculate how long it took the Earth's population to grow from (a) 1 billion to 2 billion people (b) from 2 to 3 billion people (c) from 4 to 5 billion people and (d) from 6 to 7 billion people. [4]

8.4 State by how much the high projection for population growth by 2050 differs from the low projection. [2]

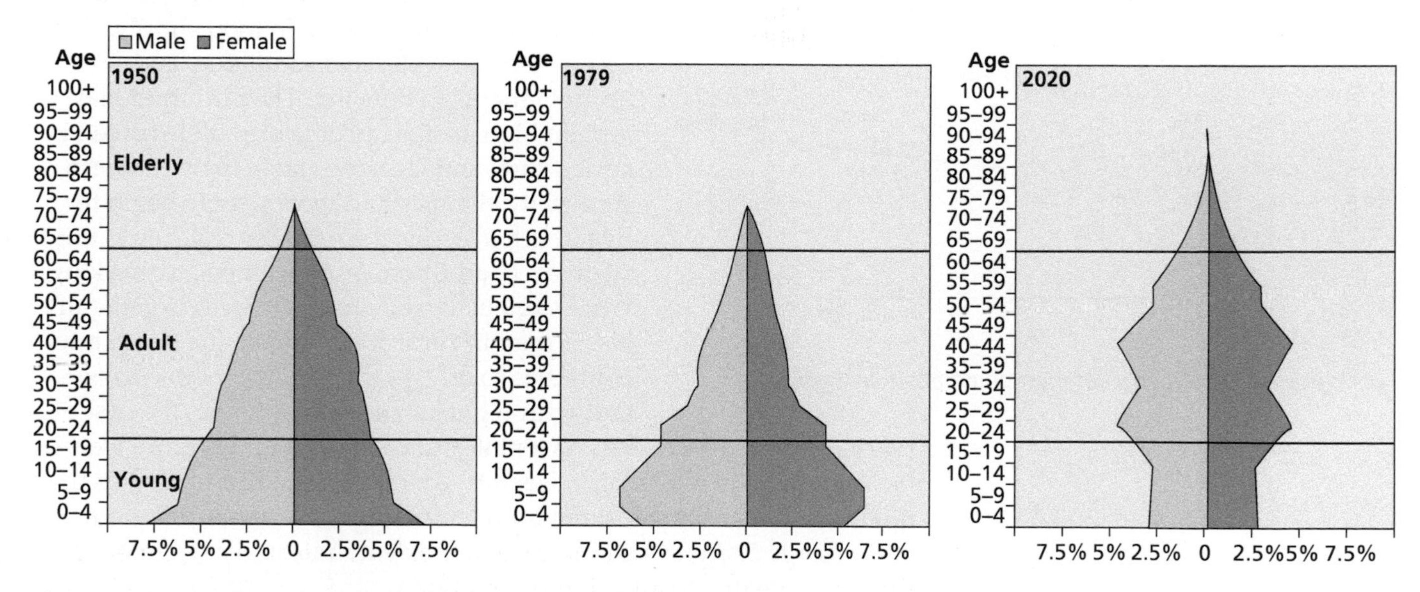

▲ **Figure 8.1.4** Population pyramids for China 1950, 1979 and 2020

Demographic transition model

The **demographic transition model** (DTM) shows the change in population structure over time (see figure 8.1.5). It shows how population changes from when a country is an LEDC to when it becomes an MEDC. When a country is an LEDC it tends to have higher birth rates and higher death rates. When a country becomes an MEDC it tends to have lower birth rates and lower death rates. This suggests that death rates fall before birth rates and that the total population expands. In stage 5 there is a higher proportion of elderly people and so the death rate rises.

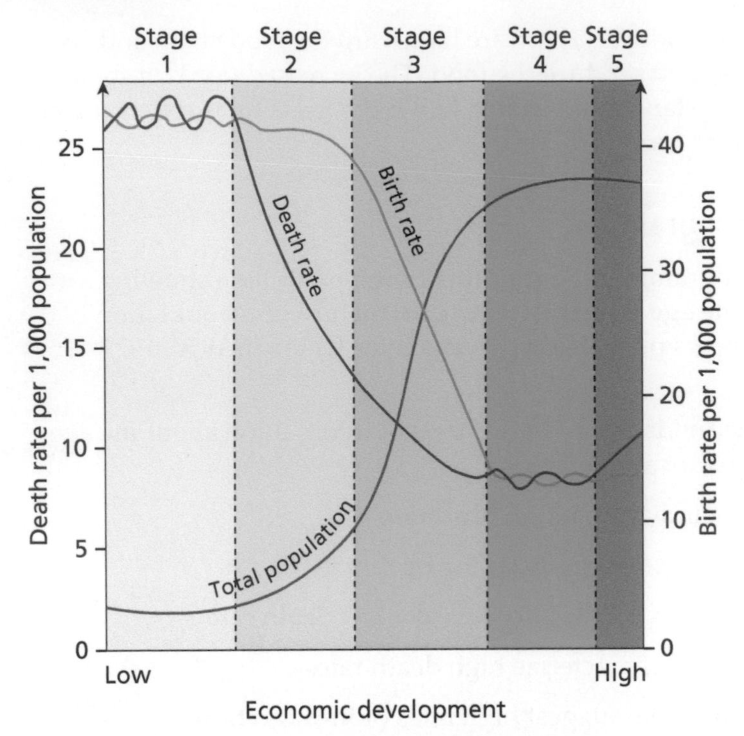

▲ **Figure 8.1.5** The demographic transition model

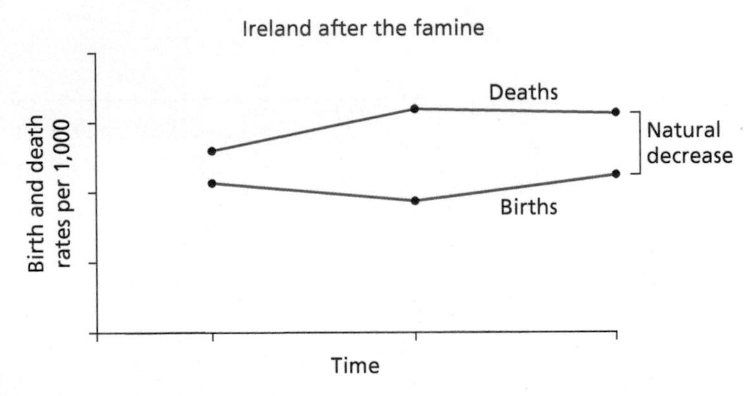

▲ **Figure 8.1.6** Ireland's demographic transition model

The fall in the death rate between stages 1 and 2 is related to improvements in food supply, access to clean water and sanitation, and improvements in housing. The birth rate may remain high as most people are still involved in agriculture, and children can help on the farm. As society changes and more people begin to live in urban areas, the need for children is not so great. In addition, more women are working. Access to health care in urban areas is generally better than in rural areas, and so the death rate continues to fall. As the country modernizes, and with higher standards of living, the cost of rearing children increases. As female education improves, more women participate in the workforce. Fertility generally falls. Eventually, as the quality of life improves, there is an increasingly elderly population, and in stage 5, there is an increase in the death rate and a decrease in the birth rate due to an aging population.

Test yourself

8.5 Describe the changes in China's population pyramids between 1950 and 2020. [3]

8.6 Suggest reasons why the birth rate declines but the death rate increases in stage 5 of the demographic transition model. [2]

The demographic transition model provides a representation of how population growth may occur. It is easy to understand. However, there are weaknesses. The DTM is based on data from just three countries, all of them European. LEDCs are taking much less time than MEDCs to progress through the stages of the demographic transition model. There are alternative demographic transition models. For example, Ireland's demographic transition (figure 8.1.6) was based on falling birth rates and rising death rates due to out-migration following the 1845–49 famine.

Influences on human population dynamics

Changes in fertility are a combination of many factors. While there may be strong correlations between these sets of factors and changes in fertility, it is impossible to prove the linkages or to prove that one set of factors is more important than another.

Cultural factors

Parents may be dependent on their children for support in their old age. This may create an incentive to have many children. Boys are more valued than girls in some cultures. This can lead to an increase in fertility so that more boys are born. For example, if girls are born the parents will try to have more children in the hope that they will have a boy.

Historical

The early stages of development are characterized by high mortality rates. Even when the death rate begins to fall, some believe that it is just another short-term fluctuation rather than a long-term trend. Hence they continue to have more children as replacements for the children they expect to die.

Religion

Religious influence on contraception can influence fertility. Some religions prohibit abortion or access to contraception. However, it is not clear-cut. Some Catholic countries, such as Mexico, have high fertility whereas others such as Ireland and Italy have relatively low fertility. However, most religions are pro families, whether their members follow their teaching or not.

Social factors

In some societies, there may be pressures on women to have a large number of children. This is more likely in rural parts of LEDCs. It is less likely with increased urbanization and industrialization.

Political factors

Most governments in LEDCs have introduced some programmes aimed at reducing birth rates. Their effectiveness is dependent on:

- focusing on family planning in general and not just on birth control
- investing sufficient finance in the schemes
- working in consultation with the local population.

Where birth controls have been imposed by government, they are less successful (except in the case of China). In the MEDCs, financial and social support for children is often available to encourage a pro-natalist approach. However, in countries where there are fears of negative population growth (as in Singapore), more active and direct measures are taken by the government to increase birth rates.

In Hungary in 2019, the prime minister announced that women with more than four children would never have to pay income tax and that those below the age of 40 who get married could obtain cheap loans, with increasing amounts written off as they have more children. They also get help buying seven-seater cars and child-care provision.

Economic prosperity

The correlation between economic prosperity and the birth rate is not total, but there are links. As GDI increases, the total fertility rate generally decreases, and as GNP/head increases so the birth rate decreases. Economic prosperity favours an increase in the birth rate, while increasing costs lead to a decline in the birth rate. Recession and unemployment are also linked with a decline in the birth rate. This is related to the cost of bringing up children. Surveys have shown that the cost of bringing up children in the UK can be over £200,000, partly through lost earnings on the mother's part. Whether the cost is real, or imagined (perceived), does not matter. If parents believe they cannot afford to bring up children, or that by having more children it will reduce their standard of living, they are less likely to have children.

National and international development policies

There are many national and international policy factors that influence human population growth. Some policies directly affect population growth and others affect it indirectly. **Population policies** can be described as **pro-natalist** when they try to increase the birth rate or **anti-natalist** when they try to reduce population growth.

Concept link 

STRATEGIES: Some governments introduce population policies in order to control their population growth. Pro-natalist population policies favour increased birth rates whereas anti-natalist policies try to reduce population growth.

The most famous anti-natalist policy is China's one-child policy. It was introduced in 1979 and limited most Chinese families to one child. It is thought that China's population would now be 400 million people larger if the one-child policy did not exist. Some critics believe that China's fertility would have reduced even without the one-child policy. They believe that urbanization and industrialization would have led to reduced population growth. In particular, they believe that improved female education and more working women would have led to reduced birth rates.

The prospect of an aging society in which one worker is left to support two parents and four grandparents led China to relax the one-child policy. In 2015, amid fears that the size of the working population was set to decline, the Chinese government changed the policy to allow couples to have two children.

Party conservatives still fear two things about loosening population controls:

- the population will grow beyond the country's capacity to feed itself
- relaxing rules too quickly will lead to a baby boom that will strain public services.

Factors other than the one-child policy have also encouraged couples to limit the number of children that they have. Education and housing costs as well as a lack of social security support make having more than one child prohibitive for many. The relaxation of the policy does not appear to have led to any change in fertility.

Singapore is an example of a country that had an anti-natalist policy and changed to a pro-natalist policy. It changed because its **fertility rate** had dropped to below 1.25 and the workforce was getting smaller. The government offered incentives to families to have three or more children if they could afford them. Singapore's fertility rate has remained low despite the incentives. Women continue to play an active role in the workforce and are choosing jobs rather than having children.

International policies that affect population growth

In 2000 the United Nations announced the Millennium Development Goals (MDGs). The aim of the Millennium Development Goals was to address issues of poverty and inequality. Many of them addressed population growth directly. These included MDG 4 Reduce child mortality and MDG 5 Improve maternal health. In 2015 the MDGs were replaced by the Sustainable Development Goals (SDGs). Goal 3 aims to ensure healthy lives and promote wellbeing for the whole population at all ages. One of its targets is improving reproductive, maternal, newborn, and child health.

National and international development policies may stimulate rapid population growth by reducing **mortality** without significantly affecting fertility. Agricultural development and improved **public health** may lead to population growth by reducing mortality without significantly affecting fertility. Birth rates often fall following urbanization. There is less demand for children and more women have jobs in the **formal** and **informal sectors**. Policies that target female education and female participation in the job market are believed to be the most effective method for reducing population pressure.

Test yourself

8.7 Identify the (a) MDGs and (b) SDGs that were created to develop maternal and child health. [2]

8.8 Suggest why China's population policy changed from anti-natalist to pro-natalist. [3]

QUESTION PRACTICE

a) With reference to Figures 1(a) and 1(b) describe the trends in Iceland's population dynamics. [2]

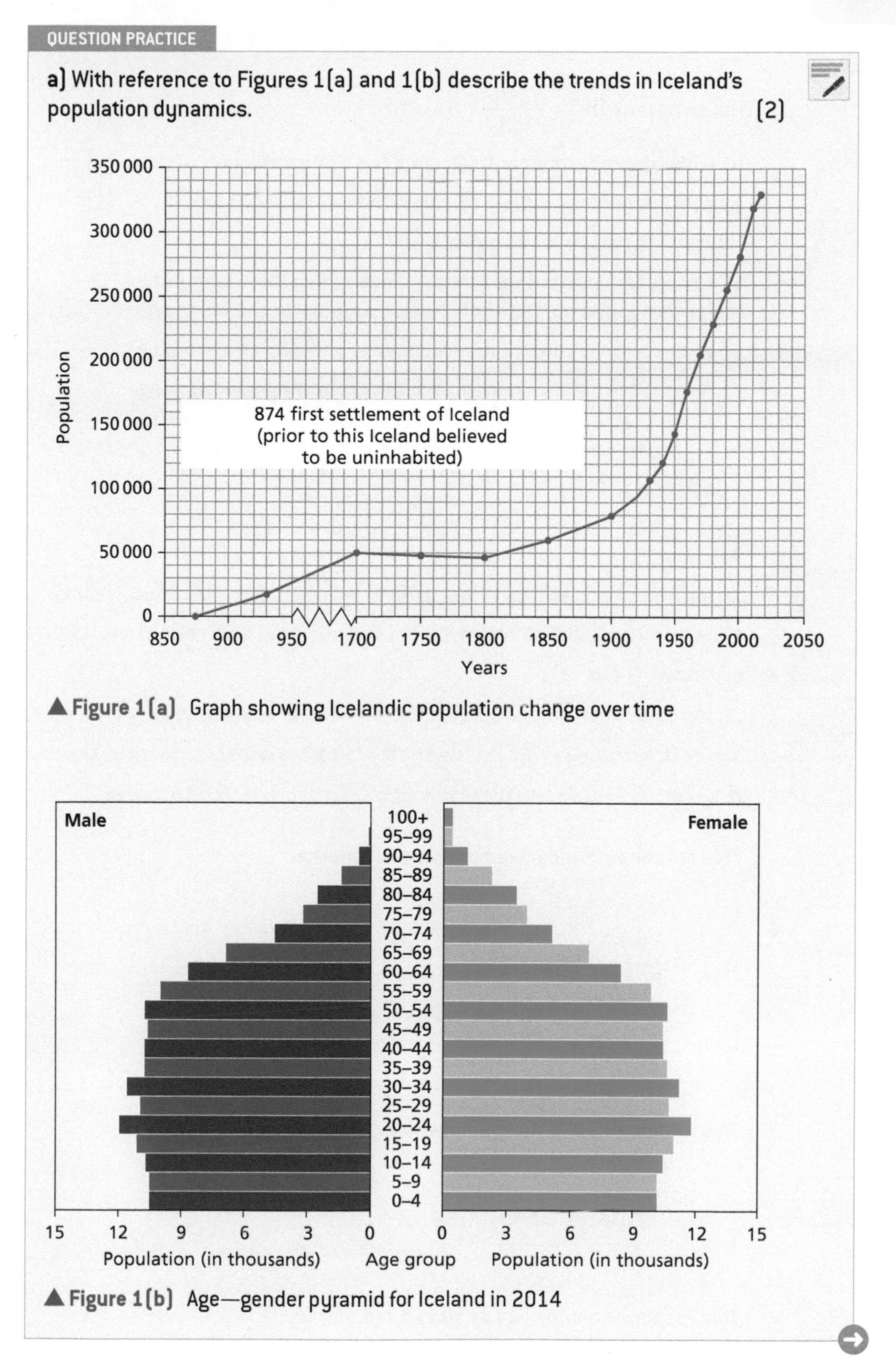

▲ **Figure 1(a)** Graph showing Icelandic population change over time

▲ **Figure 1(b)** Age—gender pyramid for Iceland in 2014

▼ **Table 1** Population data for three countries

Population variable	India	Japan	Uruguay
Crude birth rate (CBR)	21	8	14
Crude death rate (CDR)	7	10	10
Total fertility rate	2.3	1.4	1.9
Doubling time	X	–	175
Natural increase rate (NIR)	1.4	Y	10.4
Ecological footprint (Global ha per capita)	0.9	4.73	5.13

$$\text{Doubling time (DT)} = \frac{70}{\text{NIR}}$$

b) With reference to Table 1 calculate the DT for India (X). [1]

c) With reference to Table 1 calculate the NIR for Japan (Y). [1]

How do I approach these questions?

a) You are required to give a detailed account of the changes in Iceland's population. You should use/manipulate the data provided.

b) You need to obtain a numerical answer showing your working.

c) You need to obtain a numerical answer showing your working.

SAMPLE STUDENT ANSWER

a)

1. Iceland's population has grown between 874 to 1700 which then remains static followed by a dramatic increase from 1800 till now. (J curve).

2. As shown in Figure 1 (b), the birth rate is decreasing in 2014 since it has a narrow base which gives an indication that there is a slight decrease in growth rate on 1a from 2010 to 2014.

▲ Valid point, 1 mark

▲ Correct point, 1 mark

▲ Valid points, but max number of marks already reached

This response could have achieved 2/2 marks.

b)

$DT_x = \frac{70}{NIR_x}$ $NIR_x = 1.4$

$DR_x - \frac{70}{14} = 50$ years

▲ Correct calculation

This response could have achieved 1/1 marks.

c)

$NIR_y = \frac{CBR_y - CDR_y}{10} = \frac{8 - 10}{10} = -0.2$

▲ Correct calculation

This response could have achieved 1/1 marks.

8.2 RESOURCE USE IN SOCIETY

- **Natural resource** – any naturally occurring feature that provides benefits to people
- **Natural capital** – the total quantity of natural resources
- **Natural income** – the renewable/sustainable harvestable quantity generated by natural capital
- **Ecological services** – the goods and benefits provided by natural ecosystems for humanity

You should be able to show how resource use in society varies

✔ Reliable natural capital includes living species, ecosystems, and groundwater.

✔ Non-renewable natural capital can only be replaced over extremely long timescales.

✔ Renewable natural capital can be used sustainably or unsustainably.

✔ Attempts to make use of natural capital may lead to environmental degradation.

✔ Natural capital provides many goods and services.

✔ Natural capital is dynamic in nature.

If they are carefully managed, **renewable resources** can produce "**natural income**" over and over again. This income could be marketable goods such as timber and grain or it could be **ecological services** such as flood control. It also includes living species and ecosystems that require solar energy in order to function. Some non-living features, such as groundwater and the ozone layer, are also renewable resources.

However, some **natural capital** is non-renewable, for example fossil fuels, soils, and minerals. These can only be replaced or renewed over a very long timescale, longer than a human lifetime and possibly over millions of years in the case of fossil fuels.

Content link

Section 7.2 examines renewable and non-renewable energy resources.

Assessment tip

The command term "state" requires a specific name, value or other brief answer.

▼ **Table 8.2.1** Forms of natural capital

Forms of natural capital	Examples of goods	Examples of services	Time needed to be renewed
Renewable (living)	Living species such as cattle and sheep; food such as wheat and corn	Climate regulation such as the planting of trees to store water and maintain temperature; flood control	Within a human lifetime
Renewable (non-living)	Groundwater and ozone	Ozone protects against ultraviolet radiation from the sun	Within a human lifetime
Non-renewable	Fossil fuels such as oil; mineral reserves such as iron ore	Soil nutrients produced by weathering of rocks	May take millions of years to renew

Test yourself

8.9 Outline the meaning of the term "ecological services". [2]

8.10 State the timescale needed to renew fossil fuels and minerals. [1]

Renewable natural capital and the "supply chain"

Although a resource may be harvested in a sustainable way, i.e. the amount harvested is less than the annual recharge, the way in which it is extracted, transported and processed may cause environmental (and social) damage, thereby making this type of harvesting of natural capital unsustainable.

Test yourself

8.11 Suggest the resources (other than trees) needed to (a) run a saw mill and (b) produce paper products. [2]

8.12 State one environmental disadvantage of the removal of trees in lumber production. [1]

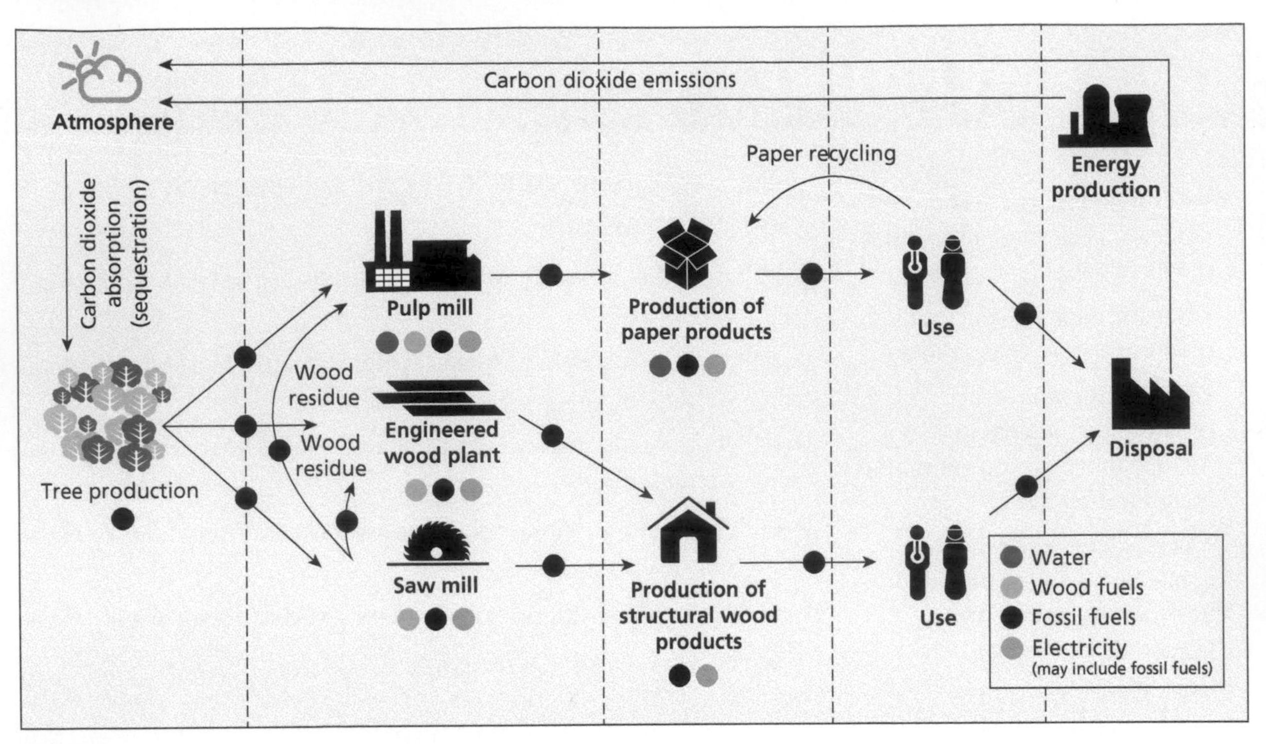

▲ **Figure 8.2.1** Environmental and social impacts in the forestry sector

For example, loggers may deprive indigenous peoples of their food source (socially unsustainable) and reduce the ecological services provided by forests. Even if the forest is logged sustainably there are still costs associated with the access roads, transport of felled trees, and processing into finished or semi-finished goods.

Concept link 

EVSs: Ecocentrics believe in the intrinsic value of the environment whereas technocentrists believe that people need to understand the environment so that they can control it.

Goods and services provided by natural capital

Natural capital produces numerous goods and services that are beneficial to societies. The value of these services takes many forms, such as economic, social, environmental, technological, cultural, aesthetic, ethical, spiritual, and/or intrinsic.

The **economic** value of an ecosystem can be quantified by the amount of money someone will pay for a good or service. The **ecological** value of an ecosystem includes **climate regulation** and **soil erosion control**, but these are difficult to quantify. The social values may be the provision of food and water for indigenous populations who live in an ecosystem. It could also be the pleasure that someone gets from being close to nature.

The **aesthetic** value of an ecosystem is the appreciation of the ecosystem for how it looks. Again, it cannot be quantified. The **intrinsic** value of an environment is the value of the environment in its own right. Ecosystems have a philosophical, spiritual, and ethical value even if they do not have an economic value.

Therefore, there are many ways of valuing the environment. Attempts are being made to give a value for benefits such as climate regulation. One advantage of climate regulation is a stable climate that allows farmers to plan for the year ahead. However, it is not possible to put a price on the benefits of climate regulation.

▲ **Figure 8.2.2** Deciduous woodland, Wytham Woods, Oxfordshire, UK

The dynamic nature of natural capital

The value of natural capital changes over time. In the Middle Ages oil was used to seal wounds, but in the 20th century it became the world's leading fuel. Similarly, there was a high demand for whale oil in the nineteenth century to make textiles easier to work with. Other resources that have changed over time include cork and uranium. Technological, economic, and cultural values all influence the value of resources.

Uranium is used in the nuclear power industry to create energy. Many factors help explain why uranium became an important resource. However, there are also reasons why the use of uranium in future may be limited.

In summary, resources change in value over time as new technology makes them more or less useful and therefore more or less valuable.

Test yourself

8.13 Identify three services provided by woodlands. [3]

8.14 Briefly **explain** why it is difficult to quantify the value of natural capital. [2]

Content link

Section 7.1 examines the use of nuclear power.

▼ **Table 8.2.2** The changing nature of the concept "resource" – uranium

Resource	Cultural factors	Economic factors	Technological factors
• Uranium has been used as a resource for less than 100 years. • It had no value until nuclear energy was developed. • Since then it has had great value.	• Awareness of the role of fossil fuels in global warming may have shifted interest in favour of nuclear energy for some societies. • There has been a change in values as the dangers of nuclear energy became clearer, e.g. Fukushima–Daiichi (Japan) explosion in 2011. • Nuclear weapons and nuclear disasters have turned some people against nuclear energy.	• Economic development led to increased demand for energy supply. • Nuclear energy was seen as clean and cheap. • Energy was created by nuclear fission. • Decommissioning of power stations is very costly. • As more countries develop, the use of nuclear energy is becoming more widespread.	• As nuclear technology has developed, the value of uranium has increased. • Fusion of hydrogen atoms has much greater energy potential than the fission of uranium. However, the technology for nuclear fusion has not been developed yet. • If nuclear fusion is developed, uranium will no longer be needed as a resource.

Test yourself

8.15 Identify how cultural factors may be (a) in favour of and (b) against nuclear power. [2]

8.16 Outline how technological developments could lead to a decline in the use of uranium. [2]

QUESTION PRACTICE

a) **Outline** four different ways in which the value of named resources has changed over time. [4]

b) **Outline** the reasons why natural capital has a dynamic nature. [4]

c) **Explain** how the inequitable distribution of natural resources can lead to conflict. [7]

How do I approach these questions?

a) You will be awarded 1 mark for each correct reason up to a maximum of 4 marks. Different kinds of influence include economic, social, political, environmental, cultural, technological, etc.

b) You will be awarded 1 mark for each correct reason and/or example, up to a maximum of 4 marks.

c) You will be awarded 1 mark for any argument/valid point, e.g. connecting unequal resource distribution to conflict; maximum of 4 marks allowed for outlining inequality in resource distribution without clear reference on how it is leading to conflict.

SAMPLE STUDENT ANSWER

a)

1. Coal has become less valuable because people have discovered that there is energy supply to last upwards of 250 years.

▼ Unclear point – over supply?

2. Oil has become more valuable because there are limited amounts left to be used.

▲ Economic reason

3. Natural gases have become less valuable because they have the ability to release harmful toxins as well as being more sustainable than a resource such as oil.

▲ Environmental reason

4. Wind energy has become more valuable because it created jobs as well as being sustainable.

▼ Needs developing. Two points made but not developed

This response could have achieved 2/4 marks.

Two more marks could have been gained by developing each of the points in detail, or by having two more creditworthy points.

b) Natural capital has a dynamic nature because it is set through human value and development. What today may seem valuable, a hundred years from now may not be as important. A great example is uranium, before the 20th century this element wasn't considered valuable or important, but since the development of nuclear weapons and energy, uranium has become essential to this development increasing its value. Natural capital is dynamic because its value changes over time depending on our needs and different human developments.

▲ Generic point – humans value resources

▲ Uranium – valid example

▼ Repetition of first point

This response could have achieved 2/4 marks.

Two more marks could have been gained by developing each of the points in detail, or by having two extra valid points

c) Resources are inequitably distributed in different parts of the world system. In water systems there exists large inequality in access to freshwater. This is worsened by the fact that many of these resources are shared, for example the Danube river is shared by 19 countries, and provides water for 81 million people. For these 2 reasons, human activity can alter access to freshwater in many ways, worsening inequality. For example, withdrawals of water at the base of the glacier that feeds the Nile makes access to water scarce in Egypt, to the point that they have to import

▲ Inequality in access to water

▲ Development of case study – the Danube with some detail

▼ Accuracy?

food because they do not have enough water to produce it. Also since water resources are common access resources, they are a good example of the Tragedy of the Commons. A conflict could be identified between individual needs and wants and society's well-being. For instance, a certain group of the population may be close to a water source (e.g. river), and may decide to extract more water, damaging water supply further below the course of the river. Moreover, some countries have larger access to fish and other food sources from marine ecosystems. This generated conflict before the United Nations passed the United Nations Convention on the Laws of the Sea, where countries enjoyed 200 nautical miles of exclusive economic zones in the seas that were found by their territories. An example of this conflict were the Cod Wars between Britain and Iceland from 1950s to 1970s.

Regarding energy resources like fossil fuels (carbon, oil, natural gas), conflicts have also arisen. Each country is concerned with energy security. Whether it has the natural resources or has to import them.

For instance, Ukraine imported gas from Russia, and continued doing so without paying. For this reason Russia cut its supply of gas to this country. Conflicts between Ukraine and Russia still exists nowadays in relation to energy security. Considering soil systems and societies, soil is a natural resource with different qualities in different regions of the world. This makes certain countries have different abilities to produce food, and it may become an issue if countries are not able to be self-sufficient. If this were the case, said country would have to import, and issues such as dumping may arise, where international institutions like the World Trade Organization has to intervene.

▲ Tragedy of the Commons – common resource

▲ Access to fish – identifies how unequal access to resources has led to conflict in the past, and how this conflict can be managed

▲ Attempts to manage access – illustrates conflict with a valid example

▲ Conflict over energy

▲ Example with some detail

▼ Does not add to the answer. Conflict? Support?

This response could have achieved 7/7 marks.

Overall, a focused answer. Refers back to the question and uses a range of resources (water, fish, oil, soil) and a range of supporting examples.

8.3 SOLID DOMESTIC WASTE

- **Solid domestic waste** – household waste/garbage—it does not include faecal waste
- **Sustainability** – the use of resources at such a rate that allows natural regeneration and/or minimizes damage to the environment
- **Landfill** – the disposal of waste onto or into the land
- **Incineration** – the burning of waste
- **Composting** – the breakdown/natural decay of organic material and its use as a fertilizer in soil

You should be able to show the range of solid domestic waste and options for its disposal

✔ There are many types of solid domestic waste.

✔ The increase in non-biodegradable waste is a major environmental problem.

✔ Waste disposal methods include composting, recycling, incineration, and landfill.

✔ Pollution management strategies for solid domestic waste include changing human activities, controlling the release of pollutants, and clean-up and restoration programmes.

Different types of solid domestic waste (SDW)

There are many types of solid domestic (or municipal) waste (rubbish or garbage). In an MEDC **solid domestic waste** usually consists of the following (typical percentage volumes are shown in brackets):

- organic waste from kitchen or garden, including waste wood (20–50%)
- paper/packaging/cardboard (20–30%)
- glass (5–10%)
- metal (less than 10%)
- plastics (5–15%)
- textiles (less than 5%)
- electrical appliances, e.g. computers/fridges etc. (less than 5%).

With increasing wealth there is:

- a decrease in organic waste/biomass waste
- an increase in paper and cardboard
- an increase in plastic waste
- an increase in metal and glass waste (figure 8.3.2).

▲ **Figure 8.3.1** The WEEE man – a robotic figure weighing over 3 tonnes and formed from waste electrical and electronic equipment

Concept link

SUSTAINABILITY: Increasing wealth is associated with increasing levels of waste.

Test yourself

8.17 Study figure 8.3.1. **State** whether the waste material is likely to come from an MEDC, NIC or LEDC (or any combination). [2]

8.18 (a) **State** how the main items of waste disposal vary between MEDCs, NICs and LEDCs. (b) **Identify** one or more features that are common to all three types of countries. [4]

Volumes of waste

Much of the world's rubbish is generated by city dwellers. According to the World Bank the potential costs of dealing with increasing rubbish are high. The world's cities currently generate around 1.3 billion tonnes of solid domestic waste a year, or 1.2 kg per city dweller per day. With increasing urbanization, this is expected to rise to 2.2 billion tonnes by 2025, or 1.4 kg per person.

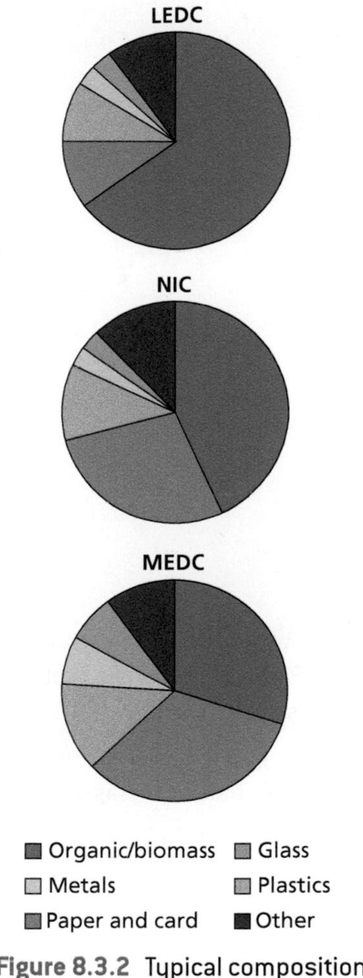

▲ **Figure 8.3.2** Typical composition of waste associated with LEDCs, NICs and MEDCs

Assessment tip

Although you are not expected to be able to identify all of the individual countries, you are expected to know the continents and use compass points/directions to locate places, e.g. southern Africa or north-west Europe.

Test yourself

8.19 Describe the global variations in the disposal of waste. [3]

8.20 Suggest why rates of SDW are projected to rise, especially in urban areas. [2]

Non-biodegradable pollution

Plastic

According to an article in *Science* in 2015, between 4.8 and 12.7 million tonnes of plastic enter the oceans each year. Much of it was concentrated in the Great Pacific Garbage Patch, a region containing as much as 100 million tonnes of plastic suspended in two separate gyres of garbage in an area twice the size of Texas. It can take centuries for plastic to decompose. Turtles, seals, and birds inadvertently eat it, and not just in the Pacific. A Dutch study of 560 fulmars (seabirds) picked up dead in countries around the North Sea found that 95% had plastic in their stomachs.

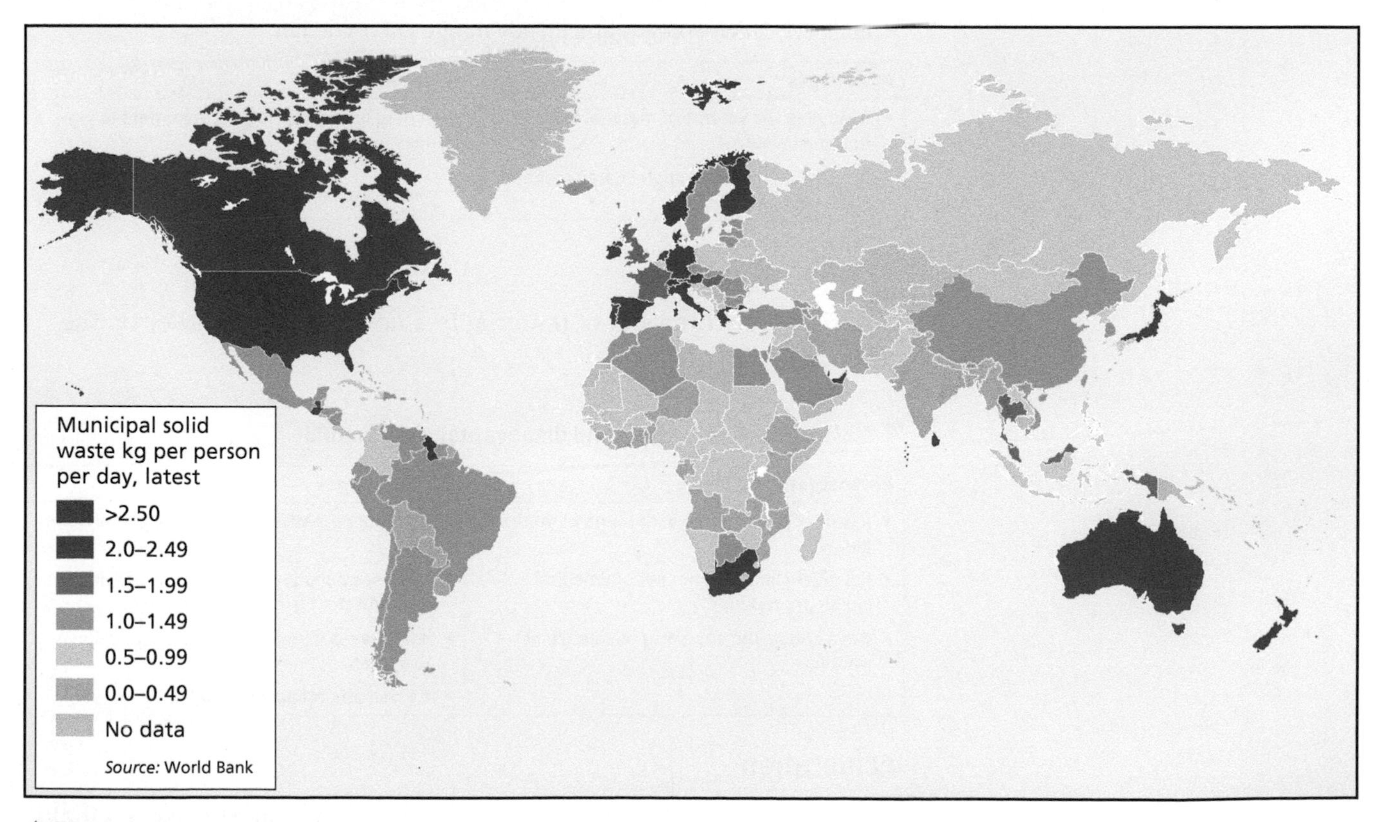

▲ **Figure 8.3.3** Volumes of the world's waste

Batteries

The increasing global demand for batteries is due, in part, to increasing consumer culture and the demand for mobile phones and laptops. Each year billions of batteries, containing toxic or corrosive materials, are disposed of. Some batteries contain lithium, lead, cadmium and/or mercury, which can be hazardous to living and non-living elements, e.g. water. Production, transport and distribution of batteries uses fossil fuels, thereby contributing to global warming. Potential risks are associated with emissions of chemicals into aquatic ecosystems. Disposal of batteries into landfill may release toxic substances into groundwater.

Electronic waste

In 2012 China generated over 11 million tonnes of e-waste, followed by the USA with around 10 million tonnes. However, per capita figures were reversed: on average, each American generated around 30 kg of e-waste, compared to less than 5 kg per person in China. The European Environment Agency estimates that between 250,000 tonnes and 1.3 million tonnes of used electrical products are shipped out of the EU every year, mostly to west Africa and Asia.

Guiyu in China has been described as the e-waste capital of the world. The industry is worth $75 million to the town each year, but Guiyu's population has high rates of lead poisoning, cancer-causing dioxins, and miscarriages. In 2018 China banned the import of e-waste.

Test yourself

8.21 Briefly **outline** the problem of waste batteries. [2]

8.22 Suggest the likely impact of China's ban on e-waste. [2]

Waste disposal options

Reduction

Reduction of waste means using fewer resources to meet the needs of the population.

▼ **Table 8.3.1** Advantages and disadvantages of reduction

Advantages	Disadvantages
• It reduces the volume of material that becomes waste. • It can be achieved through charging people e.g. for plastic bags.	• It may be difficult to get consumers to use less resources.

Landfill

Landfill is the dumping of material in a hole in the ground or on the ground.

▼ **Table 8.3.2** Advantages and disadvantages of landfill

Advantages	Disadvantages
• It can produce energy in the form of methane gas. • Relatively limited amounts of time and labour are required. • It is a cheap and easy way to dispose of waste.	• It produces methane, which is a greenhouse gas. • It causes pollution of watercourses and groundwater by leaching materials. • It increases vermin, which can cause disease to spread. • It gives off unpleasant smells.

Incineration

Incineration is the burning of household and industrial waste so that it is reduced in volume.

▼ **Table 8.3.3** Advantages and disadvantages of incineration

Advantages	Disadvantages
• It reduces the volume of waste, thereby reducing the need for landfill. • It is a way of producing energy from waste. • It produces ash that can be used in construction.	• It produces greenhouse gases. • It releases toxic chemicals. • People may object to the building of new incinerators in their neighbourhood

Recycling

Recycling is the processing of household and industrial waste so that it can be used again.

▼ **Table 8.3.4** Advantages and disadvantages of recycling

Advantages	Disadvantages
• It reduces the amount of resources used. • It reduces the amount of material in landfill sites. • It can be used to make new products.	• It involves transport of sometimes heavy goods so requires lots of energy. • It may produce toxic waste. • Cost may be a limiting factor for some countries.

Composting

Composting is the decomposition of **biodegradable** material and its use as a fertilizer in soil.

▼ **Table 8.3.5** Advantages and disadvantages of composting

Advantages	Disadvantages
• It produces organic fertilizer. • It reduces use of chemical fertilizer. • It reduces volume of waste.	• It is a slow process. • It may produce unpleasant smells. • It can attract vermin if not done properly. • It is limited to biodegradable materials.

Strategies for SDW management

The following strategies can be implemented to manage SDW:

- Altering human activity, e.g. so that consumers consider packaging when buying goods and use "bags for life" instead. Educational campaigns may lead to people changing their ways. Similarly, a government tax, e.g. on plastic bags, may lead to reduced consumption of plastic.
- Controlling the release of pollutants, e.g. governments can encourage recycling, reuse, and composting, and make facilities available for the recycling/reuse of goods. Certain goods can be reused for alternative purposes, e.g. using old car tyres to stabilize slopes and old plastic bottles to grow plants in. Alternatively, consumers may decide to **compost** their food waste.
- Reclaiming landfills and recovering some energy from waste (EfW) to produce methane (natural gas) from waste materials. In extreme cases, there may be clean-up and restoration, e.g. litter collections and removing plastic waste from beaches or from oceans.

Concept link 

STRATEGIES: There are three main strategies to manage SDW: altering human activity, controlling the release of pollutants, and clean-up/restoration strategies.

Test yourself

8.23 State which of the strategies for dealing with SDW are (a) easiest to achieve and (b) most expensive. [2]

8.24 Suggest why altering human activity may be hard to achieve. [2]

QUESTION PRACTICE

The United Nations Environment Programme (UNEP) and GRID-Arendal have suggested the hierarchy (order of preference) for waste management strategies shown in figure 1.

1) Outline two reasons why UNEP and GRID-Arendal prefer reduction over recycling as a waste management strategy. [2]

Essay

2) The management of a resource can impact the production of solid domestic waste.

To what extent have the three levels of the pollution management model been successfully applied to the management of solid domestic waste? [9]

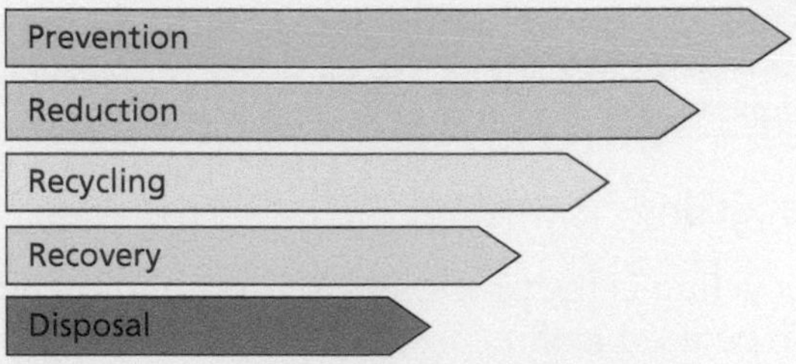

Prevention
Reduction
Recycling
Recovery
Disposal

▲ **Figure 1** Waste management hierarchy

Source: *UNEP and GRID-Arendal*

How do I approach these questions?

1) "Outline" requires you to give a brief account or summary, in this case of why reduction is better than recycling. For example, recycling is more expensive, may require a higher energy input; reduction reduces the initial use of raw materials. You will not gain extra credit for stating the opposite, e.g. recycling is more expensive, reduction is cheaper.

Essay

2) Answers should include knowledge and understanding of ESS issues and/or concepts, terminology, relevant examples, balanced analysis, and a well-supported conclusion. Answers may include:

- understanding concepts and terminology of pollution management model and its "three levels", i.e. A, altering human activity/reducing production; B, regulating/limiting release; and C, clean-up/restoration
- breadth in addressing and linking different levels with each other and with relevant management strategies to each, e.g. A, educational campaigns/legislation for reduced packaging; B, promotion of reuse/recycling, composting, waste to energy schemes; and C, landfill reclamation, litter collection
- examples of specific schemes, e.g. A, a charge for plastic bags or a tax on waste collection (e.g. Ireland); B, government waste to energy schemes, sponsored recycling schemes; and C, mining landfills to remove hazardous waste, clean-up schemes for plastic on beaches and in the ocean
- balanced analysis of the success or otherwise, i.e. relative strengths and weaknesses of a range of strategies from all three levels of the pollution management model
- a conclusion that is consistent with, and supported by, analysis and examples given.

SAMPLE STUDENT ANSWER

▲ Valid point about reduction as a strategy.

▼ Confuses recycling with re-use

(1) Reduction leads to less waste being produced because people use less packaging and less plastic bags, whereas recycling still uses the resources. Recycling of plastic bags means that you use them many times, this also reduces waste.

This response could have achieved 1/2 marks.

To get full marks the student would have to give another reason why reduction is better and/or recycling is worse e.g. the expense or the energy required.

Solid domestic waste consists of material that is no longer useful for households, so it is disposed of. A country whose waste management strategy serves as an example is Sweden. In Sweden, the three levels of waste management are employed. First, altering human activity through education has been beneficial, insofar as people separate waste before sending it to recycling plants. These plants are near residential areas so that people can take their waste themselves, increasing the efficiency of the system as less intermediaries are required (and there is a smaller chance of waste being disposed on the street as a consequence). Most importantly, cleanup is very developed in Sweden, through the use of incinerators. Through this mechanism, waste is burnt at temperatures rounding to 2000°C, and it can be later used as ashes in roadbuilding. This is so efficient that Swedes have to import waste from the UK to keep the incinerators working. This strategy, although apparently harmless, has to be controlled so that gases (greenhouse gases) released from combustion do not escape into the atmosphere. So it can be said that since incinerators always require waste, they do not encourage people to produce less waste. Yet this does not represent an issue for Sweden, and its interesting business with the UK may become a way for other EU countries to clear their landfills.

▲ Identifies a named country and the 3 levels

▲ Altering human activity

▲ Recycling plants are easy to access

▲ Cleanup through incinerators

▲ Waste burnt and then used in roadbuilding

▲ Imports waste from the UK

▲ Control of GH gases

This response could have achieved 8/9 marks.

Brief amount of text but:

- substantial knowledge and understanding
- appropriate use of terminology
- well-explained examples.

To gain the extra mark, the conclusion could have been clearer and more developed.

8.4 HUMAN POPULATION CARRYING CAPACITY

- **Carrying capacity** – the maximum number of a species that can be sustainably supported by an environment
- **Ecological footprint** – the area of land and water required to support a defined human population at a given standard of living, and to dispose of its waste materials
- **Degradation** – any decline in the quality or quantity of a resource/environment
- **Finite (resources)** – resources which can only be used once during a human lifetime

You should be able to show how the ecological footprints of societies impact on the carrying capacity of the environment

✔ The carrying capacity is the maximum population that can be supported sustainably by an environment.

✔ It is difficult to estimate the carrying capacity for human populations.

✔ Ecological footprints measure the size of area needed to support a population.

✔ Ecological footprints are used to measure the environmental demands of people.

✔ Ecological footprints vary widely.

✔ Population growth may be limited by environmental degradation and resource use.

✔ Over-use of resources may lead to population collapse.

Carrying capacity

Carrying capacity is the maximum number of a species or "load" that can be sustainably supported by a given environment.

Although it is possible to estimate the carrying capacity of an environment for a given species, based on its size and the food and water supply it contains, it is difficult to apply to human populations for many reasons. For example:

- The range of resources used by humans is usually much greater than for any other species.
- When a resource becomes scarce humans show great ingenuity in substituting one resource for another. The substitution of shale gas for oil is a good example.
- Resource requirements differ according to lifestyles. These vary from time to time and from population to population. A nomadic pastoralist, for example, uses far less resources than an urban dweller in a rich country.
- Human populations regularly import resources from outside their immediate environment. This enables them to grow beyond the carrying capacity set by their local resources. The import of food from countries such as Brazil and New Zealand into Europe is a good example. Importing resources in this way increases the carrying capacity for the local population. However, it has no influence on global carrying capacity.

All of these variables make it practically impossible to make reliable estimates of carrying capacities for human populations.

Ecological footprints

The **ecological footprint** of a population is the area of land that would be required to provide all of the population's resources and assimilate all of its wastes. It is useful as a model because it can provide a quantitative estimate of human carrying capacity. It is the inverse of carrying capacity. It refers to the area required to sustainably support a given population rather than the population that a given area can sustainably support. It is measured in global hectares (gha). The total ecological footprint for an area can be given (figure 8.4.1), or the ecological footprint per person may be given (figure 8.4.2).

The two broad categories that contribute to a person's or nation's ecological footprint are:

- land required to provide necessary resources, e.g. food, agriculture, housing, urbanization, industry, water supply
- land required for assimilation of wastes, e.g. vegetation that provides a sink for carbon waste and areas of landfill.

Ecological footprints have many advantages as a model.

- They are a useful snapshot of the **sustainability** of a population's lifestyle.
- They provide a means for individuals or governments to measure their impact and to identify potential changes in lifestyle.
- They are a symbol for raising awareness of environmental issues.

However, there are many disadvantages.

- Ecological footprints do not include all information on the environmental impacts of human activities.
- They are only a model so they are a simplification and lack precision.
- They use approximation of actual figures which cannot be accurately calculated.

Test yourself

8.25 State the relationship between carrying capacity and ecological footprint. [2]

Assessment tip

The world's biocapacity may increase as more land comes into production and yields per hectare increase.

Variations by country and over time

Figure 8.4.1 shows how the world's total ecological footprint increased between 1961 and 2014, and how its composition changed. Figure 8.4.2 shows global variations in ecological footprint per person.

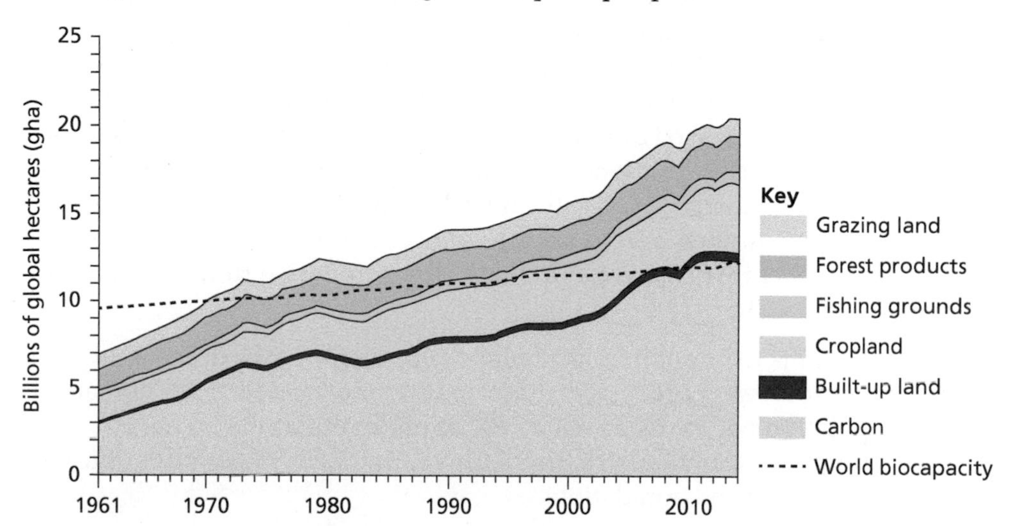

NB Biocapacity is the amount of biologically productive land, total hectares

▲ **Figure 8.4.1** The world's ecological footprint, 1961–2014

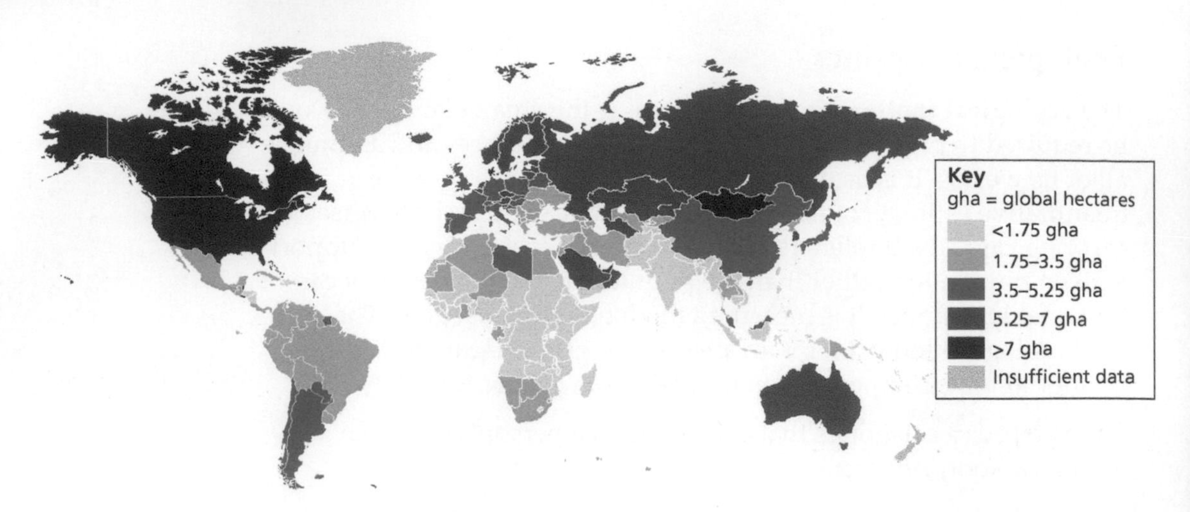

▲ **Figure 8.4.2** Global variations in ecological footprint per person

Test yourself

8.26 State the unit of measure for ecological footprints in figures 8.4.1 and 8.4.2. [1]

8.27 Describe the main changes in global ecological footprints, as shown in figure 8.4.1. [3]

8.28 Outline the main variations in ecological footprints per head, as shown in figure 8.4.2. [3]

LEDCs tend to have smaller ecological footprints than MEDCs because of their much smaller rates of resource consumption. In MEDCs, people have more disposable income, leading to greater demand for and consumption of energy resources. The resource use of MEDCs is often wasteful and MEDCs produce far more waste and pollution. People in LEDCs, by contrast, have less to spend on consumption, and the informal economy in LEDCs is responsible for recycling many resources. However, as LEDCs develop, the size of their ecological footprint increases. Countries with high rates of primary productivity (due to climate and undisturbed vegetation) need less land to absorb a given quantity of carbon waste, thus reducing their ecological footprint.

Unsustainable footprints

As countries develop and improve, their resource consumption increases. The use of technology reinforces the consumption of resources. In MEDCs and in NICs the use of cars and electrical goods has increased. This increases the demand for energy resources. MEDCs account for about 20% of the world's population but consume over 50% of the world's resources.

The world has been described as a "globalized consumer culture". Resources are extracted and manufactured into goods. They are then transported and stored in warehouses and shops. These goods are sold and eventually thrown away. The demand for consumer goods has increased dramatically since the 1960s, putting the world's resources under great pressure. Many believe that the world is reaching its carrying capacity. Indeed, many have argued that the over-consumption of resources, pollution, and **degradation** of the environment will lead to a decline in the population from around 2040 (figure 8.4.3).

Concept link 

SUSTAINABILITY: The ecological footprint in some LEDCs is sustainable, whereas the ecological footprint in many MEDCs is unsustainable.

The Limits to Growth model has a pessimistic view of population and resources. It predicts that the limits to the growth of the human population will be reached by 2100. It predicts that the human population would outstrip the ability of the Earth to provide sufficient resources for the population.

However, many people believe that human carrying capacity can be increased through technological developments. They argue that the

Earth can hold more people if we learn to use energy and resources more efficiently. Technological improvements include irrigation and fertilizers. Crops could be grown in nutrient-enriched water. This is known as hydroponics. High-yielding varieties (HYVs) of plants could be used.

On the other hand, many of the "solutions", e.g. HEP and nuclear energy, require vast amounts of fossil fuels for their construction, and many forms of sustainable development are very expensive. Governments and oil companies will continue to take the easy option while it is cheaper to extract fossil fuels than develop alternative technologies. Governments face many pressures and calls on their resources. Investing in technologies that might not be profitable for many decades is not seen as a vote winner by politicians.

Concept link

EVSs: Technocentrists believe that technological developments, such as hydroponics and genetically modified crops, will allow carrying capacity to be increased.

Content link

Section 7.1 examines HEP and nuclear power in more detail.

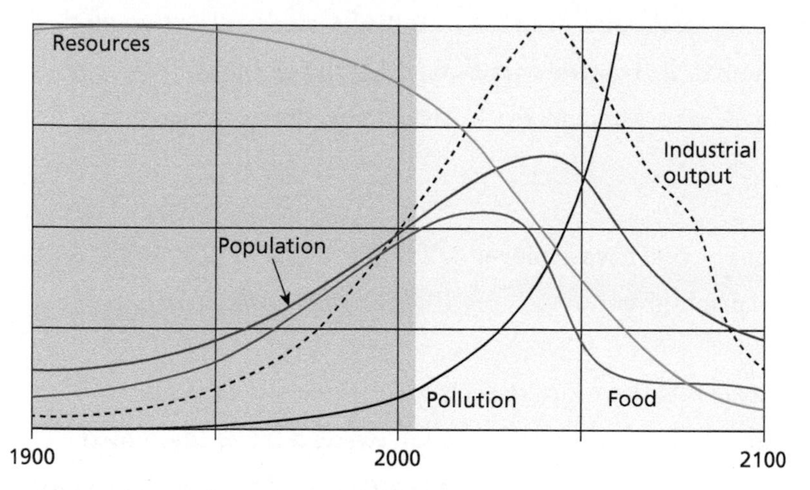

▲ **Figure 8.4.3** The Limits to Growth model

▼ **Table 8.4.1** The composition of ecological footprints in the USA and Bangladesh (size in gha/rank in world out of 149 countries)

	USA	Bangladesh
Carbon	4.87 gha/4	0.15 gha/112
Grazing	0.19 gha/82	0.01 gha/114
Forest	0. 86 gha/10	0.07 gha/137
Fishing	0.09 gha/49	0.02 gha/96
Cropland	1.09 gha/16	0.32 gha/127
Built-up land	0.07 gha/48	0.07 gha/48
Total	7.17 gha/5	0.63 gha /144

Source: *Living Planet Report 2012*

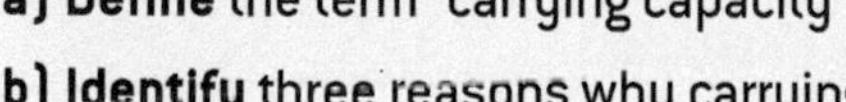

QUESTION PRACTICE

a) Define the term "carrying capacity". [1]

b) Identify three reasons why carrying capacity can be difficult to estimate. [3]

c) Identify two reasons why Uruguay has the biggest ecological footprint. [2]

▼ **Table 1** Ecological footprint for three countries

Country	India	Japan	Uruguay
Ecological footprint (Global ha per capita)	0.9	4.73	5.13

How do I approach these questions?

a) The definition needs to make clear that carrying capacity is the maximum population that can be supported sustainably in an environment.

b) You will receive 1 mark for each correct reason identified, up to a maximum of 3 marks, e.g. changing environmental conditions, dependence on a range of resources, variations in lifestyles/levels of wealth, technological developments, etc.

c) You will receive 1 mark for each correct reason identified, up to a maximum of 2 marks, e.g. high standard of living, high dependence on fossil fuel, limited vegetation/ecosystems to absorb waste CO_2, meat-rich diet, etc.

Test yourself

8.29 Compare the compositions of the USA's ecological footprint with that of Bangladesh (Table 8.4.1). [4]

8.30 Describe the main features of the Limits to Growth model (figure 8.4.3). [4]

SAMPLE STUDENT ANSWER

▲ Correct definition

a) Carrying capacity is the maximum number of organisms of a certain species that the environment can sustain with its resources.

This response could have achieved 1/1 marks.

▲ New resources

▲ Import resources

▲ New technology

b) In the case of humans, it is difficult to estimate because we are creative, and find new resources that meet the same needs as the ones we currently use. Also, importing/exporting allows us to consume goods that aren't available in our ecosystem, and improvements in technology may allow us to extract resources that were once unavailable to humans, increasing carrying capacity.

This response could have achieved 3/3 marks.

Three valid points given.

▲ Diet

▲ Non-renewable energy

▼ Would not affect gha/person

▼ The maximum number of marks have been achieved already, so this last point will not be taken into account.

c) Uruguay has the biggest ecological footprint of 5.13 as it might have a meat heavy diet which needs a lot of land and water. Secondly, it may have the biggest ecological footprint due to the high consumption of non-renewable energy. Moreover, it has more births than deaths, leading to an increasing population leading to a large ecological footprint. Moreover, it is an LEDC with not many renewable energy sources.

This response could have achieved 2/2 marks.

Two valid points given.

INTERNAL ASSESSMENT

The internal assessment (IA) is an individual investigation that counts for 25% of your final grade in ESS.

The purpose of the IA is to explore an environmental issue through a personal investigation. The investigation is recorded as a written report which should be 1,500 to 2,250 words long.

The IA will be marked by your teacher and moderated by an external examiner. Marks will only be awarded for the first 2,250 words, and appendices will not be read. You may include screen shots of survey questions and these will not be included in the word count, but it is not appropriate to include all of the raw, untabulated results of a survey.

The IA is marked according to six criteria, each examining a different aspect of the investigation (see table 1).

The IA needs to have a focused research question that arises from a broad environmental issue. Your IA needs to:

- develop a methodology that successfully investigates the environmental issue
- evaluate the success of the research process and the findings of your study
- discuss the extent to which your study applies to the environmental issue.

You also need to suggest realistic solutions to the environmental issue, either based on the findings of your study or from secondary sources. This discussion (the "application" of your study) should lead you to develop creative thinking and novel solutions, or to inform current political and management decisions relating to the issue. For example, if you carry out a study on the impact of wind turbines erected in the vicinity of your school, you may suggest solutions for the erection of wind turbines in other areas based on your findings.

▼ **Table 1** Mark criteria for the IA, with the percentage of total marks available shown in brackets

Criterion	Maximum marks (percentage of IA total marks)
Identifying the context	6 (20%)
Planning	6 (20%)
Results, analysis and conclusion	6 (20%)
Discussion and evaluation	6 (20%)
Applications	3 (10%)
Communication	3 (10%)
Total	30 (100%)

Assessment tip

In your IA you must engage in practical research, with either primary or secondary data, in order to fulfil the requirements of the IA criteria. A general essay will normally achieve very few marks.

Identifying the context

You need to clearly specify an environmental issue that you will be addressing in your investigation, for example habitat disturbance through activities such as logging, water pollution, increased sedimentation due to deforestation, and so on. The ESS syllabus provides many opportunities for investigating both local and global environmental issues. The environmental issue provides context for the research question. The research question that you develop from the environmental issue must also be clearly stated. It is not sufficient to simply carry out a traditional practical—your investigation must address an environmental issue. For example, it would not be sufficient to carry out a study examining the effect of light intensity on photosynthesis in pondweed (*Elodea*) because there is no associated environmental issue—a more appropriate investigation for your ESS IA would be to look at the effect of increased siltation and turbidity on photosynthetic rates as a result of deforestation. Usually a research question includes an independent variable (i.e. the variable that varies

Assessment tip

The best IA reports tend to come from a personal connection between you and your research, especially when this includes some direct investigation—field, lab-based or supported by surveys. Secondary sources can be used but it is actually more difficult to do this well.

in a study) and a dependent variable (i.e. the variable that is being measured). The research question must be precise and give a clear understanding of the aim of the investigation. A research question that simply asks "What is the effect of pollution on plants?", for example, would be too general and not give a clear idea of what was being investigated. If the environmental issue is increased salinization of soils due to vegetational changes, the issue can be addressed by studying the effects of increased salt concentration on one species of plant that is easily grown in the laboratory, such as broad beans, and the research question would be "What is the effect of salt concentration on the germination of broad-bean seedlings?".

>> Assessment tip

The writing of a good research question is essential for the development of the rest of the practical. Your research question needs to be focused enough for adequate treatment in the given word limit, and accessible in the context of the IA.

>> Assessment tip

A common mistake in the identifying of the context criterion is a failure to clearly identify an environmental issue. A research question investigating, for example, how differently textured soils (e.g. compost, sand-based, clay-based and loam) affect plant growth does not link to an obvious and meaningful environmental issue. Marks may also be lost in framing a research question—these are often imprecise.

>> Assessment tip

The question should not mistake the model for the reality. For example, if what is being researched is the amount of suspended solids in water that has been allowed to flow over samples of soil with and without plant cover, the question should not be, "What is the impact of soil erosion on agriculture?", because this was not being directly investigated.

The connection between the environmental issue and research question must be clearly made, and this then leads on to the methodology.

Planning

Your methodology should provide sufficient data for a robust analysis—if you do not record sufficient data, it is difficult to score highly in the subsequent criteria. If carrying out a statistical test, you need to ensure that sufficient data is gathered for the specific test you will be using. Repeats should be considered to ensure reliability of your data, and so that anomalies can be identified. Repeated data also allows variation in data to be considered—something that you can refer to when evaluating the procedure in subsequent criteria.

You need to justify your sampling technique, not simply describe it. This means that you need to justify how subjects for a survey were selected and why this method, and no other, was used. If sampling a local forest with quadrats, you should indicate why these are being placed randomly instead of along a transect. If quadrats are being placed randomly within a set of axes (e.g. by putting down two tape measures at right angles to describe a sampling area), how was the location of those axes determined? You are also required to address ethical considerations and risk assessments where appropriate. These will be appropriate for most practicals, even those using secondary data, where the relevant ethical consideration will be to ensure that you are not engaged in purposeful bias. If carrying out a survey-based questionnaire, you need to discuss the need to guarantee anonymity to subjects, and indicate that when approaching strangers in order to ask them to fill out a survey; this is more safely done in the company of others in your class. When working in laboratories you would be expected to explain the use of standard safety equipment, gloves, aprons, goggles, and so on. Fieldwork can also be hazardous, for example working in the intertidal zone can be dangerous, especially in wave-exposed areas, and these risks should be addressed.

Assessment tip

In this criterion you need to justify the choice of sampling strategy used, i.e. establish why your method is being chosen over another, or why samples are being taken "here" and not "there". If your investigation involves taking a survey, you must indicate how participants are selected and why. Why have they been asked the questions that are included in the survey? Central to this criterion is knowledge of how to recognize bias and minimize it.

Results, analysis and conclusion

You need to fully analyse your data and explain all trends, patterns or relationships in the data. The analysis must be coherent with the research question. Your conclusion should link back to your research question and environmental issue.

Trend line generating software (Excel for example) can be used to add a trend line to scatter plots. It is possible that a curve fits the data better and this can be explored using these same tools and the corresponding R^2 value calculated (this is the squared correlation coefficient, which indicates the strength of the relationship between the two variables in a scatter-plot). Care should be taken when fitting curves to data—the curve needs to have some logical explanation within the context of the investigation. These tools are not required but if used will be assessed for correct application and interpretation.

Assessment tip

If an investigation monitors the growth of a plant over five weeks and records the length of the plant five times in week 1, week 2, and so on, these five values cannot be averaged because they represent different points in a time sequence. When monitoring a continuously changing variable, a mean cannot be calculated. In this case, five replicates should have been made at each point in the time sequence (week 1, week 2, and so on) so that a mean of these replicates can be calculated. Replicates also allow standard deviation to be calculated, which can be drawn on graphs as error bars—non-overlapping error bars indicate that mean results are significantly different from one another, whereas overlapping error bars indicate no significant difference.

The processed data upon which a graph is built should be present in your report. A well-constructed graph of raw data may be considered under the first aspect of this criterion; however, it is not considered to be analysis. You should carefully check calculations to ensure they are correct—your external moderator will do this and it is a pity if you lose points for careless mistakes.

The conclusion should identify clear trends, interpret, and explain the data. Make sure you do not over-interpret data: you need to focus on what your data actually show. Avoid over-ambitious generalizations based on a very small study—limit yourself to what the data actually say.

Discussion and evaluation

In this criterion you need to:

- Evaluate your conclusion in the context of the environmental issue—you are expected to see how your conclusion measures up against what might be known regarding the issue. For example, if you conclude that salinity does impact germination rates in broad bean seeds, how is this related to the original problem? Is the conclusion in keeping with the literature? If it is in keeping with the literature, is this a trustworthy conclusion? Much can be justified with the simple calculation of variation in the data (i.e. standard deviation).
- Discuss the strengths, weaknesses and limitations of the methodology. This should not just be a description—a discussion is required. A table of strengths and weaknesses may provide a very

Assessment tip

Indicating that "all pertinent safety measures were followed" is not sufficient alone in a risk assessment. Work that relies exclusively on secondary data may not require a risk assessment or ethical considerations, but you should explain why this is the case in your report.

Assessment tip

Repeat readings are essential. If a research question is, for example, how a mine spillage has affected pH at different depths in surrounding rivers, it would not be sufficient to take only one reading at each depth in each of six locations measured, because little can be done with these data (i.e. standard deviations and mean values at each depth cannot be calculated because there is only one reading at each depth). It would be inappropriate to calculate the mean pH (i.e. combining all depths) and standard deviation at each site because this is not a relevant calculation in light of the research question. It is also inappropriate to calculate a mean value if working with a time series.

Assessment tip

This criterion requires you to evaluate the conclusion in the context of the environmental issue. How well did your investigation address the environmental issue? What were the strengths and limitations of your study in this regard?

useful scaffold although it is difficult to achieve full marks with a table that is not accompanied by text.

- Suggest modifications and further areas of research. Further areas of research should be qualitatively different from what has been done before, and not just more of the same.

Assessment tip

It is difficult to evaluate and discuss your conclusion in one brief paragraph or in a table format. Make sure you read the mark descriptors in the ESS subject guide carefully. Sometimes the connector between various characteristics in the descriptor is an "or" and other times an "and" is used, and this affects how the criterion is assessed.

You need to address possible modifications and further areas of research in the discussion and evaluation criterion. You need to suggest something that is qualitatively different in order to score full marks. For example, doing more of the same would not be considered further research. Researching percentage cover of plants after having done a study of diversity would be considered further research.

Applications

This criterion measures your ability to apply your research and consider potential solutions to the environmental issue you are exploring. To achieve full marks for the criterion, one well-thought-out application with an evaluation of pros and cons is all that is needed. You need to provide an evaluation of the strategy you are proposing—what are its potential strengths and limitations? It is important to consider that the solution does not need to be based directly on the data generated by the study—the examiner will appreciate that your results may be unclear, and this should not have a knock-on effect on your success in this criterion. Information from the literature or secondary sources can be used to help you develop your solution.

Assessment tip

Do not give three or four solutions; this is unnecessary and does not achieve more marks. One solution or application needs to be justified, and this needs to be evaluated. If a candidate lists a number of good and relevant solutions, in the absence of an evaluation, they are likely to be scored with zero marks.

Assessment tip

Low marks in this criterion are the result of only a very general discussion and failure to include any sort of evaluation. For example, if writing about how to deal with acid rain, more is required than simply stating that governments should pass legislation making it illegal. If this is to be the suggestion then examples from successful application of legislation should be cited and referenced, and the success of the solution should be evaluated.

Communication

The failure to follow some conventions may result in lowered marks in this criterion, for example poor titles in tables, inclusion of units in individual cells, lack of horizontal/vertical axes labels in graphs. Raw data must be tabulated; a list is not appropriate.

Assessment tip

While the report would be expected to be correctly referenced, you will not be penalized under the Communication criterion for a lack of bibliography or other form of citation. It is likely, however, that such an omission would be treated under the IB Diploma Programme academic honesty policy.

If you make very specific statements regarding some issue in the context or discussion and evaluation criterion, these should be accompanied with a citation. Failure to do so may result in a loss of marks in communication. There is no required citation format in the IB and you should not lose marks for using the wrong style of citation.

Examiners are instructed to stop reading when they reach the word limit. All work beyond this limit will not be considered. Examiners will also not read appendices. You may include screen shots of survey questions and these will not be included in the word count, although only a reasonable quantity should be included (i.e. 50 extra pages of survey results would not be appropriate!).

PRACTICE EXAM PAPERS

At this point, you will have re-familiarized yourself with the content from the topics of the IB ESS syllabus. Additionally, you will have picked up some key techniques and skills to refine your exam approach. It is now time to put these skills to the test; in this section you will find practice examination papers 1 and 2 with the same structure as the external assessment you will complete at the end of the DP course. Additional guidance to these papers is available at www.oxfordsecondary.com/ib-prepared-support

Paper 1: Resource booklet

The scenery of the Killarney area, including the National Park, is world-renowned. It is a major attraction and the area is one of the most visited tourist venues in Ireland. Over a million visitors travel to Kerry each year, bringing an estimated £160 million to the area. Of these the majority visit Killarney, a town with a resident population of 14,000 and over 4,000 tourist rooms! Under legislation in Ireland, where conflict arises between the need for tourism and the need for conservation, the protection of the natural heritage takes precedence over other considerations.

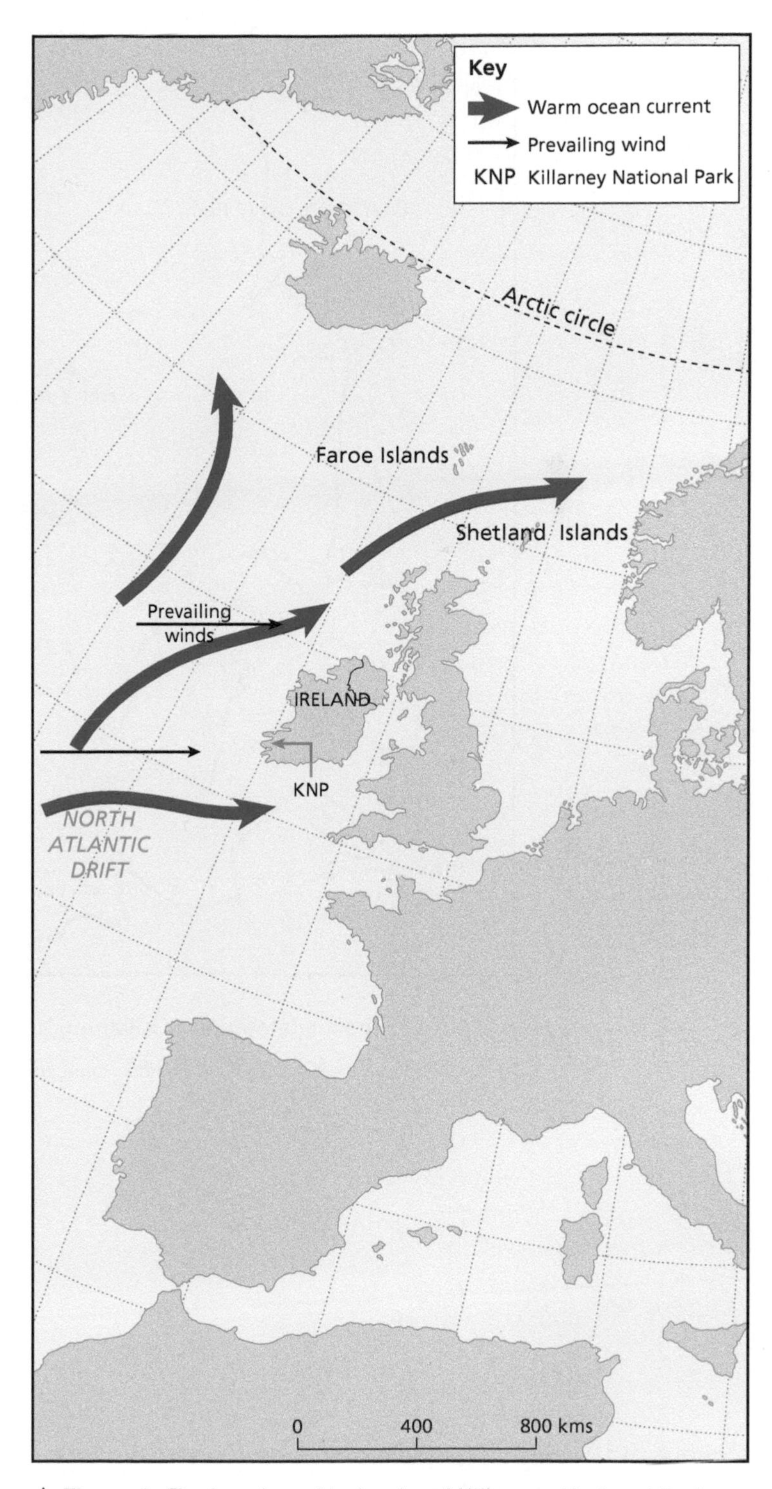

▲ **Figure 1** The location of Ireland and Killarney National Park, Ireland

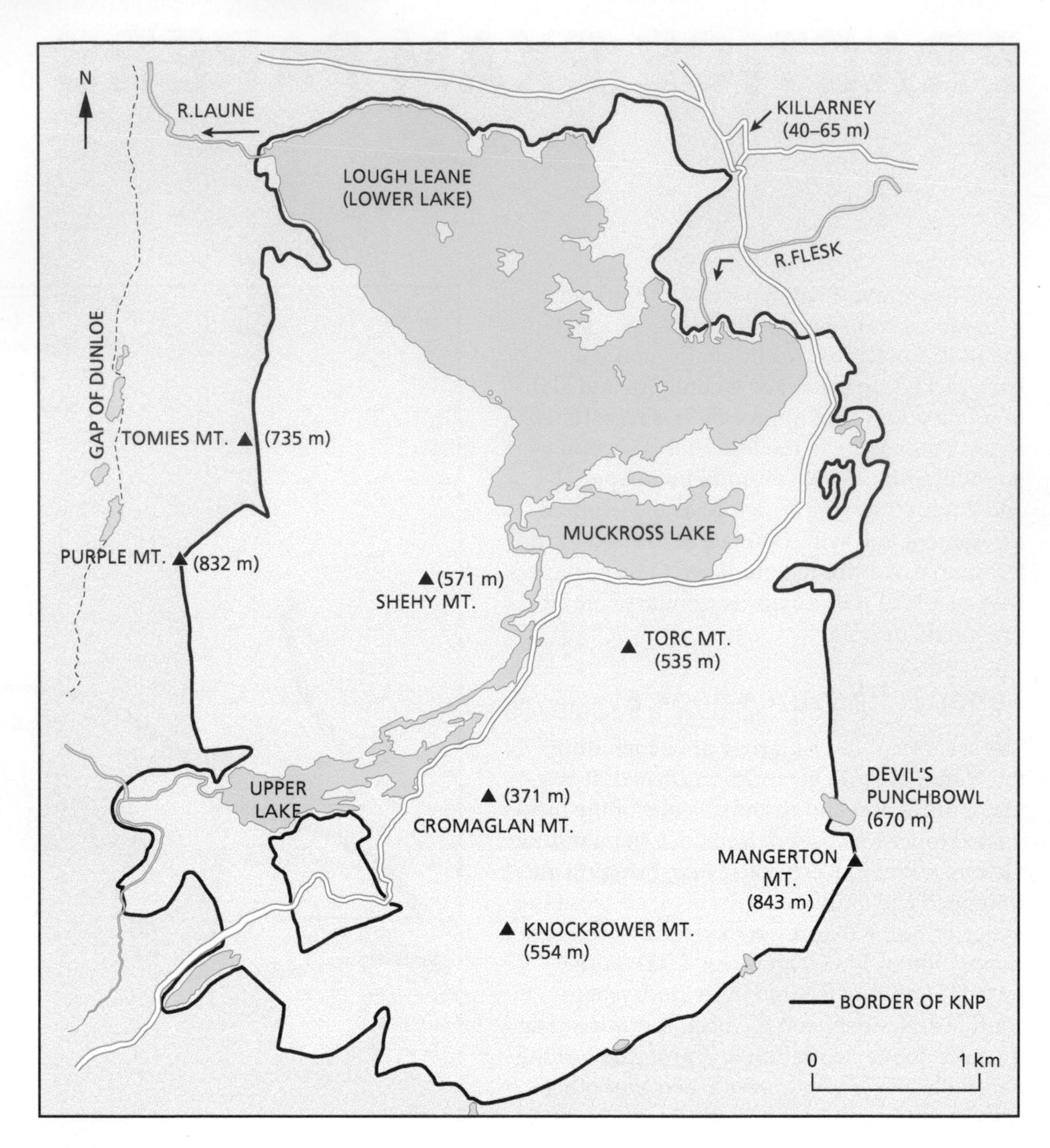

▲ **Figure 2** The main features of Killarney National Park

Source: Adapted from *Killarney National Park Management Plan, 1990*, The Office for Public Works

Limestone rocks by Muckross Lake

Boat on Lough Leane

Dinis Bridge

Upper lake from Ladies View

Ross Castle

Lough Leane (Lower lake)

Muckross House – the main cultural attraction

Busy street scene from Killarney

Visitors admiring the view of the Lakes

▲ **Figure 3** Pictures of Killarney National Park

▼ **Table 1** Result of a survey of fish populations in Lough Leane in 2014

Fish species	***n***	**(*n* – 1)**	***n*(*n* – 1)**
Perch *Perca fluviatilis*	331	330	109,230
Brown trout *Salmo trutta*	179	178	31,862
Rudd *Scardinius erythrophthalmus*	50	49	2,450
Eel *Anguilla anguilla*	27	26	702
Killarney shad *Alosa fallax killarnensis*	36	35	1,260
Flounder *Platichthys*	19	18	342
Tench *Tinca tinca*	6	5	30
Salmon *Salmo salar*	4	3	12
Arctic char *Salvelinus alpinus*	1	0	0
	$\sum$		$\sum =$

Source: http://wfdfish.ie/wp-content/uploads/2011/11/Leane_report_2014.pdf

Factfile 1: Invasive species: the rhododendron

- The shrub, *rhododendron pontium*, is a native of Spain, Portugal, and around the Black Sea.
- It was introduced into the British Isles in 1763, and was widely used as a decorative plant and as a cover for game.
- It was planted in the Muckross and Kenmare Estates in the 19th century, and now affects almost 40% of Killarney National Park's area.
- It grows vigorously on the acid soil, and thrives in the mild, ice-free, damp environment of Killarney.
- The plant can survive fire and frost.
- In some woodlands it has replaced the holly understorey and shaded out the ground cover.
- It does not get grazed as its leaves are toxic.
- Efforts to eradicate rhododendron have been in place since the 1960s.

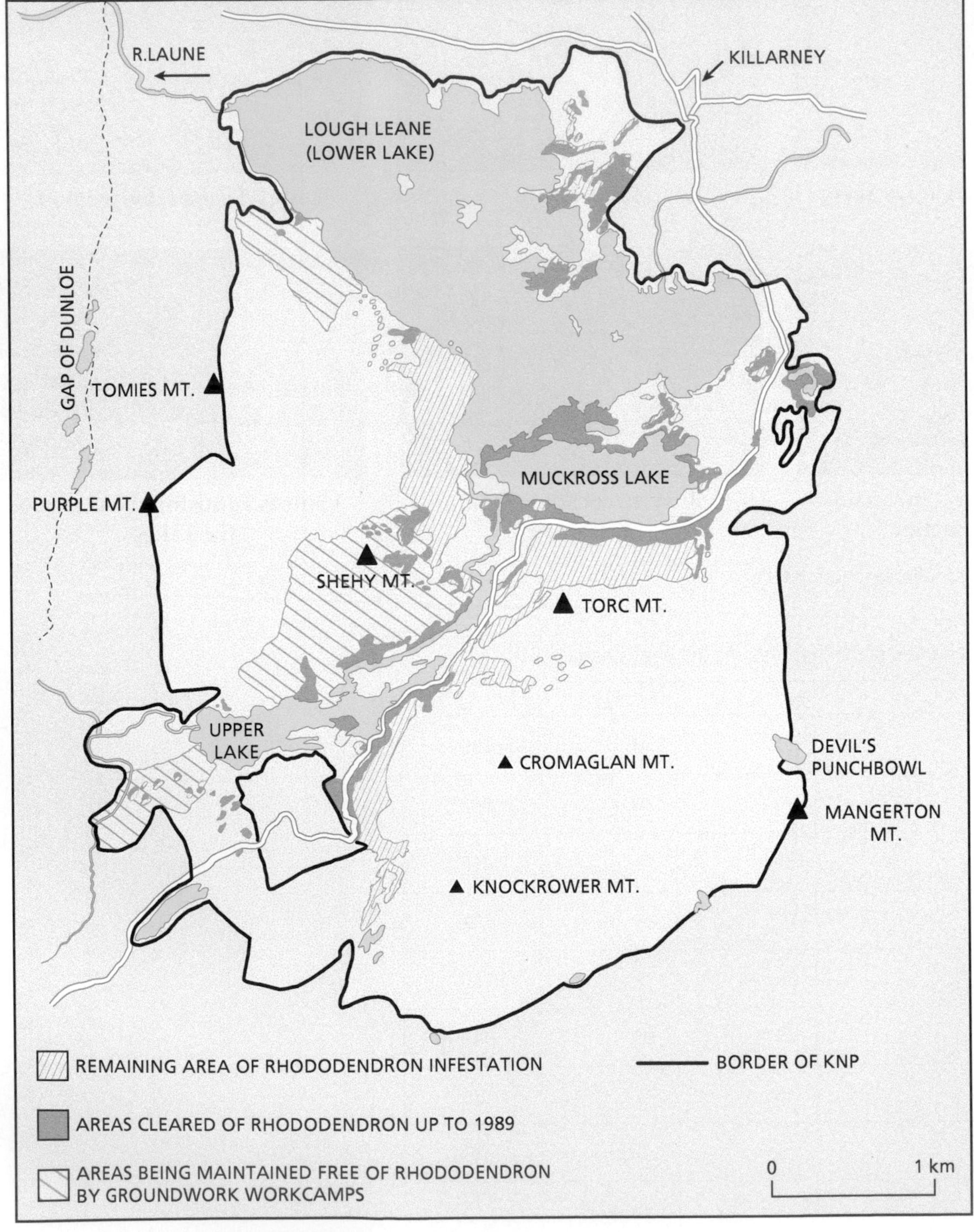

▲ **Figure 4a** Rhododendron infestation and clearance up until 1989

Source: *Killarney National Park Management Plan, 1990*, The Office of Public Works

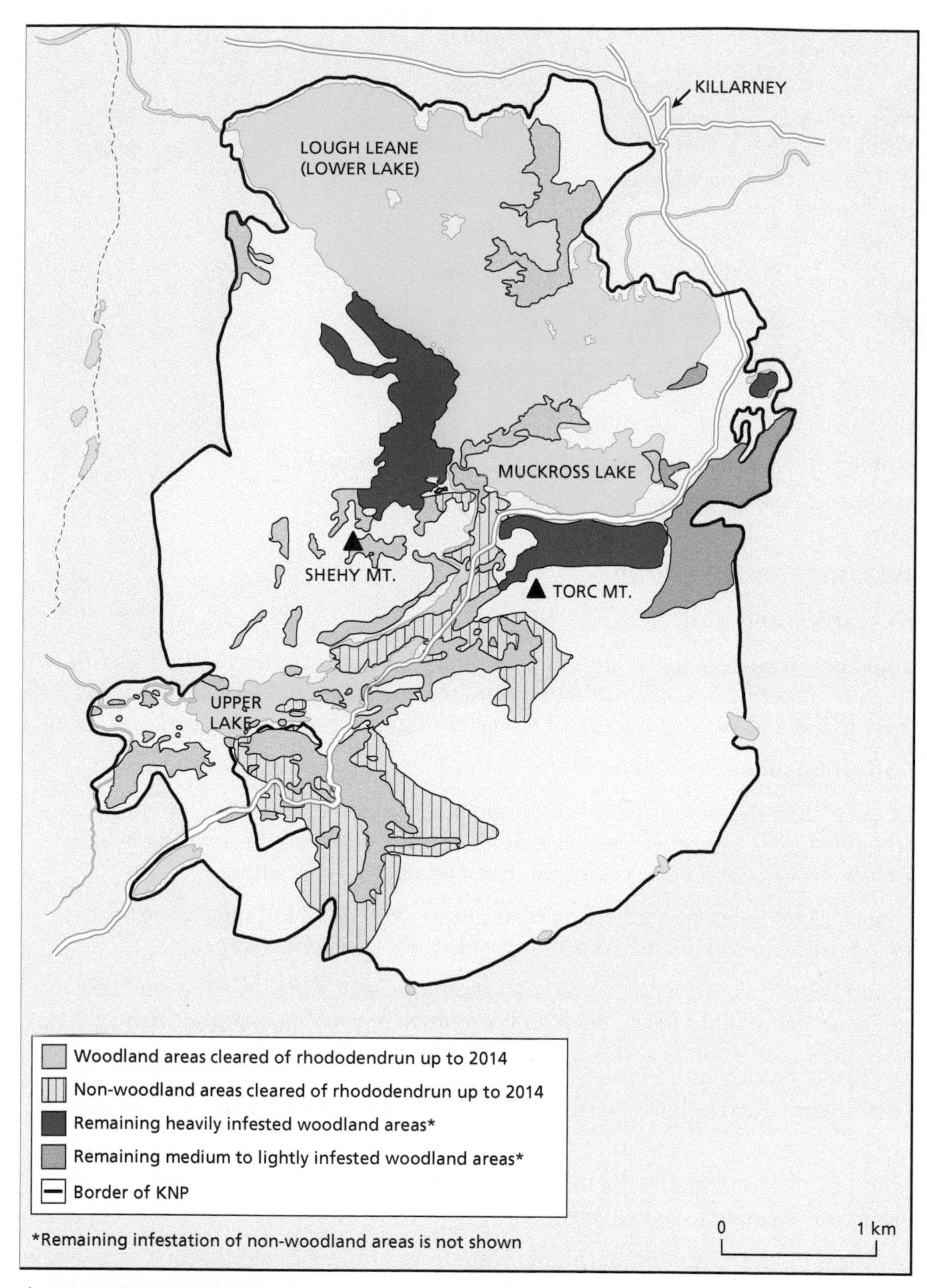

▲ **Figure 4b** Rhododendron infestation and clearance up until 2014

Source: Adapted from *UNESCO, Killarney National Park Biosphere Reserve, 2017*, Department of Arts, Heritage, Regional, Rural and Gaeltacht Affairs

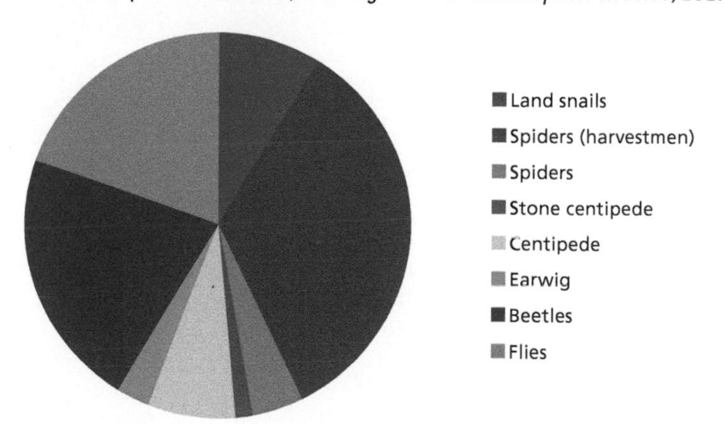

▲ **Figure 5a** Composition of invertebrate community in rhododendron stands at Killarney National Park

Source: https://www.killarneynationalpark.ie/about-us-killarney/biodiversity/

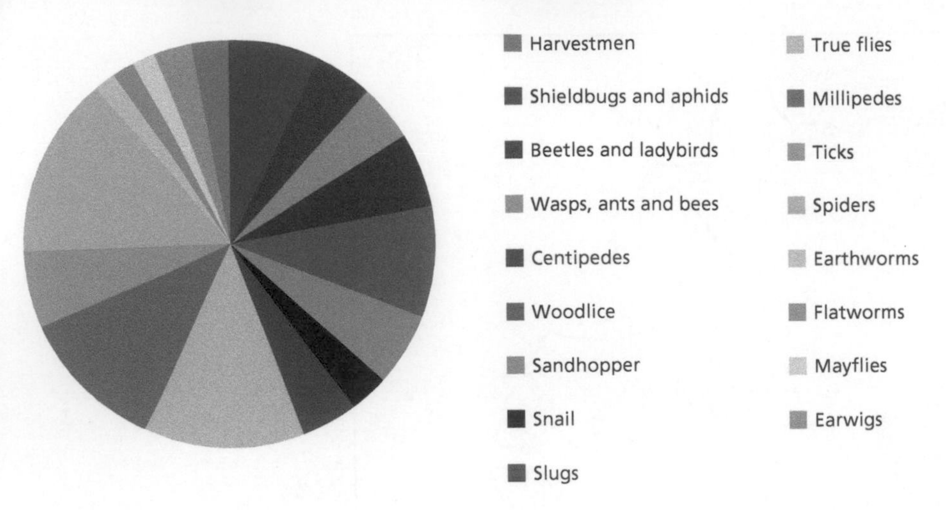

▲ **Figure 5b** Composition of invertebrate community in native woodland at Killarney National Park

Source: https://www.killarneynationalpark.ie/about-us-killarney/biodiversity/

Factfile 2: Methods of treating rhododendron

- **Remove regrowth from cut stems and treat.**

  Rhododendron stems should be cut as close to the ground as possible. Low-cut stumps produce less growth and therefore minimize the use of herbicides. After 18–24 months, regrowth is removed and herbicide applied. The method is easy to use and reduces the risk of herbicide being carried into surrounding areas.

- **Stem and plant base treatment using a chain saw**

  The rhododendron is cut as close to the base as possible. Larger stems may be cut a number of times. Herbicide is applied immediately to fresh saw cuts, and a dye is used to show which areas have been treated. Plant death normally occurs within one year. Leaf death takes longer in winter.

  Advantages include – can be used on most rhododendron plant types; very good kill rate; cost-effective; low volume of herbicide required; no soil disturbance; effective in light showery weather.

  Disadvantages include – second large-scale work phase needed if standing dead plants are to be removed; potential fire risk; standing dead plants and subsequent litter may inhibit regeneration and site management.

- **Stem treatment**

  This requires making a downward cut. Herbicide should be applied. Leaf die-back begins 3-4 weeks after treating the plant.

  Advantages include – a good kill rate; minimal herbicide use; no soil disturbance; minimal risk of herbicide drift; limited equipment needed; cost-effective.

  Disadvantages include – requires mainly dry weather; all plant stems require treatment; second large-scale work phase required if dead plants are to be removed; standing dead plants and litter pose a fire risk.

  This method is most effective for single-stem plants.

▼ **Table 2** The value of ecosystem services provided by native woodlands

Amenity (non-market value)	€65 million per year
Tourism expenditure	€60 million per year
Health	€4 million per year
Biodiversity utility value	€60 million per year
Water quality, flood and erosion control	€3 million per year
Carbon storage and sequestration	€45 million per year
Timber and wood fuel	€37 million per year

Source: Bullock, C. and Hawe, J., *The natural capital values of Ireland's native woodland*, Woodlands of Ireland, 2014.

Factfile 3: Sika deer in Inisfallan Island, Killarney National Park

- Introduced into Ireland in 1860.
- Two females and a male were introduced to Killarney in 1864.
- Maximum density 50 deer per 100 ha.
- In 2010 a group of sika deer swam 1.5 km to colonize Inisfallan Island (8.5 ha).
- In 2018 four sika deer were found to have starved on the island and a further 27 were culled. Fifteen deer remained on the island.
- Trees had been stripped bare by the deer and much of the ground vegetation destroyed.
- The forest has been unable to regenerate due to the sika deer grazing/browsing.
- Deer are protected under the 1976 Wildlife Act.
- The only natural predators of deer, the wolf and the Golden Eagle, are now extinct.

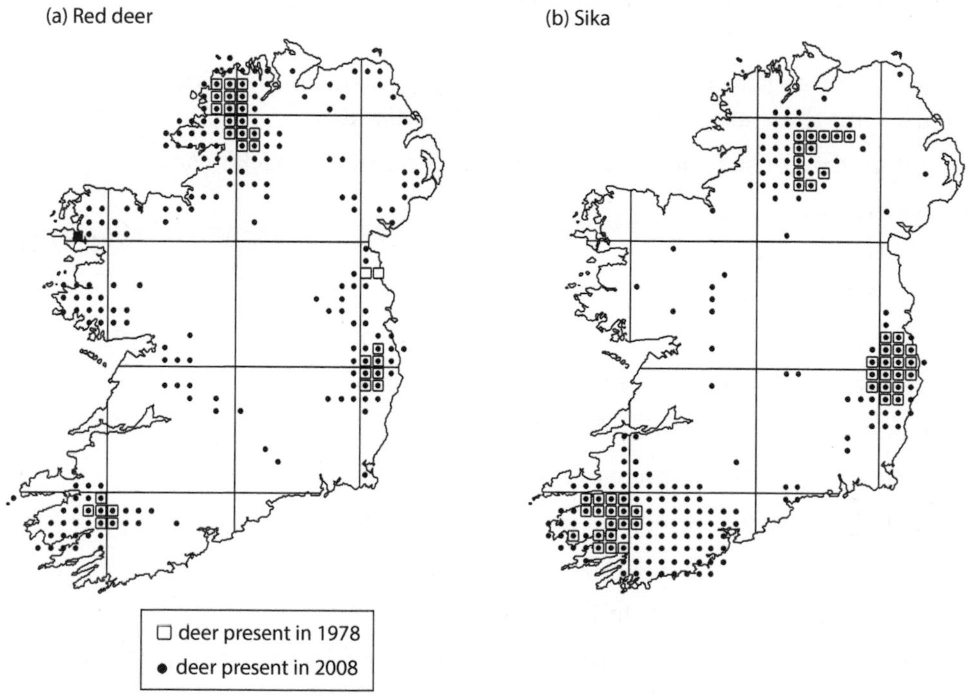

▲ **Figure 6** Distribution of red deer and sika deer in Ireland

Source: Carden, R. F. et al, 2011, Distribution and range expansion of deer in Ireland, *Mammal Review*, 2011, Vol. 41, 4, 313–325.

▼ **Table 3** Expansion of deer population in Ireland, 1978–2008

Species	Number of squares occupied in 1978	Number of squares occupied in 2008	Total increase in range 1978–2008
Red deer	31	206	564.5%
Sika	47	213	353.2%

Source: Carden, R. F. et al, 2011, Distribution and range expansion of deer in Ireland, *Mammal Review*, 2011, Vol. 41, 4, 313–325.

Paper 1

The resource booklet provides information on Killarney National Park, Ireland. Use the resource booklet and your own studies to answer the following.

1. Using Figures 1 and 2, identify **two** reasons why Killarney has a wet and relatively warm climate. [1]
2. Using Figures 2 and 3, identify two different types of attractions of Killarney National Park for tourists. [2]
3. Using data in Table 1, calculate the Simpson diversity index for fish in Lough Leane in 2014.

   Simpson diversity index $= \dfrac{N(N-1)}{\sum n(n-1)}$ [2]

4. In 2008 the Simpson diversity index for the fish population in Lough Leane was 2.81 and in 2011 it was 3.13. Suggest possible reasons for the change in diversity index between 2008 and 2014. [2]
5. Using Factfile 1 and Figures 4a and 4b, compare and contrast the area of rhododendron infestation in 1989 with that of 2014. [3]
6. Using Figures 5a and 5b, outline the differences in the composition of organisms in rhododendron stands in Killarney National Park with those associated with native woodland. [2]
7. **a)** Using Factfile 2, evaluate two methods used for the control of rhododendron in Killarney National Park. [5]

   b) Define the term 'natural capital'. [1]

   c) Explain the meaning of the term 'ecosystem service'. [2]

   d) Using Table 2, calculate the value of ecosystem services provided by native forests in Ireland. [1]
8. **a)** Using Factfiles 1 and 3, suggest a definition for the term 'invasive species'. [1]

   b) **i)** Using Figure 6, compare and contrast the distribution of sika deer with that of red deer in 2008. [3]

   ii) Using Table 3, calculate the increase in the range of sika deer between 1978 and 2008. [1]

   iii) Suggest two contrasting reasons for the expansion of the range of sika deer after 1978. [2]

   c) Define the term 'carrying capacity'. [1]

   d) Using Factfile 3, calculate the density of sika deer on Inisfallan Island. [2]

   e) Contrast the number of deer on the island with the maximum carrying capacity of 50 deer/100 ha. [1]

   f) Explain three consequences of too many deer in an environment. [3]

Paper 2

Section A

1. Figure 1 shows total fertility rate and gross domestic product for countries in 2016.

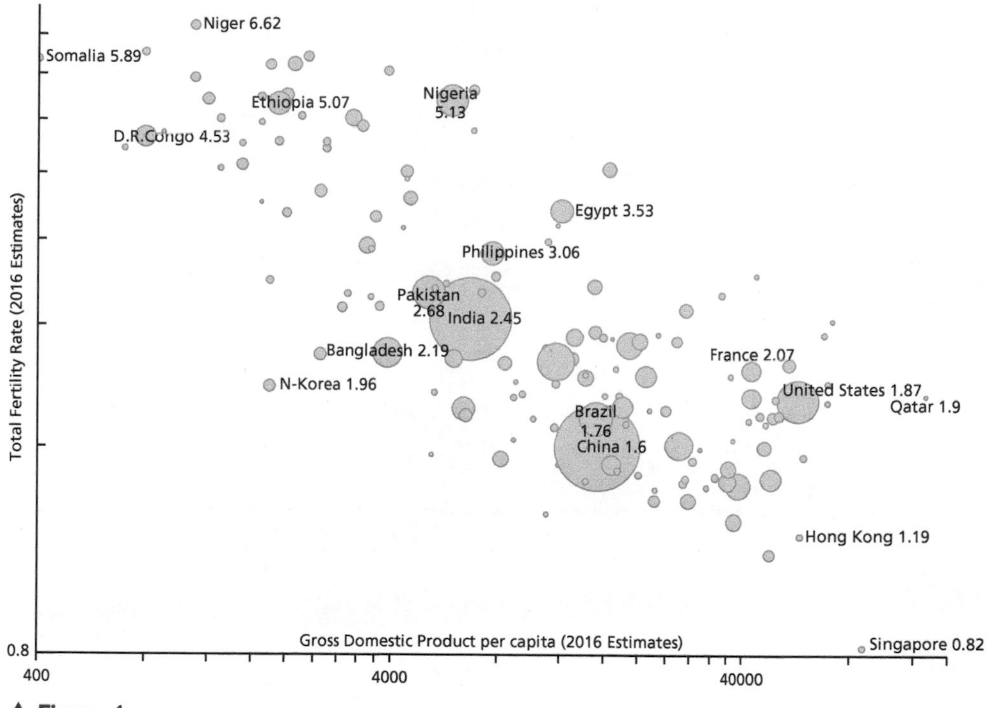

▲ **Figure 1**

Source: Based on: https://commons.wikimedia.org/wiki/File:CIA_WFB_TotFertilityRate-GDP-Population_-_Simplified_2016.png

a) Define 'total fertility rate'. [1]

b) Identify the country with the highest total fertility rate. [1]

c) Identify the country with the lowest total fertility rate. [1]

d) Calculate the range from the highest total fertility rate to the lowest total fertility rate. [1]

e) Estimate the GDP of (i) Bangladesh and (ii) France. [2]

f) Describe the relationship between total fertility rate and gross domestic product. [3]

g) Suggest why a log-log graph is an appropriate method to show the data. [2]

2. Figure 2 shows changes in the pH of rainwater 1993–2007 near Oxford, a city in an MEDC.

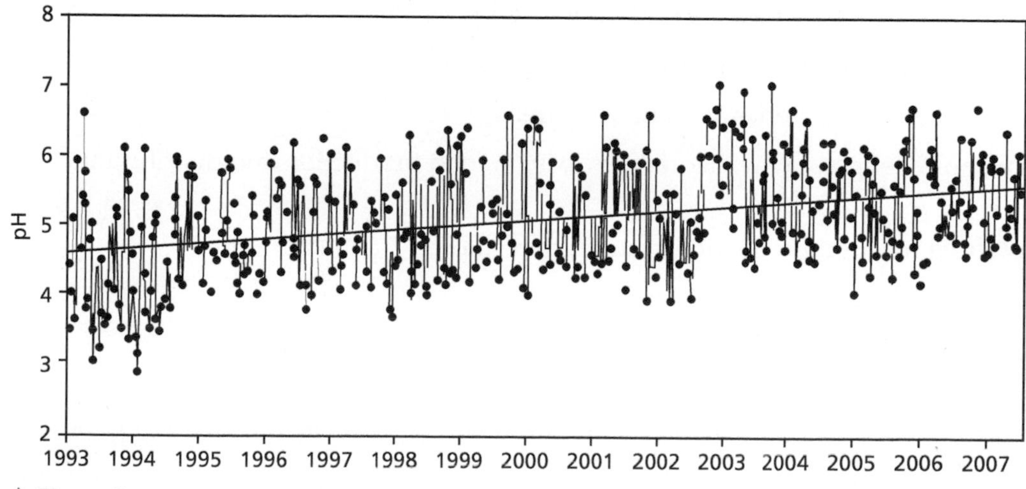

▲ **Figure 2**

Source: Savill, P.S., et al., 2010, *Wytham Woods: Oxford's Ecological Laboratory*, Oxford University Press.

a) Identify **two** trends in the pH of rainwater, as shown in Figure 2. [2]

b) Calculate the difference between mean pH in 1993 and 2007. [1]

c) Suggest reasons why the trend in pH, as shown, is changing. [3]

3. Figure 3 shows carbon stores and flows in an undisturbed tropical rainforest and ten years after deforestation.

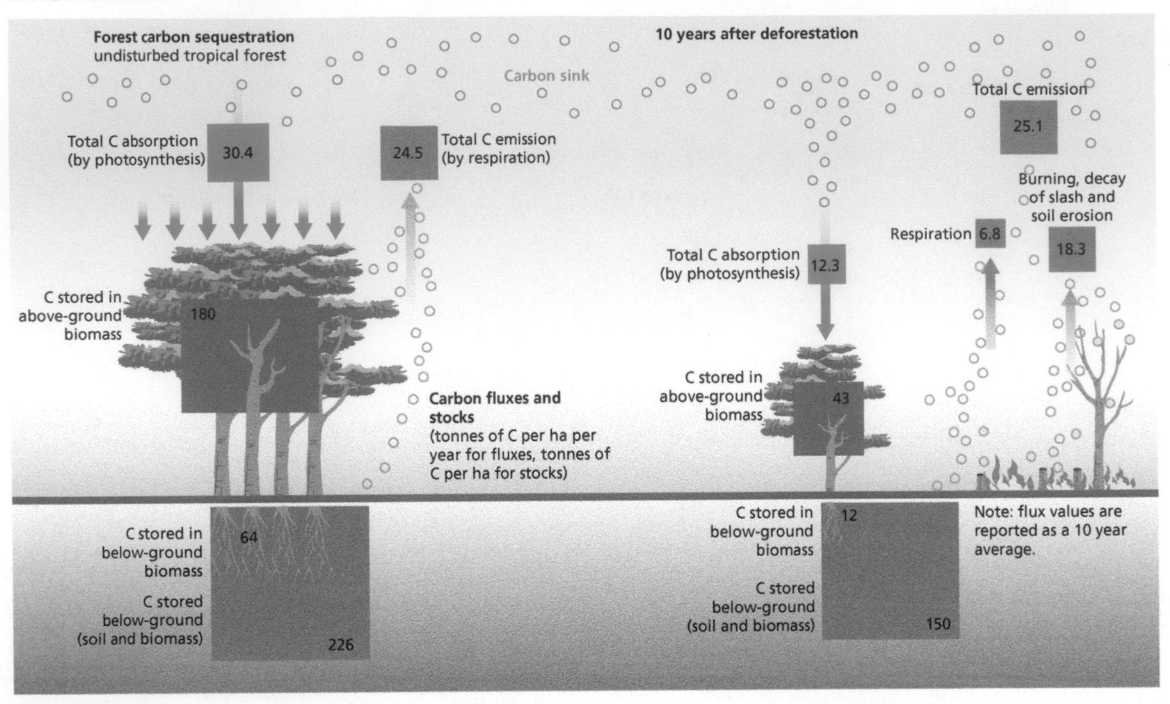

▲ Figure 3

a) Identify i) the largest store of carbon in the undisturbed forest [1]

ii) the largest flow of carbon in the undisturbed forest. [1]

b) Calculate the decline in carbon stored in biomass 10 years after deforestation. [2]

c) Identify the largest relative decrease in carbon flow between the undisturbed forest and following deforestation. [1]

d) Outline how deforestation leads to an increase in atmospheric carbon. [3]

Section B Essays

4. a) Describe how urbanization can lead to changes in surface runoff and infiltration. [4]

b) Explain how water supplies may be increased in rich and poor countries. [7]

c) Evaluate the success of different strategies to manage pollution. [9]

5. a) Distinguish between capture fisheries and aquaculture. [4]

b) Using named examples, explain why inequalities exist in food production and distribution around the world. [7]

c) Examine the potential for food production systems to be more sustainable. [9]

6. a) Outline the working of the greenhouse effect. [4]

b) Explain the role of positive and negative feedback mechanisms associated with global climate change. [7]

c) Discuss the view that most countries' attempts to achieve energy security are damaging the world's environment. [9]

7. a) Outline the dynamic nature of natural capital. [4]

b) Explain the impact of acid deposition on soils, water and living organisms. [7]

c) Discuss the view that waste management strategies fail to protect the environment. [9]

Index

Page numbers in **bold** refer to key terms boxes.